PROGRESS IN

Nucleic Acid Research and Molecular Biology

Volume 10

PROGRESS IN
Nucleic Acid Research and Molecular Biology

edited by

J. N. DAVIDSON
Department of Biochemistry
The University of Glasgow
Glasgow, Scotland

WALDO E. COHN
Biology Division
Oak Ridge National Laboratory
Oak Ridge, Tennessee

Volume 10

1970

ACADEMIC PRESS
New York and London

ACADEMIC PRESS, INC.
111 Fifth Avenue, New York, New York 10003

United Kingdom Edition published by
ACADEMIC PRESS, INC. (LONDON) LTD.
Berkeley Square House, London W1X 6BA

LIBRARY OF CONGRESS CATALOG CARD NUMBER: 63-15847

PRINTED IN THE UNITED STATES OF AMERICA

List of Contributors

Numbers in parentheses refer to the pages on which the authors' contributions begin.

M. R. ATKINSON (87), *School of Biological Sciences, Flinders University of South Australia, Bedford Park, South Australia*

ORIO CIFERRI* (121), *Department of Genetics, University of Pavia, Pavia, Italy*

DAPHNE C. ELLIOTT (87), *School of Biological Sciences, Flinders University of South Australia, Bedford Park, South Australia*

ROSS H. HALL (57), *Department of Biochemistry, McMaster University, Hamilton, Ontario, Canada*

TAMIKO KANO-SUEOKA (23), *Department of Biology and Program in Biochemical Sciences, Princeton University, Princeton, New Jersey*

ALAN H. MEHLER (1), *Department of Biochemistry, Marquette School of Medicine, Milwaukee, Wisconsin*

A. W. MURRAY (87), *School of Biological Sciences, Flinders University of South Australia, Bedford Park, South Australia*

BRUNO PARISI (121), *Department of Genetics, University of Pavia, Pavia, Italy*

ALEXANDER RICH (183), *Department of Biology, Massachusetts Institute of Technology, Cambridge, Massachusetts*

ZOE A. SHABAROVA (145), *Department of Chemistry, Laboratory of Bio-organic Chemistry, Moscow State University, USSR*

NOBORU SUEOKA (23), *Department of Biology and Program in Biochemical Sciences, Princeton University, Princeton, New Jersey*

DONALD VOET† (183), *Department of Biology, Massachusetts Institute of Technology, Cambridge, Massachusetts*

* Present address: Botanical Institute, University of Cagliari, Cagliari, Italy.
† Present address: Department of Chemistry, University of Pennsylvania, Philadelphia, Pennsylvania.

Preface

In this volume of *Progress in Nucleic Acid Research and Molecular Biology* the contributions follow our usual pattern of attempting to present "essays in circumscribed areas" in which recent developments in particular aspects of the field of nucleic acids and molecular biology are discussed by workers provided with an opportunity for more personal expression of points of view that are individualistic and perhaps even controversial. We have not attempted to define or restrict any author's approach to his chosen subject, and have confined our editing to ensuring maximum clarity to the reader, whom we envisage to be a person himself active in or concerned with the general field of nucleic acids or molecular biology. Needless to say, we do not necessarily share all the opinions or concepts of all the authors and accept no responsibility for them. We seek rather to provide a forum for discussion and debate, and we will welcome further suggestions from readers as to how this end may best be served. Indeed, we should like again to remind readers that we wish them to write to us with their comments.

Abbreviations and symbols used for nucleic acids and their derivatives are now well established by the authority of the International Union of Biochemistry and the International Union of Pure and Applied Chemistry. Those pertinent to our subject are not listed at the beginning of each chapter, but will be found on the following pages.

J.N.D.
W.E.C.

November, 1969

Abbreviations and Symbols

All contributors to this Series are asked to use the terminology (abbreviations and symbols) as formulated by the IUPAC–IUB Commission on Biochemical Nomenclature (CBN) and approved by IUPAC and IUB, and the Editors endeavor to assure conformity. These Rules have been published in many journals (*1*) and compendia (*2*) in four languages and are available in reprint form from the NAS–NRC Office of Biochemical Nomenclature (OBN) (*3*), as stated in each publication, and are therefore considered to be generally known. Those used in nucleic acid work, as set out in section 5 of the above Rules (*1*) and recently revised and expanded (*2, 3*) are given in condensed form (I–V) below for the convenience of the reader.

I. Bases, Nucleosides, Mononucleotides

1. *Bases* (in tables, figures, equations, or chromatograms) are symbolized by Ade, Gua, Hyp, Xan, Cyt, Thy, Oro, Ura; Pur = any purine, Pyr = any pyrimidine, Base = any base. The prefixes S–, H_2, F–, Br, Me, etc., may be used for modifications of these.

2. *Ribonucleosides* (in tables, figures, equations, or chromatograms) are symbolized, in the same order, by Ado, Guo, Ino, Xao, Cyd, Thd, Ord, Urd (Ψrd), Puo, Pyd, Nuc. Modifications may be expressed as indicated in (1) above. Sugar residues may be specified by the prefixes r (optional), d (=deoxyribo), a, x, l, etc., to these, or by two three-letter symbols, as in Ara-Cyt (for aCyd) or dRib-Ade (for dAdo).

3. *Mono-, di-, and triphosphates of nucleosides* (5′) are designated by NMP, NDP, NTP. The N (for "nucleoside") may be replaced by any one of the nucleoside symbols given in II-1 below. 2′-, 3′-, and 5′- are used as prefixes when necessary. The prefix d signifies "deoxy." [Alternatively, nucleotides may be expressed by attaching P to the symbols in (2) above.]

II. Oligonucleotides and Polynucleotides

1. Ribonucleoside Residues

(a) Common: A, G, I, X, C, T, O, U, Ψ, R, Y, N (in the order of I-2 above).

(b) Base-modified: sI or M for thioinosine = 6-mercaptopurine ribonucleoside; sU or S for thiouridine; brU or B for 5-bromouridine; hU or D for 5,6-dihydrouridine; i for isopentenyl; f for formyl. Other modifications are similarly indicated by appropriate *lower-case* prefixes (in contrast to I-1 above) (*2, 3*).

(c) Sugar-modified: prefixes are d, a, x, or l as in I-2 above; alternatively, by italics or boldface type (with definition) unless the entire chain is specified by an appropriate prefix. The 2′-*O*-methyl group is indicated by *suffix* m (e.g., -Am- for 2′-*O*-methyladenosine, but -mA- for *N*-methyladenosine) (*2, 3*).

(d) Locants and multipliers, when necessary, are indicated by superscripts and subscripts, respectively, e.g., -m^6_2A- = 6-dimethyladenosine; -s^4U- or -^{4}S- = 4-thiouridine; -ac^4Cm- = 2′-*O*-methyl-4-acetylcytidine (*2, 3*).

(e) When space is limited, as in two-dimensional arrays or in aligning homologous sequences, the prefixes may be placed *over the capital letter*, the suffixes *over the phosphodiester symbol* (see ref. *2*, H62–63).

2. Phosphoric Acid Residues [left side = 5′, right side = 3′ (or 2′)]

(a) Terminal: p; e.g., pppN . . . is a polynucleotide with a 5′-triphosphate at one end; Ap is adenosine 3′-phosphate; C>p is cytidine 2′:3′-cyclic phosphate (*1, 2, 3*).

(b) Internal: hyphen (for known sequence), comma (for unknown sequence); unknown sequences are enclosed in parentheses. E.g., pA-G-A-C(C_2,A,U)A-U-G-C>p is a sequence with a (5′) phosphate at one end, a 2′:3′-cyclic phosphate at the other, and a tetranucleotide of unknown sequence in the middle. (Only codon triplets are written without some punctuation separating the residues.)

3. Polarity, or Direction of Chain

The symbol for the phosphodiester group (whether hyphen or comma or parenthesis, as in 2b) represents a 3′-5′ link (i.e., a 5′ . . . 3′ chain) unless otherwise indicated by appropriate numbers. "Reverse polarity" (a chain proceeding from a 3′ terminus at left to a 5′ terminus at right) may be shown by numerals or by right-to-left arrows. Polarity in any direction, as in a two-dimensional array, may be shown by appropriate rotation of the (capital) letters so that 5′ is at left, 3′ at right when the letter is viewed right-side-up.

4. Synthetic Polymers

The complete name or the appropriate group of symbols (see II-1 above) of the repeating unit, enclosed in parentheses if complex, is either (a) preceded by "poly," or (b) followed by a subscript "n" or appropriate number.

The conventions of II-2b are used to specify known or unknown (random) sequence, e.g.,

polyadenylate = poly A or A_n, a simple homopolymer;

poly(3 adenylate, 2 cytidylate) = poly (A_3C_2) or $(A_3,C_2)_n$, a *random* copolymer of A and C in 3:2 proportions;

poly(deoxyadenylate-deoxythymidylate) = poly d(A-T) or $d(A\text{-}T)_n$, an *alternating* copolymer of dA and dT;

poly(adenylate, guanylate, cytidylate, uridylate) = poly (A,G,C,U) or $(A,G,C,U)_n$, a random assortment of A, G, C, and U residues, proportions unspecified.

The prefix copoly or oligo may replace poly, if desired. The subscript "n" may be replaced by numerals indicating actual size.

III. Association of Polynucleotide Chains

1. *Associated* (e.g., H-bonded) chains, or bases within chains, are indicated by a *center dot* (not a hyphen or a plus sign) separating the *complete* names or symbols, e.g.:

poly A·poly U or $(A)_n\cdot(U)_m$

poly A·2 poly U or $(A)_n\cdot 2(U)_m$

poly(dA-dC)·poly(dG-dT) or $(dA\text{-}dC)_n\cdot(dG\text{-}dT)_m$;

also, "the adenine·thymine base-pair" or "A·T base-pair" in text.

2. *Nonassociated* chains are separated by the plus sign, e.g.:

$$2[\text{poly A}\cdot\text{poly U}] \xrightarrow{\Delta} \text{poly A}\cdot 2\text{ poly U} + \text{poly A} \quad \text{(II-4a)}$$

or $$2[A_n\cdot U_m] \xrightarrow{\Delta} A_n\cdot 2U_m + A_n \quad \text{(II-4b)}.$$

3. Unspecified or unknown association is expressed by a comma (again meaning "unknown") between the completely specified residues.

Note: In all cases, each chain is completely specified in one or the other of the two systems described in II-4 above.

IV. Natural Nucleic Acids

RNA	ribonucleic acid or ribonucleate
DNA	deoxyribonucleic acid or deoxyribonucleate
mRNA; rRNA; nRNA	messenger RNA; ribosomal RNA; nuclear RNA
D-RNA; cRNA	"DNA-like" RNA; complementary RNA
mtDNA	mitochondrial DNA
tRNA	transfer (or acceptor or amino acid-accepting) RNA; replaces sRNA, which is not to be used for any purpose
aminoacyl-tRNA	"charged" tRNA (i.e., tRNA's carrying aminoacyl residues); may be abbreviated to AA-tRNA
alanine tRNA or $tRNA^{Ala}$, etc.	tRNA normally capable of accepting alanine, to form alanyl-tRNA
alanyl-tRNA or alanyl-$tRNA^{Ala}$	The same, with alanyl residue covalently attached. [*Note:* fMet = formylmethionyl; hence $tRNA^{fMet}$ or $tRNA^{Met}_{f}$]

Isoacceptors are indicated by appropriate subscripts, i.e., $tRNA^{Ala}_{1}$, $tRNA^{Ala}_{2}$, etc.

V. Miscellaneous Abbreviations

P_i, PP_i	inorganic orthophosphate, pyrophosphate
RNase, DNase	ribonuclease, deoxyribonuclease

Others listed in Table II of Reference 1 may also be used without definition. No others, with or without definition, are used unless, in the opinion of the editors, they increase the ease of reading.

Enzymes

In naming enzymes, the recommendations of the IUB Commission on Enzymes, approved by IUB in 1964 (*4*), are followed as far as possible. At first mention, each enzyme is described *either* by its systematic name *or* by the equation for the reaction catalyzed, followed by its EC number in parentheses. Subsequent mention may use a trivial name. Enzyme names are not to be abbreviated except when the substrate has an approved abbreviation (e.g., ATPase, but not LDH, is acceptable).

References

1. *J. Biol. Chem.* **241,** 527 (1966); *Biochemistry* **5,** 1445 (1966); *Biochem. J.* **101,** 1 (1966); *Arch. Biochem. Biophys.* **115,** 1 (1966), **129,** 1 (1969); and elsewhere.

2. "Handbook of Biochemistry" (H. A. Sober, ed.), Chemical Rubber Co., Cleveland, Ohio, 1968, pp. A8–9, G3–8, H14–19, H62-5.

3. In press; available, as are all CBN Rules, from the Office of Biochemical Nomenclature (W. E. Cohn, Director), Biology Division, Oak Ridge National Laboratory, Box Y, Oak Ridge, Tennessee, 37830, USA.

4. "Enzyme Nomenclature," Elsevier Publ. Co., New York, 1965.

Contents

List of Contributors v

Preface vii

Abbreviations and Symbols ix

Contents of Previous Volumes xvii

Some Articles Planned for Future Volumes xxiii

Induced Activation of Amino Acid Activating Enzymes by Amino Acids and tRNA

Alan H. Mehler

I. Introduction 1
II. Reactions of Amino Acid Activating Enzymes 3
III. Consequences of Amino Acid Structure 4
IV. Association of tRNA and Amino Acid Activating Enzymes . . . 11
V. Interactions of Ligands 15
VI. Binding of Nucleotides 18
VII. Speculations 18
References 21

Transfer RNA and Cell Differentiation

Noboru Sueoka and Tamiko Kano-Sueoka

I. Introduction 23
II. Control of Protein Synthesis Mediated by Transfer RNA . . . 24
III. Chemical Basis for Adaptor Alteration 28
IV. Specific Biological Systems 35
V. Conclusion 49
References 51

N^6-(Δ^2-Isopentenyl)adenosine: Chemical Reactions, Biosynthesis, Metabolism, and Significance to the Structure and Function of tRNA

Ross H. Hall

I. Introduction 57
II. Occurrence of N^6-(Δ^2-Isopentenyl)adenosine and Its Derivatives in Nature 60
III. N^6-(Δ^2-Isopentenyl)adenosine and Related Compounds in the Free State 61
IV. Synthesis of N^6-(Δ^2-Isopentenyl)adenosine 62
V. Synthesis of N^6-(Δ^2-Isopentenyl)-2-methylthioadenosine 64

VI. Chemical Properties of N^6-(Δ^2-Isopentenyl)adenosine and Derivatives . 64
VII. Chemical Reactions of N^6-(Δ^2-Isopentenyl)adenosine 64
VIII. Role of N^6-(Δ^2-Isopentenyl)adenosine in the Function of tRNA . . 67
IX. Biosynthesis of N^6-(Δ^2-Isopentenyl)adenosine 70
X. Enzymatic Degradation of N^6-(Δ^2-Isopentenyl)adenosine 75
XI. Biological Activity of N^6-(Δ^2-Isopentenyl)adenosine 76
XII. Discussion of Possible Significance of N^6-(Δ^2-Isopentenyl)adenosine to the Control of tRNA Function 80
Appendix 82
References 84

Nucleotide Biosynthesis from Preformed Purines in Mammalian Cells: Regulatory Mechanisms and Biological Significance

A. W. Murray, Daphne C. Elliott, and M. R. Atkinson

I. Introduction 87
II. Catalytic and Regulatory Properties of Enzymes Involved in Formation and Utilization of Purine Nucleosides and Bases 90
III. Supply of Purine Nucleosides and Bases 99
IV. Chemotherapy and Control of Growth 103
V. General Conclusions 111
References 113

Ribosome Specificity of Protein Synthesis *in Vitro*

Orio Ciferri and Bruno Parisi

I. Introduction 121
II. Role of Ribosomes in the Synthesis of Protein 122
III. Protein Synthesis *in Vitro* by Heterologous Systems 125
IV. Attempts to Characterize the Sites Responsible for Ribosome Specificity 129
V. Ribosome Specificity, Cellular Organelles, and the Regulation of Protein Synthesis 135
VI. Ribosome-Specific Inhibitors of Protein Synthesis 137
VII. Conclusions 139
References 141
Note Added in Proof 144

Synthetic Nucleotide-peptides

Zoe A. Shabarova

I. Introduction 145
II. Nomenclature of Nucleotide-peptides 147
III. Synthesis of Nucleotide-(P-N)-peptide Phosphoramidates and Nucleotide-(P-O)-seryl(or threonyl)-peptide Phosphoesters 148
IV. Properties of Nucleotide-peptide Bonds in Synthetic Models . . . 151

V. Nucleotide-(P-N)-peptide Phosphoramidates as Models of "Active Nucleotides" . . . 170
VI. Conclusion . . . 178
References . . . 179

The Crystal Structures of Purines, Pyrimidines and Their Intermolecular Complexes

DONALD VOET AND ALEXANDER RICH

I. X-Ray Crystallography and the Nucleic Acids . . . 183
II. The Experimental X-Ray Data . . . 186
III. What Are the Molecular Structures? . . . 196
IV. How Do the Purines and the Pyrimidines Interact? . . . 205
V. Summary and Conclusions . . . 243
Appendix Tables . . . 245
References . . . 260

AUTHOR INDEX . . . 267

SUBJECT INDEX . . . 286

Contents of Previous Volumes

Volume 1

"Primer" in DNA Polymerase Reactions
F. J. Bollum

The Biosynthesis of Ribonucleic Acid in Animal Systems
R. M. S. Smellie

The Role of DNA in RNA Synthesis
Jerard Hurwitz and J. T. August

Polynucleotide Phosphorylase
M. Grunberg-Manago

Messenger Ribonucleic Acid
Fritz Lipmann

The Recent Excitement in the Coding Problem
F. H. C. Crick

Some Thoughts on the Double-Stranded Model of Deoxyribonucleic Acid
Aaron Bendich and Herbert S. Rosenkranz

Denaturation and Renaturation of Deoxyribonucleic Acid
J. Marmur, R. Rownd, and C. L. Schildkraut

Some Problems Concerning the Macromolecular Structure of Ribonucleic Acids
A. S. Spirin

The Structure of DNA as Determined by X-Ray Scattering Techniques
Vittorio Luzzati

Molecular Mechanisms of Radiation Effects
A. Wacker

Author Index—Subject Index

Volume 2

Nucleic Acids and Information Transfer
Liebe F. Cavalieri and Barbara H. Rosenberg

Nuclear Ribonucleic Acid
Henry Harris

Plant Virus Nucleic Acids
Roy Markham

The Nucleases of *Escherichia coli*
I. R. Lehman

Specificity of Chemical Mutagenesis
David R. Krieg

Column Chromatography of Oligonucleotides and Polynucleotides
Matthys Staehelin

Mechanism of Action and Application of Azapyrimidines
J. Skoda

The Function of the Pyrimidine Base in the Ribonuclease Reaction
Herbert Witzel

Preparation, Fractionation, and Properties of sRNA
G. L. Brown

Author Index—Subject Index

Volume 3

Isolation and Fractionation of Nucleic Acids
K. S. Kirby

Cellular Sites of RNA Synthesis
David M. Prescott

Ribonucleases in Taka-Diastase: Properties, Chemical Nature, and Applications
Fujio Egami, Kenji Takahashi, and Tsuneko Uchida

Chemical Effects of Ionizing Radiations on Nucleic Acids and Related Compounds
Joseph J. Weiss

The Regulation of RNA Synthesis in Bacteria
Frederick C. Neidhardt

Actinomycin and Nucleic Acid Function
E. Reich and I. H. Goldberg

De Novo Protein Synthesis *in Vitro*
B. Nisman and J. Pelmont

Free Nucleotides in Animal Tissues
P. MANDEL

AUTHOR INDEX—SUBJECT INDEX

Volume 4

Fluorinated Pyrimidines
CHARLES HEIDELBERGER

Genetic Recombination in Bacteriophage
E. VOLKIN

DNA Polymerases from Mammalian Cells
H. M. KEIR

The Evolution of Base Sequences in Polynucleotides
B. J. MCCARTHY

Biosynthesis of Ribosomes in Bacterial Cells
SYOZO OSAWA

5-Hydroxymethylpyrimidines and Their Derivatives
T. L. V. ULBRICHT

Amino Acid Esters of RNA, Nucleosides, and Related Compounds
H. G. ZACHAU AND H. FELDMANN

Uptake of DNA by Living Cells
L. LEDOUX

AUTHOR INDEX—SUBJECT INDEX

Volume 5

Introduction to the Biochemistry of D-Arabinosyl Nucleosides
SEYMOUR S. COHEN

Effects of Some Chemical Mutagens and Carcinogens on Nucleic Acids
P. D. LAWLEY

Nucleic Acids in Chloroplasts and Metabolic DNA
TATSUICHI IWAMURA

Enzymatic Alteration of Macromolecular Structure
P. R. SRINIVASAN AND ERNEST BOREK

Hormones and the Synthesis and Utilization of Ribonucleic Acids
J. R. TATA

Nucleoside Antibiotics
JACK J. FOX, KYOICHI A. WATANABE, AND ALEXANDER BLOCH

Recombination of DNA Molecules
CHARLES A. THOMAS, JR.

Appendix I. Recombination of a Pool of DNA Fragments with Complementary Single-Chain Ends
G. S. WATSON, W. K. SMITH, AND CHARLES A. THOMAS, JR.

Appendix II. Proof That Sequences of A, C, G, and T Can Be Assembled to Produce Chains of Ultimate Length Avoiding Repetitions Everywhere
A. S. FRAENKEL AND J. GILLIS

The Chemistry of Pseudouridine
ROBERT WARNER CHAMBERS

The Biochemistry of Pseudouridine
EUGENE GOLDWASSER AND ROBERT L. HEINRIKSON

AUTHOR INDEX—SUBJECT INDEX

Volume 6

Nucleic Acids and Mutability
STEPHEN ZAMENHOF

Specificity in the Structure of Transfer RNA
KIN-ICHIRO MIURA

Synthetic Polynucleotides
A. M. MICHELSON, J. MASSOULIÉ, AND W. GUSCHLBAUER

The DNA of Chloroplasts, Mitochondria, and Centrioles
S. GRANICK AND AHARON GIBOR

Behavior, Neural Function, and RNA
H. HYDÉN

The Nucleolus and the Synthesis of Ribosomes
ROBERT P. PERRY

The Nature and Biosynthesis of Nuclear Ribonucleic Acids
G. P. GEORGIEV

Replication of Phage RNA
CHARLES WEISSMANN AND SEVERO OCHOA

AUTHOR INDEX—SUBJECT INDEX

Volume 7

Autoradiographic Studies on DNA Replication in Normal and Leukemic Human Chromosomes
FELICE GAVOSTO

Proteins of the Cell Nucleus
LUBOMIR S. HNILICA

The Present Status of the Genetic Code
CARL R. WOESE

The Search for the Messenger RNA of Hemoglobin
H. CHANTRENNE, A. BURNY, AND G. MARBAIX

Ribonucleic Acids and Information Transfer in Animal Cells
A. A. HADJIOLOV

Transfer of Genetic Information During Embryogenesis
MARTIN NEMER

Enzymatic Reduction of Ribonucleotides
AGNE LARSSON AND PETER REICHARD

The Mutagenic Action of Hydroxylamine
J. H. PHILLIPS AND D. M. BROWN

Mammalian Nucleolytic Enzymes and Their Localization
DAVID SHUGAR AND HALINA SIERAKOWSKA

AUTHOR INDEX—SUBJECT INDEX

Volume 8

Nucleic Acids—The First Hundred Years
J. N. DAVIDSON

Nucleic Acids and Protamine in Salmon Testes
GORDON H. DIXON AND MICHAEL SMITH

Experimental Approaches to the Determination of the Nucleotide Sequences of Large Oligonucleotides and Small Nucleic Acids
ROBERT W. HOLLEY

Alterations of DNA Base Composition in Bacteria
G. F. GAUSE

Chemistry of Guanine and Its Biologically Significant Derivatives
ROBERT SHAPIRO

Bacteriophage ϕX174 and Related Viruses
ROBERT L. SINSHEIMER

The Preparation and Characterization of Large Oligonucleotides
GEORGE W. RUSHIZKY AND HERBERT A. SOBER

Purine *N*-Oxides and Cancer
GEORGE BOSWORTH BROWN

The Photochemistry, Photobiology, and Repair of Polynucleotides
R. B. SETLOW

What Really Is DNA? Remarks on the Changing Aspects of a Scientific Concept
ERWIN CHARGAFF

Recent Nucleic Acid Research in China
TIEN-HSI CHENG AND ROY H. DOI

AUTHOR INDEX—SUBJECT INDEX

Volume 9

The Role of Conformation in Chemical Mutagenesis
B. SINGER AND H. FRAENKEL-CONRAT

Polarographic Techniques in Nucleic Acid Research
E. PALEČEK

RNA Polymerase and the Control of RNA Synthesis
JOHN P. RICHARDSON

Radiation-Induced Alterations in the Structure of Deoxyribonucleic Acid and Their Biological Consequences
D. T. KANAZIR

Optical Rotatory Dispersion and Circular Dichroism of Nucleic Acids
JEN TSI YANG AND TATSUYA SAMEJIMA

The Specificity of Molecular Hybridization in Relation to Studies on Higher Organisms
P. M. B. WALKER

Quantum-Mechanical Investigations of the Electronic Structure of Nucleic Acids and Their Constituents
BERNARD PULLMAN AND ALBERTE PULLMAN

The Chemical Modification of Nucleic Acids
N. K. KOCHETKOV AND E. I. BUDOWSKY

AUTHOR INDEX—SUBJECT INDEX

Some Articles Planned for Future Volumes

Transcription and Translation in Mitochondria
W. E. Barnett

Uptake of Nucleic Acids by Mammalian Cells
P. M. Bhargava and G. Shanmugan

The Ribosomal Cistrons of Higher Organisms
M. L. Birnstiel

The Relaxed Control Phenomenon
E. Borek

Nucleic Acid Maturation in Animal Cells
R. H. Burdon

Three-dimensional Structure of tRNA
F. Cramer

Induction of Interferon by Natural and Synthetic Polynucleotides
C. Colby

Molecular Aspects of Genetic Recombination
C. I. Davern

DNA Ligases
J. Hurwitz and A. Becker

Recent Studies Relevant to the Genetic Code
T. H. Jukes

The Principles and Practices of Nucleic Acid Hybridization
D. Kennell

X-Ray Diffraction Studies of Nucleic Acids
R. Langridge

Synthesis and Biochemical Properties of Peptidyl Transfer RNA
Y. Lapidot and N. de Groot

Supercoiled DNA Molecules in Biological Structures
M. F. Maestre

Mechanisms of Polypeptide Synthesis on Ribosomes
H. Matthaei and E. Bermek

Deoxycytidylate Aminohydrolase in DNA Synthesis
E. Scarano

Liporibonucleoprotein Complexes as an Integral Part of Animal Cell Membranes
V. S. Shapot

Initiation of Protein Synthesis
R. E. Thach

Induced Activation of Amino Acid Activating Enzymes by Amino Acids and tRNA[1]

ALAN H. MEHLER

Department of Biochemistry,
Marquette School of Medicine,
Milwaukee, Wisconsin

I. Introduction 1
II. Reactions of Amino Acid Activating Enzymes 3
III. Consequences of Amino Acid Structure 4
A. Reactions with Analogs of Amino Acids 4
B. Competitive Inhibition of Amino Acids 6
C. Reactions with Aminoacyl Adenylates 8
IV. Association of tRNA and Amino Acid Activating Enzymes . . 11
A. Recognition of tRNA 11
B. Effects of tRNA on Enzyme Activity 13
C. Binding of tRNA 13
V. Interactions of Ligands 15
A. Effects of tRNA on Binding of Amino Acid 15
B. Effects of Amino Acids on Binding of tRNA 16
C. Effect of tRNA on Catalysis 17
VI. Binding of Nucleotides 18
VII. Speculations 18
A. Variability and Similarity 18
B. The Basis for Specificity 19
References 21

I. Introduction

A characteristic of protein synthesis is the absolute specificity with which the correct amino acids are incorporated into each position of the polypeptide chain. This requires at least the same degree of specificity in the ability of amino acid activating enzymes to discriminate both among potential amino acid substrates and among the many species of tRNA that exist within each cell. An explanation of the mechanism that permits precise selection of amino acids and their specific carriers may be properly

[1] The current research on which this review has been based was supported by National Institutes of Health grant GM 13037 and Atomic Energy Commission contract AT(11-1)-1498.

considered an important component of the answer to the mystery of life. Unlike the partial answer given by the elucidation of the structure of DNA, the question of enzyme specificity cannot be explained by a structure; instead, a more elaborate description of a process is required.

Enzyme mechanisms in general are poorly understood and detailed explanations are only now beginning to be proposed for some enzymes. The most completely analyzed enzymes are hydrolytic, and the success in correlating chemical mechanisms with the structures in the cases of lysozyme, ribonuclease, chymotrypsin, and a few others has resulted in the popular acceptance of a general model of enzymes. Enzymes that function in an extracellular environment tend to have characteristics different from those of intracellular catalysts. One difference is size; the enzymes best known are from one-half to one-tenth the size of most intracellular enzymes. Another difference is the role of cysteine residues; extracellularly these are usually oxidized to disulfide bridges that are required for enzymatic activity, whereas intracellular proteins contain few, if any, disulfide bonds and often function best in reducing environments that prevent oxidation of sulfhydryl groups. These differences permit the consideration of the variety of enzymes that catalyze the diverse reactions of intermediary metabolism as flexible molecules that can undergo more extensive conformational changes than can the smaller compact, hydrolytic enzymes.

The amino acid activating enzymes are a family of catalysts related to a single biological phenomenon, protein synthesis (*1*) [although it may be noted in passing that some aminoacyl-tRNA's have been found to participate in other biosynthetic systems (*2*, *3*)] and to a single formal reaction (*4*), which may be written:

$$\text{Amino acid} + \text{ATP} + \text{Enzyme} \rightleftharpoons \text{enzyme–aminoacyl adenylate} + \text{PP}_i$$

$$\text{Enzyme-aminoacyladenylate} + \text{tRNA} \rightleftharpoons \text{aminoacyl-tRNA} + \text{AMP} + \text{enzyme}$$

Relatively little has been done to characterize these enzymes further. The information that has been obtained indicates a greater diversity of detailed properties than might have been anticipated. This essay is devoted to a survey of those properties that may give insight into the mechanism of action of individual activating enzymes, which includes the chemical basis for specific recognition. The thesis will be proposed that the combination of an enzyme with a specific substrate results in conformational changes that may result in the formation of a catalytic site, in the binding site for another substrate, or in the approximation of one site to another. Individual differences among enzymes of this group suggest that a unitary mechanism cannot be applied in detail to all members, although certain features may be expected to be found to be common to all amino acid activating enzymes.

II. Reactions of Amino Acid Activating Enzymes

The activity of aminoacyl-tRNA synthetases has been assayed by several methods. The formation of amino acid hydroxamates, which was the first assay used, is still used but was recently shown by Hirsh and Lipmann (*5*) to give rates not proportional to the rates of other reactions catalyzed by the activating enzymes. Loftfield and Eigner (*6*) showed that the kinetics of hydroxamate formation are second order with respect to hydroxylamine and that many other amines also activate in a similar fashion. The activations generally require high concentrations of amines, which inhibit at still higher concentrations. These properties have great potential value in revealing components of the catalytic mechanism, but they discourage the use of hydroxylamine in assays designed to study other aspects of the enzyme.

The two assays generally used are amino acid-dependent exchange of radioactive pyrophosphate into ATP and esterification of a radioactive amino acid to tRNA. The exchange reaction is sometimes considered to measure the activation process, and the esterification (binding, transfer) reaction is considered to be the sum of activation and a subsequent transfer. This interpretation is derived from the evidence with some representative enzymes that aminoacyladenylates are formed as firmly bound complexes with the enzyme (*7–10*) and is strengthened by the observations (Table I) that the exchange rate is never slower than the rate of binding and is

TABLE I
RATES OF REACTION OF AMINO ACYL-tRNA SYNTHETASES

Synthetase	Rate[a] of exchange	Transfer	Ref.
Tyrosine (*E. coli*)	56	2.6	(*11*)
Tyrosine (*B. subtilis*)	36	5.8	(*11*)
Valine (*E. coli*)	42	6	(*13a*)
Valine (*E. coli*)	574	26.4	(*4*)
Lysine (*E. coli*)	21	0.05	(*12*)
Isoleucine (*E. coli*)	460	3.3	(*4*)
Leucine (*E. coli*)	120	2.8	(*4*)
Methionine (Yeast)	22.1	0.007	(*4*)
Methionine (*E. coli*)	244	3	(*4*)
Methionine (*Sarcina lutea*)	3.65	0.037	(*70*)
Glutamate (*E. coli*)	10	5.4	(*57*)
Glutamate (*E. coli*)	11	5.5	(*57*)
Arginine (*E. coli*)	1	1	(*28*)
Proline (*E. coli*)	11.5	6.8	(*35*)

[a] The rates of all reactions were recalculated when necessary to express the values as micromole per minute per milligram protein.

often much faster (*11*). An important difference between the two reactions is seen in many cases when substrate analogs are used; the specificity of the transfer reaction is greater than that of the exchange.

III. Consequences of Amino Acid Structure

A. Reactions with Analogs of Amino Acids

Studies on enzyme specificity require highly purified enzymes for unambiguous results. A preparation of the lysine activating enzyme from *Escherichia coli* did not react with other naturally occurring amino acids and in preliminary physical studies appeared to be homogeneous (*12*). This enzyme reacted with thiosine, the analog of lysine with a sulfur atom in place of the methylene carbon in position 4. The K_m of this analog is greater than that of lysine but the V_{max} is identical in the exchange reaction. In the transfer reaction, in contrast, the maximum rate of esterification of thiosine is significantly less than the rate of lysine esterification. δ-Hydroxylysine, which also has a high K_m, is similarly effective in causing an exchange reaction with the same V_{max}, but it is not esterified to a measurable extent.

The elegant and elaborate studies of Norris (Baldwin) and Berg (*9, 13*) on the isoleucine enzyme of *E. coli* include the one well characterized case of a second naturally occurring amino acid reacting with an activating enzyme. Valine has a high K_m but reacts effectively to promote pyrophosphate exchange and, in the absence of pyrophosphate, to form valyl adenylate bound to the enzyme. However, under no circumstances is valine transferred to tRNA by this enzyme. Analyses of the kinetics of the activation reactions led Loftfield and Eigner (*13a*) to propose a rigid model for the site of binding of the amino acid to the enzyme and to conclude that activity is not induced.

Analogs are not necessarily poorer substrates than the amino acids that are believed to be the natural substrates, those that are found in proteins. In the case of canavanine, with an oxygen in place of carbon-5 of arginine, and a highly purified arginine-activating enzyme from *E. coli* (*14*), the rates of both exchange and transfer are about twice the corresponding rates with arginine, although the K_m for arginine is 100 times smaller.

A somewhat different situation is found with the proline enzyme of *E. coli* (*15*). A large element of the specificity of this enzyme is conferred by the requirement for a secondary amine, which excludes all the usual amino acids except hydroxyproline. The trans hydroxyl group at position 4 compeltely eliminates binding of this amino acid to the enzyme; it not only is not a substrate, it is not a competitive inhibitor at concentrations as high as 0.1 M. The isomer, allohydroxyproline, which does not occur naturally, is a substrate. Other substrates include sarcosine, N-methylalanine, N-ethylglycine, and N-propylglycine. Each of these amino acids has a K_m near 10^{-2} M and gives a V_{max} very similar to that of proline. For technical reasons careful kinetics have not been carried out with the transfer reaction, but it has been found that each of the analogs that is activated is transferred to tRNA, at rates comparable to that of proline esterification.

It is clear that no simple generalization can explain the nature of reaction of the activating enzymes with various analogs of the amino acids that are the natural substrates. It is possible, however, to offer a qualitative explanation for the kinetic differences seen with various amino acids. The exchange of pyrophosphate is evidence for the activation of the carboxyl group, which may be considered to exist as part of an anhydride in aminoacyl adenylate. The anhydride formed with an analog is reasonably viewed as being formed and bound to the same part of the enzyme as the natural substrate. The tRNA that is the acceptor of the amino acid is the same molecule for all reactions of a given enzyme. Nevertheless, the rates of transfer from the activated complex to tRNA vary in many cases, and in some cases analogs are not transferred. Since the only structural difference is the side chain of the amino acid, this structure must determine the rate of transfer. This is independent of any role it plays in binding the amino acid to the enzyme in the case of substrates that give equal rates of pyrophosphate exchange. It is conceivable that the side chain of the amino acid participates itself in the catalytic mechanism. A more satisfying explanation is that the side chain induces a conformational change in the enzyme. This is a mechanism that can be applied in general to all amino acids, since all the side chains must be bound to the specific recognition sites of the family of enzymes. It is much more difficult to invent mechanisms for direct involvement of many of the side chains in the catalysis of transfer.

Although a detailed description of conformational changes cannot be made at this time, it may be concluded that extremely small differences in side-chain structure can cause changes in the positioning of groups in the catalytic site. The resulting structures are responsible for catalysis; if the necessary functional groups are at appropriate positions, catalysis is efficient, but if the distances are modified (as a result of an alteration in the side chain of the substrate), the rate of transfer is decreased. In extreme cases, there may be no transfer.

Proline activation appears to involve an exception to the generalization derived from kinetic studies with other activating enzymes. The proline enzyme behaves as though it contained a rigid binding site for the amino acid substrate, one that cannot accommodate a number of analogs for steric reasons (e.g., hydroxyproline) but that can accept secondary amino acids whose total bulk is not greater than that of proline. If the side chains were important in inducing an active site through a conformational change, it would not be expected that similar conformations would be induced by proline and open-chain compounds. The similarity of the rates of activation of these analogs and the transfer of all of the activated amino acids to tRNA lead to the conclusion that the side chains are unimportant in the formation of an active catalyst and in influencing the rates of the catalytic reaction.

B. Competitive Inhibition of Amino Acids

Since amino acid activating enzymes react specifically with three quite different substrates (amino acid, ATP, and tRNA), there are opportunities for at least three types of competitive inhibitors, whose structures can give insight into the nature of binding of the corresponding substrate. Relatively little is known about competition for the RNA site; apparently the structures of tRNA's vary sufficiently that there is little, if any, competition by nonaccepting species for a given activating enzyme of a homologous species. The observations of Loftfield and associates (*16, 17*) on heterologous systems suggest that these might be useful in determining the basis for selective binding of tRNA's, but at this time there is insufficient physical or chemical characterization of tRNA's for such an analysis.

Not much has been learned about the requirements for ATP binding. Substitution of any of the common bases for adenine effectively eliminates binding in those cases studied (*18*). dATP has a higher K_m than ATP in those cases examined. Kinetic studies of transfer reactions with dATP show the same V_{max} as with ATP but different rates of pyrophosphate exchange were observed. The exchange kinetics have not been further analyzed. The analog ADP(CH_2)P (*19*), which closely resembles ATP, is a very powerful competitive inhibitor of ATP with the proline enzyme (*15*). Very recently this analog has been found to support the transfer reaction of the lysine enzyme of *E. coli*, although it failed to react in the arginine system. Detailed analyses of the reactions of ADP(CH_2)P and its presumed product, P(CH_2)P, may be of great value in elucidating the nature of the interaction of nucleotides and activating enzymes.

Although many more opportunities are available for studying amino acid analogs, only a few cases have been explored. Owens and Bell (*20*) used several analogs of valine as competitive inhibitors of the valine-dependent pyrophosphate exchange to evaluate the roles of various parts of the amino acid in binding to the valine enzyme of *E. coli*. They found that the carboxyl group contributes little or nothing to the binding since isobutylamine has a K_i equivalent to the K_m of L-valine. Both α-aminobutyric acid and alanine are bound to the enzyme, but the corresponding amines are bound more tightly. Although D-amino acids are not substrates for this enzyme, they are inhibitors. D-Valine has a K_i about 50 times higher than the K_m of L-valine but D-α-aminobutyric acid and D-alanine are bound at least as well as the L-isomers. These findings were interpreted as indicating a steric effect of the carboxyl group in valine that is minimized by a decrease in the size of the aliphatic side chain.

The carboxyl group of the amino acid must be free to accept the phosphate of adenylic acid in the process of activation. The indication above, that the carboxyl group is not involved in binding the amino acid to the enzyme, is strengthened by observations in several other systems. Calendar and Berg (*11*) found that tyrosinamide and tyrosinol both have K_i values very similar to the K_m of tyrosine with the enzyme from *E. coli*. Rouget and Chapeville (*30*) in similar studies with the leucine enzyme of *E. coli* found that the K_i of leucinol is close to the K_m of leucine. Proline amide has a K_i about twice the K_m of proline (*15*). These examples indicate that in general no strong interaction (such as a salt linkage) exists between the carboxyl group to be activated and any group of the enzyme. The function of the carboxyl group cannot be simply passive, however, since the corresponding alcohols, which are more easily acylated and give more stable products, do not react. The environment into which the carboxyl group is inserted can accommodate groups that occupy more space than the carboxyl group, but such compounds as the methyl and ethyl esters have larger K_i values (*11*); the decreased energy of binding may be a measure of the energy of the conformational change in the protein needed to provide room for the bulkier groups.

Contrary to the report of Atherly and Bell (*21*) that pyrrolidine (decarboxylated proline) does not inhibit the proline activating enzyme, Papas and Mehler (*15*) found competitive inhibition by this compound, tetrahydrofuran, and tetrahydrothiophene. Several other analogs, including Δ-3-pyrroline and tetrahydrofurfuryl alcohol, are also inhibitors with K_i values 10 to 100 times larger than the K_m of proline. Analogs with larger rings, L-pipecolic acid and 4,5-dehydropipecolic acid, and with carbon in place of nitrogen, cyclopentane carboxylic acid, are not inhibitors of this enzyme.

In various cases, D-amino acids are recognized quite differently. The proline enzyme is completely indifferent to D-proline. The arginine enzyme of *E. coli* also does not bind D-arginine significantly (*22*). In contrast, the tyrosine enzyme of *E. coli* activates D-tyrosine and transfers it to tRNA (*23*).

The binding of analogs of amino acids studied with several enzymes indicates that the α-amino group and the entire side chain of the amino acid participate in binding the amino acid to the enzyme. If the carboxyl group is not bound, both D- and L-isomers can occupy the same site unless a portion of the protein prevents the attachment of the unnatural form by imposing a steric restriction on the space needed to accommodate the carboxyl group. Among the few enzymes studied, qualitative differences are seen in this respect, ranging from binding and activation of D-tyrosine through the binding of D-amino acids as inhibitors of the valine enzyme to failure of the D-amino acids to be bound, as in the cases of proline and arginine. Even in these latter cases the corresponding amines, pyrrolidine and agmatine, are bound tightly. It is not difficult to visualize a steric reason for the failure of amino acids to be bound when their side chains are bulkier than those of the natural substrate. It is less obvious to see how the nitrogen of the α-amino group fulfills an essential role. Rings with sulfur or oxygen in place of nitrogen are bound to the proline enzyme as competitive inhibitors, and the hetero atoms (because of unpaired electrons?) are essential for this property, since the carbon analog is not bound. A question for future investigation is to determine why analogs such as α-hydroxy acids are not substrates for amino acid activating enzymes.

C. Reactions with Aminoacyl Adenylates

Part of the evidence for the two-step reaction mechanism generally written to describe the action of amino acid activating enzymes is the reaction with the presumed intermediate, aminoacyl adenylate. A partially purified leucine enzyme was found by DeMoss *et al.* to catalyze the pyrophosphorolysis of synthetic L-leucyl adenylate (*24*). Surprisingly, slow reactions were also observed with the derivatives of D-leucine and alanine, although these amino acids did not stimulate an ATP-pyrophosphate exchange. Berg showed that the methionine enzyme of yeast also is less specific in its reactions with aminoacyl adenylates than with free amino acids (*25*). The tryptophan-activating enzyme of animal pancreas (*26*) was also shown to pyrophosphorolyze the adenylates of D-tryptophan and other amino acids that are inert in the forward reaction (*28*). The relative lack of specificity of the tryptophan enzyme may be a reflection of a mechanism more similar to the proline enzyme of *E. coli* than to those enzymes that seem to require precise conformational changes for catalytic activity. Sharon and Lipmann (*27*) found that a series of tryptophan analogs react with this enzyme, as substrates in several cases and as inhibitors when methyl or other "bulky" groups were substituted for hydrogens. In all cases in which activation was seen (hydroxyamate formation), there was evidence for transfer to tRNA. The requirements in the partial back reaction are not simply for an acid anhydride of AMP. In the few cases studied, differences in K_m were noted; the methionine enzyme exhibited a lower K_m for methionyl adenylate than for seryl adenylate (*25*).

Not all activating enzymes react nonspecifically with aminoacyl adenylates. The arginine, lysine, and glutamine enzymes of *E. coli* were found to react with their corresponding aminoacyl adenylates but not to cross-react with the substrates for the other enzymes (*28*).

The difference in K_m values noted for methionyl and seryl adenylates with the yeast methionine enzyme demonstrates that in this case the amino acid side chain influences the binding of the intermediate to the enzyme. It is probable that those enzymes that do not react nonspecifically do not bind appreciably the adenylates of amino acids that are not substrates.

Several aminoacyl adenylate–enzyme complexes have been isolated (*8, 9, 10, 29*). In general, these react with pyrophosphate or tRNA to give the substrates or products of the normal reaction. In some cases the intermediate complex reacts with hydroxylamine to yield the amino hydroxamic acid. In the case of the threonine enzyme of *E. coli*, Hirsh (*31*) found that hydroxylamine caused only hydrolysis of the complex except when a mercurial was added. In this case, the transfer to tRNA was inhibited, but the reaction with hydroxylamine resulted in the formation of threonine hydroxamate. It is necessary to conclude that the effect of the mercurial was not simply to modify a binding group or to inhibit tRNA binding sterically. The granting of access to hydroxylamine must be the result of a conformational change in the protein, which might also be the cause of the failure to transfer. It will be of interest to learn whether the inhibition involves tRNA binding or catalysis.

The variability shown by the activating enzymes with regard to specificity makes it difficult to fit a simple model to the different cases. The nature of the reaction, however, permits some rationalization. The question may be phrased, "How can an enzyme catalyze a reversible reaction in only one direction?" From prior considerations of reactions of amino acid analogs, it was concluded that the side chain of the amino acid plays a critical role in the formation of a catalytic mechanism. If the appropriate side chain is not present in the aminoacyl adenylate, clearly the same catalytic site cannot be formed to catalyze the reverse reaction. It is wrong, however, to consider that there must be a rigid structure that is the "catalyst." Instead, it must be considered that changes in the protein occur as a result of the partial enzymatic reaction. Therefore, a conformation that must be induced to catalyze a reaction between an amino acid and ATP may not be the conformation required to catalyze a reaction between an aminoacyl adenylate and pyrophosphate. The conformation induced by any aminoacyl adenylate, in the cases of low specificity, may permit a facile reaction with pyrophosphate and the very reactive anhydride, although a specific side chain may still be required to induce the further conformational change required to permit transfer to tRNA.

It might appear that a thermodynamic dilemma is presented by the situation in which one aminoacyl adenylate is formed and another is not, yet both are degraded by a given enzyme. Two considerations resolve this problem. One, the fact that aminoacyl adenylates do not dissociate from activating enzymes establishes a difference in the two systems that are being compared; the pyrophosphorolysis of exogenous aminoacyl adenylates is not the reverse of the reactions in which enzyme–aminoacyl adenylate complexes are formed. Two, the Briggs-Haldane relationship of enzyme rates to chemical equilibrium contains a K_m factor. Since K_m can be a complex constant, affected by the rates of partial reactions and since the binding of one substrate can affect the binding of another, this factor can compensate for any discrepancy in the rates of reaction in the two directions.

A mechanism for studying the binding of aminoacyl adenylates has been made available by the synthesis by Sandrin and Boissonnas (*32*) and by Cassio *et al.* (*33*) of analogous structures, the esters of the corresponding amino alcohols with AMP. These compounds are inert as substrates and inhibit the enzymes that activate the corresponding amino acids. In a few cases, valinyl[2] adenylate with the isoleucine enzyme and isoleucinyl[2] and alanyl adenylates with the valine enzyme, there was weak inhibition with high concentrations (10^{-4} *M*) of the aminoalkyl adenylates. The inhibitors are competitive with amino acid. It is suggested by Cassio *et al.* that the relatively great specificity of the inhibitors compared with the lesser specificity of certain enzymes in the pyrophosphorolysis of aminoacyl adenylates is explained by the presence of free amino acids. However, the fact that the more specific enzymes were from *E. coli* suggests that the nonspecific interactions were only those that correspond to nonspecific activation of amino acids and that binding of aminoacyl adenylates and aminoalkyl adenylates requires specific interaction of the side chain of the substrate or inhibitor with the enzyme.

The aminoalkyl adenylate corresponding to methionine protects a fraction of the sulfhydryl groups in the methionine enzyme from *p*-hydroxymercuribenzoate (*34*). The mercurial causes complete loss of activity in the absence of the adenylate but eliminates only the transfer reaction in the presence of methioninyl adenylate. [The dissociation of the second step has been noted also in the case of the reversible cold inactivation of the proline enzyme, which does not involve the activation step (*35*).] It has not been determined whether the protection requires the adenylate or whether methionine alone can protect the enzyme against *p*HMB. Many other questions of protein conformation may be studied profitably with the analogs of the putative intermediate.

[2] The amino alcohols corresponding to amino acids have been given trivial names such as histidinol, tyrosinol, etc., to show this relationship. Valinyl adenylate in this nomenclature is the ester of adenylic acid and the amino alcohol corresponding to valine.

The ability of aminoacyl adenylates to protect some sulfhydryl groups in activating enzymes has been demonstrated with the tryptophan enzyme of beef pancreas (*36*) and the valine enzyme of *E. coli* (*37*). Recently Kuo and DeLuca (*38*) have described an effect of *p*-mercuribenzoate on the isoleucyl-tRNA synthetase of *E. coli* that is very similar to that seen by Cassio in the methionine system. In these experiments, the presence of isoleucine and ATP limited the reaction with the mercurial so that the transfer reaction was lost while the activation was not affected. The roles of the adenylic acid and amino acid components of this system also remain to be elucidated. Iaccarino and Berg (*38a*) have shown that one sulfhydryl group in the isoleucine enzyme reacts with *N*-ethyl maleimide much more rapidly than other sulfhydryl groups in this protein and that it is the reactive group that is protected by the presence of isoleucine and ATP. The reaction with the maleimide decreases the rate of pyrophosphate exchange to less than 1% of the rate catalyzed by the unmodified enzyme but does not alter the rate or extent of binding of tRNA.

The presence of a sulfhydryl compound, mercaptoethanol, was found to dissociate the activities of the tryptophan activating enzyme of rat liver. Parin and Kisselev (*38b*) have described a reversible alteration of this enzyme in which a prior incubation with mercaptoethanol eliminates the transfer reaction without affecting the formation of tryptophan hydroxamate.

IV. Association of tRNA and Amino Acid Activating Enzymes

A. Recognition of tRNA

Since the original experiments of Preiss *et al.* (*39*) on the independent acceptance of amino acids by tRNA, it has been known that each activating enzyme recognizes its own specific nucleic acid among a family of similar 4 S molecules. The finding of Hecht *et al.* (*40*) that the entire family have identical 3′-terminal compositions, the sequence -C-C-A, restricted the specificity, and the analyses of Singer and Cantoni (*41*), which show that most of the 5′-termini are pGp, further restricted the structures to be considered as recognition sites. The complete primary structure of a tRNA elucidated by Holley and associates (*42*) has had many important consequences, including the discovery of several rare bases, the determination of the primary structures of almost 20 (at this writing) other species of tRNA, and the possibility of a common secondary structure based on the "cloverleaf" pattern of base-pairing. Cramer and associates (*43*) have described a further folding of the cloverleaf into a compact three-dimensional structure, supported by model-building and restricted chemical reactivity of most of the bases. These descriptions of the structures of tRNA have until now given little insight into the chemical basis of recognition by specific enzymes.

It cannot be assumed that the various activating enzymes recognize corresponding parts of their tRNA substrates. On the other hand, no evidence at this time indicates that a specific portion of a tRNA molecule is functional in one case but not in another. It is, therefore, possible that a collage composed of fragments derived from studies on different enzymes and tRNA's may describe the properties of all such systems.

Many investigations have shown that alteration or removal of the terminal adenosine eliminates the ability of tRNA to accept amino acids. In one study, however, Daniel and Littauer (*44*) prepared rat liver tRNA with only one cytidylic acid residue adjacent to the adenylic acid (R-C-A instead of R-C-C-A) and have shown that the modified tRNA retained the ability to accept several amino acids, although the rate of transfer was less than when unmodified tRNA was used.

The one structure known to be characteristic of each tRNA is the anticodon, located in each known tRNA structure in the loop opposite the stem of the cloverleaf. This has been considered to be a possible site of recognition, although only two of the three bases in the anticodon are identical in the iso-accepting species that can be esterified by individual activating enzymes. That this portion of the tRNA is not important for effective reaction with enzyme has been demonstrated most completely for the yeast $tRNA^{Val}$. Bayev and associates (*45*) used ribonuclease T_1 to hydrolyze specifically the phosphodiester bond linking the anticodon region to the 3′ half of the unfractionated yeast tRNA. The resulting halves, when recombined, accepted valine, tyrosine, and serine (the only amino acids tested). Removal of the newly exposed nucleotides by stepwise elimination does not prevent association of the halves or acceptance of amino acid by the altered tRNA. Similarly, the unidentified base, originally termed "Y," was eliminated from $tRNA^{Phe}$ by Zachau and associates (*46*), who then split the nucleic acid at the point of elimination, adjacent to the anticodon, and found the modified tRNA to accept phenylalanine.

Another type of evidence was used by Sundharadas *et al.* (*46a*) to show that the anticodon does not participate in recognition or binding of tRNA to the activating enzyme for serine from *E. coli*. They obtained three fractions that responded to the codons UCU or UCC, AGU or AGC, and UCA or UCG, respectively, and showed the same rate of reaction with the enzyme.

An alteration of tRNA that does cause loss of activity was described by Schulman and Chambers (*47*). They correlated the loss in activity with modification by ultraviolet irradiation of uracil in the stem of the cloverleaf. Their generalization, that the stem is a site of recognition, has not been widely accepted, since several pairs of nucleic acids that react with the same enzyme have been found to differ in the structure of their stems. Indeed, a comparison of the structures of $tRNA^{Met}$ and $tRNA^{fMet}$ (*48*), which are charged with methionine by a single enzyme, and of $tRNA^{Val}$ from yeast and *E. coli* (*49*), both of which accept valine from the enzyme from both organisms, fails to reveal conspicuous similarities that could provide a basis for specific recognition.

B. Effects of tRNA on Enzyme Activity

The activation of amino acids by most of the synthetases is demonstrated in the absence of tRNA. Hele (*50*) demonstrated some stimulation by tRNA of the pyrophosphate exchange reaction by enzymes of rat liver. Complete dependence on the specific acceptor tRNA was shown by Ravel and associates (*51*) for the activation of glutamine. They also showed a similar requirement for tRNA for the activation of glutamate at usual concentrations, although very high (0.1 *M*) concentrations of glutamate were found to support an exchange of pyrophosphate in the absence of tRNA. Further studies (*52, 53*) showed that these requirements for tRNA are characteristic of the enzymes of yeast and liver as well as of the *E. coli* enzyme used originally.

The arginine activating enzyme of *E. coli* also has an absolute requirement for tRNA in the pyrophosphate exchange reaction (*54*). This requirement appears to be an activation without ester formation. The failure to form arginyl-tRNA in the presence of pyrophosphate was demonstrated by an isotope-trapping experiment. Subsequently a tRNA requirement for a reaction with hydroxylamine was shown. Finally, tRNA was needed to permit pyrophosphorolysis of synthetic arginyl adenylate (*29*).

The mechanism by which tRNA serves as an activator could be by increasing the binding of the amino acid or ATP or by modifying the catalytic site. In the case of the glutamate enzyme, it is not necessary to propose a function other than a decrease in the K_m for glutamate, since the rate of pyrophosphate exchange in the presence of very high concentrations of glutamate (>0.1 *M*) is not increased by the presence of tRNA. Evidence presented below shows that tRNA has only a modest effect on the binding of arginine. Therefore, it is apparent that more than one mechanism may be involved in activation by tRNA.

C. Binding of tRNA

Interaction of tRNA and activating enzymes in the absence of a catalytic reaction has been demonstrated in several ways. Yarus and Berg (*55*) developed a method in which the enzyme is adsorbed to nitrocellulose filters and showed that $tRNA^{Ile}$ bound to the isoleucine enzyme was also retained by the filter. Lagerkvist *et al.* (*56*) used gel filtration to demonstrate binding of yeast $tRNA^{Val}$ to the corresponding yeast enzyme. Ohta *et al.* (*57*) showed a change in the circular dichroism of the tyrosine enzyme of yeast when its tRNA was added. Yaniv and Gros (*58*) employed sucrose gradient centrifugation to measure the binding of *E. coli* $tRNA^{Val}$ to its synthetase. James *et al.* (*59*) demonstrated several cases of protection of yeast activating enzymes from thermal denaturation by tRNA.

All the methods listed above have been used to examine the role of the 3′-terminus in binding. Yarus and Berg found essentially no change in affinity for tRNA when the terminal adenosine was oxidized with periodate or when the terminal adenylic and cytidylic acid residues were removed with snake venom phosphodiesterase. James *et al.* and Yaniv and Gros also found no difference between intact and periodate-oxidized tRNA. In contrast, the reports of Lagerkvist *et al.* indicated that oxidized tRNA was not bound, and Ohta *et al.* did not observe an effect on circular dichroism when oxidized tRNA was used.

An examination of the *E. coli* arginine system by the measurement of protection of the enzyme against thermal denaturation has shown independent roles for elements of tRNA in binding and in stabilizing the enzyme (*22*). This enzyme is stabilized by $tRNA^{Arg}$ and does not bind other species of tRNA. When strength of binding was measured by the concentration of RNA required to give a given degree of protection (compared with the maximum protection obtained with that RNA under given conditions of temperature, pH, and time), no change was observed after periodate oxidation or removal of three terminal nucleotides. The degree of protection also was not affected by periodate treatment but venom diesterase treatment caused a reduction in the protection given to the enzyme. Stepwise degradation of the tRNA showed that the change occurred on removal of the penultimate phosphate residue. The protective ability of the RNA was restored when cytidylic acid was replaced by the -C-C-A enzyme. Replacement of the terminal adenylic acid caused no further change.

A reverse approach was used by Yaniv and Gros (*58*) to describe the binding of $tRNA^{Val}$ to the valine activating enzyme. They subjected the enzyme·tRNA complex to digestion by ribonuclease T_1 and observed protection of part of the tRNA by the enzyme. Fragments of the digested tRNA sedimented with the enzyme in the ultracentrifuge. Preliminary analyses suggest that the anticodon region is not protected while the dihydrouracil loop and the 5′-terminus are bound to the enzyme. The formation of a 1:1 complex between the valyl-tRNA synthetase of yeast and two species of $tRNA^{Val}$ from yeast was described by Lagerkvist and Rymo (*60*). They also showed that the tRNA in the complex was completely resistant to digestion by ribonuclease T_1 and partially resistant to snake venom phosphodiesterase.

Oxidation of the terminal adenosine could alter the interaction of tRNA with an enzyme in several ways. One could involve the positioning of the adenine after the sugar is cleaved. Another could result from aldehyde groups replacing secondary alcohols. The experiments of Cramer *et al.* (*61*) suggest that, in the case of yeast phenylalanine tRNA, it is the loss of alcohol groups that is important. They reduced the dialdehyde with borohydride and recovered a tRNA that served as a substrate, accepting phenylalanine from a purified activating enzyme. This is not an universal finding; the dialcohol formed by reducing the dialdehyde of $tRNA^{Arg}$ is completely inactive in catalytic reaction (*62*).

The sum of the studies to date has given only negative results; certain components of tRNA have been eliminated as participants in binding to the enzyme. Identification of the groups that do interact with the protein is of greatest importance to understanding the fundamental mechanisms of enzyme-nucleic acid recognition. A complete answer to this problem requires determination of the complementary groups in both protein and nucleic acid. Preliminary or partial evidence must be interpreted cautiously to discriminate between the possibilities of direct participation of a given group and indirect effects, such as conformational changes.

A few examples have been described of modifications in tRNA that affect catalysis but not binding. As was discussed about amino acid side-chain functions, it is possible that the groups in tRNA that are modified participate directly in catalysis, but the more likely possibility, that these groups induce conformational changes in the protein and enable a catalytically active structure to be formed, must be examined in each case.

V. Interactions of Ligands

A. Effects of tRNA on Binding of Amino Acid

The finding of Ravel *et al.* (*51*) that the glutamate and glutamine enzymes require tRNA for the activation reaction could be explained either by an effect on the catalytic mechanism or by an effect on the binding of the other substrates. In the case of the glutamate enzyme, high concentrations of the amino acid eliminate the need for tRNA, and it may be concluded that the activator does increase the affinity of the enzyme for glutamic acid and that it does not influence the rate of the activation once the amino acid is bound. Since the glutamine and arginine enzymes remain inactive in the presence of very high concentrations of their respective amino acids, the same function for tRNA may not apply. It is necessary to use an alternative technique to measure binding of the amino acids to determine whether it is modified by tRNA.

Amino acids as well as tRNA have been demonstrated to protect activating enzymes from thermal denaturation. Chuang, Atherly, and Bell (*63*) have measured the binding of several substrates (the natural amino acids and some analogs) to the proline and valine enzymes of *E. coli* and found binding constants somewhat smaller than K_m values. Mitra *et al.* (*22*) found that arginine confers greater stability than tRNA on the activating enzyme. The greater resistance to heat in the presence of amino acid made it possible to use this technique to investigate an effect of tRNA on the binding of arginine.

In the absence of tRNA, a curve of protection (enzyme activity remaining after heating) against arginine concentration has a conventional hyperbolic shape and shows a requirement for high concentrations of the amino acid. In the presence of $tRNA^{Arg}$ the concentrations of arginine required to stabilize the enzyme to a given extent is reduced. The effect of $tRNA^{Arg}$ is called synergistic protection. Other tRNA species have no effect on this system.

It may be considered that the same conformational change induced by tRNA that results in stabilization of the enzyme also increases the affinity for arginine. If this were so, the terminal adenosine would play no role in synergistic protection. However, modification of this group by periodate oxidation completely eliminates synergistic protection. Although the adenylic acid residue does not contribute significantly to the strength of binding of the tRNA and does not affect the stability of the enzyme, it is essential for the effect of tRNA on the binding of arginine.

The synergistic protection of the arginine enzyme requires an intact -C-C-A terminus of the tRNA but it does not require all the tRNA structure (*62*). Hydrolysis by small amounts of ribonuclease T_1 at low temperature results in only a few breaks in $tRNA^{Arg}$. This treatment results in rapid and complete loss of all catalytic activity in both the pyrophosphate exchange and the esterification reactions. The catalytically inactive tRNA digest retains a large fraction of the ability to protect the arginine enzyme against thermal denaturation and also to exhibit synergistic protection. A fragment that appears to be the 3′-half of $tRNA^{Arg}$ has been separated from the T_1 digest; this fragment acts synergistically with arginine.

B. Effects of Amino Acids on Binding of tRNA

Degradation of the 3′ end of tRNA has been seen to produce variable effects on the interactions with enzymes. Some of these interactions have also been studied with aminoacyl-tRNA, the product of a reversible reaction.

Direct binding studies by Yarus and Berg (*55*) and by Yaniv and Gros (*58*) show very little effect of the amino acid ester on the binding of tRNA by the isoleucine and valine enzymes, respectively. Similarly, the protection afforded the arginine enzyme by tRNA is not modified by aminoacylation (*22*). It might have been imagined that the amino acid ester would bind to both the amino acid and tRNA binding sites simultaneously and produce the stabilization characteristic of synergistic protection. Instead, no more protection was given than by tRNA alone. The presence of the aminoacyl group is not without effect; it interferes with free amino acid binding to the enzyme so that synergistic protection is not seen.

The nitrocellulose adsorption technique has been ingeniously exploited by Yarus and Berg (*64*) to measure not only strength of binding, but also the rate of dissociation of tRNA from its complex with the isoleucine enzyme. In a simple system the dissociation is surprisingly slow, requiring minutes. The addition of amino acid causes a great increase in the rate of dissociation.

C. Effect of tRNA on Catalysis

The activation of the arginine enzyme by tRNA cannot be explained by the effect on arginine binding, since arginine is not activated in the absence of tRNA even when present at concentrations that fully protect the enzyme against denaturation. It must be concluded that the tRNA serves an additional function. A possibility that has not been tested is the binding of ATP. Another possibility is attractive as the explanation of the observations of Baldwin and Berg on the reactions of the isoleucine enzyme (*13*).

The isoleucine enzyme was isolated as an intermediary complex with either isoleucyl adenylate or valyl adenylate. The transfer of isoleucine to its acceptor tRNA occurred on addition of the tRNA. Valine was not transferred, either to $tRNA^{Ile}$ or to $tRNA^{Val}$. The addition of the substrate $tRNA^{Ile}$ resulted, however, in the hydrolysis of the enzyme-valyladenylate complex to free valine and AMP. This experiment definitively shows an effect of tRNA on the catalytic process, which can be explained either by direct participation of a group within the tRNA or by a conformational change in the protein as a consequence of tRNA binding.

A similar phenomenon was studied by Bluestein *et al.* (*65*) with the serine activating enzyme of yeast. In this case the enzyme–seryl adenylate complex transferred only a portion of the serine to $tRNA^{Ser}$, but all of the complex was broken down when the tRNA was added. When $tRNA^{Ser}$ lacking the 3′-terminal adenosine was used in place of intact tRNA, no transfer occurred but the breakdown of the complex was complete.

VI. Binding of Nucleotides

Complexes of the leucine activating enzyme of *E. coli* with various substrates were examined by Rouget and Chapeville (*66*) by means of gel filtration. This method did not indicate any binding of leucine in the absence of ATP, but ATP formed a demonstrable complex in the absence of amino acid. Makman and Cantoni (*67*) found that ATP contributed to the stability of the yeast serine enzyme at elevated temperatures in the presence of Mg^{2+}. In contrast, the arginine enzyme is not stabilized by ATP alone or in the presence of arginine or tRNA. It should be noted that the binding of ATP is highly dependent upon the adenine component. None of the common bases can replace adenine in the activation of amino acids. The ribose also plays a role in binding the nucleotide to the enzyme. In a random series of enzymes, all were found to react with dATP at rates similar to those with ATP, but in each case the K_m for the deoxynucleotide was much higher than the K_m for ATP (*18, 53*).

Differential effects of ATP and amino acid on the conformation of a purified tryptophan activating enzyme from beef pancreas were observed by Lemaire, Dorizzi, and Labouesse (*67a*). They found that tryptophan alone protected the enzyme against thermal denaturation whereas ATP had very little ability to protect and did not increase the protection by tryptophan. Partial protection against *p*-hydroxymercuribenzoate is also given by tryptophan alone and not by ATP, but in this system the two substrates together give complete protection, probably through formation of the enzyme-aminoacyladenylate complex. When proteolysis by trypsin was used to probe complex formation, ATP alone gave protection, as did tryptophan, and the two together were more effective than either alone.

VII. Speculations

A. Variability and Similarity

There is reason to believe that every living cell contains specific enzymes for activating the twenty amino acids that compose most proteins. In several cases, distinguishable enzymes are found in mitochondria and the soluble fraction of the cell. The homologous enzymes from each species studied are also distinguishable from each other. Of the myriads of these enzymes, only a few individual representatives have been studied in any detail, but these few demonstrate extensive differences in most properties. The preceding sections describe examples of enzyme interaction with tRNA, amino acids, and ATP. Each type of substrate is capable of inducing changes in certain enzymes that result in altered physical properties and in some cases altered catalytic properties. However, none of these effects is general. The diversity of responses to tRNA was described by Kukhanova, Favorova, and Kiselev (*68*), who studied the effects of photooxidation and thermal denaturation on rat liver synthetases in the presence of ATP, amino acids, or tRNA. tRNA protects the valine enzyme against thermal denaturation but labilizes the lysine enzyme and has no effect on the glycine enzyme. A different pattern was seen when photooxidation was studied: tRNA protected the glycine enzyme, had no effect on the phenylalanine enzyme, and increased the sensitivity of the enzymes that activated leucine, valine, and methionine. The effects of ATP and amino acid were similarly variable.

The individuality of the amino acid activating enzymes is reason for extreme caution in proposing general mechanisms to explain any property shared by the group. It may be that some structures will be found in all enzymes of this type. One such structure including a histidine residue, was suggested by the finding of similar curves of K_m of tRNA *vs.* pH with several enzymes (*22*). Another similarity is the specificity for ATP and dATP (*18*). On the other hand, caution in generalizing is invoked by the absence of a sensitive sulfhydryl group in the lysine enzyme of *E. coli*, in contrast to all of the other activating enzymes of this organism (*69*). Resistance to *p*-chloromercuribenzoate was also found in studies of the methionine activating enzyme of *Sarcina lutea* (*70*).

B. The Basis for Specificity

With the reservation above, the generalization may be made that the interaction of an amino acid activating enzyme with each of its substrates results in conformational changes, some of which result in the formation of a catalytic site. This induced fit is proposed as a general mechanism for the extreme specificity of these enzymes. Loftfield and Eigner analyzed the forces that might bind alkyl side chains in the cases of the isoleucine and valine enzymes and concluded that the variable K_m's and relatively constant V_{max}'s could best be explained by rigid binding sites (*13a*). The extreme specificity of the transfer reaction in these cases suggests that the binding site for amino acid is not rigid, but that it responds to the structure of the amino acid to facilitate or prevent transfer. In the experiments of Baldwin and Berg, the tRNA combines with an enzyme that already bears an activated amino acid (*13*). However, this complex does not react spontaneously with water or other reagents except powerful nucleophiles, such as hydroxylamine. The presence of $tRNA^{Ile}$ induces the catalytic structure for transfer, and isoleucine is transferred. However, when the side chain is deficient in one methylene group, valine transfer is abortive. A model that may be considered is that the isoleucine structure is required to bring the activated carboxyl group into approximation to the adenosine group of the tRNA. When valine is the activated amino acid, the approximation is incomplete and a water molecule intervenes and receives the valine. More subtle changes in side-chain structure, such as the replacement of a methylene of lysine by sulfur to give thiosine, do not change the nature of the reaction but alter the rate, which can also be explained by small differences in the distance between reacting groups. The modification of a side chain does not necessarily impair the catalytic process; canavanine, with a high K_m, reacts more rapidly than arginine, the natural amino acid with a low K_m, in both pyrophosphate exchange and transfer to tRNA.

The participation of the amino acid side chain in conferring specificity is attractive as a generalization, but it does not appear to apply to the proline enzyme, which transfers all the amino acids that are activated. In this case specificity is provided by the requirement for a secondary amine, which is an infrequent structure, and by the steric factors that prevent the most prevalent analog, 4-hydroxyproline, from binding at all.

The role of tRNA is complex, as befits a macromolecular substrate. It influences the binding of amino acids by at least two enzymes, the glutamate and arginine synthetases. It activates the activation of arginine. It activates the isoleucine and serine enzymes for transfer in addition to serving as the acceptor for the amino acid. In each case, only the specific tRNA binds to the enzyme to a measurable extent. The chemical basis for specific recognition remains to be determined. It may be predicted that the groups on the enzyme that interact with a given tRNA shift to accommodate the nucleic acid and in doing so induce other changes that are essential for enzyme activity. A disturbing question has been raised with the recent elucidation of the structures of isoaccepting tRNA's that show no more obvious structural similarity to each other than to unrelated tRNA's. An answer may be found not in the binding of single groups but in the cooperative binding of several groups that induce appropriate conformational responses in the enzyme. Several different enzyme-tRNA structures may result in the formation of similar catalytic sites as a consequence of patterns induced by combinations of groups in the tRNA. This proposal is made as an alternative to the search for common groups that might be expected to "fit" a rigid binding site on an enzyme. Although the hypothesis does not encourage a simple explanation of specific recognition through the identification of the points of interaction, it is amenable to experimental analysis by detailed study of the roles of groups on the tRNA in binding and in altering the conformation of the protein.

Acknowledgments

I am indebted to Drs. P. Berg, M. Yaniv, and M. DeLuca for sending me copies of manuscripts before publication.

REFERENCES

1. M. B. Hoagland, E. B. Keller, and P. C. Zamecnik, *J. Biol. Chem.* **221**, 45 (1956).
2. M. Matsuhashi, C. P. Dietrich, and J. L. Strominger, *Proc. Natl. Acad. Sci. U.S.* **54**, 587 (1965).
3. W. J. Lennarz, J. A. Nesbitt, III, and J. Reiss, *Proc. Natl. Acad. Sci. U.S.* **55**, 934 (1966).
4. P. Berg, F. Bergmann, E. J. Ofengand, and M. Dieckmann, *J. Biol. Chem.* **236**, 1726 (1961).
5. D. I. Hirsh and F. Lipmann, *J. Biol. Chem.* **243**, 5724 (1968).
6. R. B. Loftfield and E. A. Eigner, *Biochemistry* **7**, 1100 (1968).
7. H. S. Kingdon, L. T. Webster, Jr., and E. W. Davie, *Proc. Natl. Acad. Sci. U.S.* **44**, 757 (1958).
8. P. R. Krishnaswamy and A. Meister, *J. Biol. Chem.* **235**, 408 (1960).
9. A. T. Norris and P. Berg, *Proc. Natl. Acad. Sci. U.S.* **52**, 330 (1964).
10. C. C. Allende, J. E. Allende, M. Gatica, J. Celis, G. Mora, and M. Matamola, *J. Biol. Chem.* **241**, 2245 (1966).
11. R. Calendar and P. Berg, *Biochemistry* **5**, 1690 (1966).
12. R. Stern and A. H. Mehler, *Biochem. Z.* **342**, 400 (1965).
13. A. N. Baldwin and P. Berg, *J. Biol. Chem.* **241**, 839 (1966).

13a. R. B. Loftfield and E. A. Eigner, *Biochim. Biophys. Acta* **130**, 426 (1966).

14. S. K. Mitra and A. H. Mehler, *J. Biol. Chem.* **242**, 5490 (1967).
15. T. S. Papas and A. H. Mehler, submitted to *J. Biol. Chem.*
16. R. B. Loftfield and E. A. Eigner, *Acta Chem. Scand.* **17**, S117 (1963).
17. R. B. Loftfield, E. A. Eigner, and J. Nobel, *Biol. Bull.* **135**, 181 (1968).
18. S. K. Mitra and A. H. Mehler, *Europ. J. Biochem.* **9**, 79 (1969).
19. G. M. Kosalopoff, "Organophosphorus Compounds." Wiley, New York, 1950.
20. S. L. Owens and F. E. Bell, *J. Mol. Biol.* **38**, 145 (1968).
21. A. G. Atherly and F. E. Bell, *J. Elisha Mitchell Soc.* **80**, 156 (1964).
22. S. K. Mitra, K. Chakraburtty, and A. H. Mehler, *J. Mol. Biol.* in press.
23. R. Calendar and P. Berg, *J. Mol. Biol.* **26**, 39 (1967).
24. J. A. DeMoss, S. M. Genuth, and G. D. Novelli, *Proc. Natl. Acad. Sci. U.S.* **42**, 325 (1956).
25. P. Berg, *J. Biol. Chem.* **233**, 601 (1958).
26. E. W. Davie, V. V. Koningsberger, and F. Lipmann, *Arch. Biochem. Biophys.* **65**, 21 (1956).
27. N. Sharon and F. Lipmann, *Arch. Biochem. Biophys.* **69**, 219 (1957).
28. A. H. Mehler and S. K. Mitra, *J. Biol. Chem.* **242**, 5495 (1967).
29. U. Lagerkvist and J. Waldenström, *J. Biol. Chem.* **240**, PC2264 (1965).
30. P. Rouget and F. Chapeville, *Europ. J. Biochem.* **4**, 305 (1968).
31. D. I. Hirsh, *J. Biol. Chem.* **243**, 5731 (1968).
32. E. Sandrin and R. A. Boissonnas, *Helv. Chim. Acta* **49**, 76 (1966).
33. D. Cassio, F. Lemoine, J.-P. Waller, E. Sandrin, and R. A. Boissonnas, *Biochemistry* **6**, 827 (1967).
34. D. Cassio, *Europ. J. Biochem.* **4**, 222 (1968).
35. T. S. Papas and A. H. Mehler, *J. Biol. Chem.* **243**, 3767 (1968).
36. M. DeLuca and W. D. McElroy, *Arch. Biochem. Biophys.* **116**, 103 (1966).
37. H. George and A. Meister, *Biochim. Biophys. Acta* **132**, 165 (1967).
38. T. Kuo and M. DeLuca, *Biochemistry* **8**, 4762 (1969).

38a. M. Iaccarino and P. Berg, *J. Mol. Biol.* **42**, 151 (1969).

38b. A. V. Parin and L. L. Kiselev, *Mol. Biol.* (*Russ.*) **3,** 639 (1969).
39. J. Preiss, P. Berg, E. J. Ofengand, F. Bergmann, and M. Dieckmann, *Proc. Natl. Acad. Sci. U.S.* **45,** 319 (1958).
40. L. Hecht, M. Stephenson, and P. Zamecnik, *Biochim. Biophys. Acta* **29,** 460 (1958).
41. M. F. Singer and G. L. Cantoni, *Biochim. Biophys. Acta* **39,** 182 (1960).
42. R. W. Holley, J. Apgar, G. A. Everett, J. T. Madison, M. Marquisee, S. H. Merrill, J. R. Penswick, and A. Zamir, *Science* **147,** 1462 (1965).
43. F. Cramer, H. Doepner, F. v.d. Haar, E. Schlimme, and H. Seidel, *Proc. Natl. Acad. Sci. U.S.* **61,** 1384 (1968).
44. V. Daniel and U. Z. Littauer, *J. Mol. Biol.* **11,** 692 (1965).
45. I. I. Chuguev, V. D. Axelrod, and A. A. Bayev, *Biochem. Biophys. Res. Commun.* **34,** 348 (1969).
46. P. Philippsen, R. Thiebe, W. Wintermeyer, and H. G. Zachau, *Biochem. Biophys. Res. Commun.* **33,** 922 (1968).
46a. G. Sundharadas, J. R. Katze, D. Söll, W. Konigsberg, and P. Lengyel, *Proc. Natl. Acad. Sci. U.S.* **61,** 693 (1968).
47. L. H. Schulman and R. W. Chambers, *Proc. Natl. Acad. Sci. U.S.* **61,** 308 (1968).
48. S. K. Dube, K. A. Marker, B. F. C. Clark, and S. Cory, *Nature* **218,** 232 (1968).
49. M. Yaniv and B. G. Barrell, *Nature* **222,** 278 (1969).
50. P. Hele, *Biochim. Biophys. Acta* **87,** 449 (1964).
51. J. M. Ravel, S.-F. Wang, C. Heinemeyer, and W. Shive, *J. Biol. Chem.* **240,** 432 (1965).
52. L. W. Lee, J. M. Ravel, and W. Shive, *Arch. Biochem. Biophys.* **121,** 614 (1967).
53. M. P. Deutscher, *J. Biol. Chem.* **242,** 1132 (1967).
54. S. K. Mitra and A. H. Mehler, *J. Biol. Chem.* **241,** 5161 (1966).
55. M. Yarus and P. Berg, *J. Mol. Biol.* **28,** 479 (1967).
56. U. Lagerkvist, L. Rymo, and J. Waldenström, *J. Biol. Chem.* **241,** 5391 (1966).
57. T. Ohta, I. Shimada, and K. Imahori, *J. Mol. Biol.* **26,** 519 (1967).
58. M. Yaniv and F. Gros, *J. Mol. Biol.* **44,** 1 (1969).
59. H. L. James, J. C. Morrison, R. N. Shiflet, T. C. Trass, W. D. Whybrew, and E. T. Bucovaz, *Biochem. Biophys. Res. Commun.* **33,** 574 (1968).
60. U. Lagerkvist and L. Rymo, *J. Biol. Chem.* **244,** 2476 (1969).
61. F. Cramer, F. v.d. Haar, and E. Schlimme *FEBS Letters* **2,** 136 (1968).
62. K. Chakraburtty and A. H. Mehler, unpublished experiments.
63. H. Y. K. Chuang, A. G. Atherly, and F. E. Bell, *Biochem. Biophys. Res. Commun.* **28,** 1013 (1967).
64. M. Yarus and P. Berg, *J. Mol. Biol.* **42,** 171 (1969).
65. H. G. Bluestein, C. C. Allende, J. E. Allende, and G. L. Cantoni, *J. Biol. Chem.* **243,** 4693 (1968).
66. P. Rouget and F. Chapeville, *Europ. J. Biochem.* **4,** 310 (1968).
67. M. H. Makman and G. L. Cantoni, *Biochemistry* **5,** 2246 (1966).
67a. G. Lemaire, M. Dorizzi, and B. Labouesse, *Biochim. Biophys. Acta* **132,** 155 (1967).
68. M. K. Kukhanova, O. O. Favorova, and L. L. Kiselev, *Biokhimiya* **33,** 493 (1968).
69. R. Stern, M. DeLuca, A. H. Mehler, and W. D. McElroy, *Biochemistry* **5,** 126 (1966).
70. G. A. Hahn and J. W. Brown, *Biochim. Biophys. Acta* **146,** 264 (1967).

Transfer RNA and Cell Differentiation

NOBORU SUEOKA AND
TAMIKO KANO-SUEOKA

*Department of Biology and
Program in Biochemical Sciences,
Princeton University,
Princeton, New Jersey*

I. Introduction 23
II. Control of Protein Synthesis Mediated by Transfer RNA . . . 24
A. Modulation Hypothesis 24
B. Adaptor Modification Hypothesis 25
C. Abundancy Hypothesis 26
D. New Synthesis 27
E. Inhibitor Hypothesis 28
III. Chemical Basis for Adaptor Alteration 28
A. Chain Cleavage 28
B. Conformational Changes 29
C. Methylation 29
D. Oxidation and Reduction of Thiolated Bases 31
E. Terminal-CCA Addition 32
F. Minor Nucleosides 32
G. Aminoacyl-tRNA Synthetase 33
IV. Specific Biological Systems 35
A. Virus–Host 35
B. Growth Conditions 40
C. Different Tissues and Organs 43
D. Differentiation and Development 45
E. Neoplasia 47
V. Conclusion 49
References 51

I. Introduction

Cell differentiation often involves drastic changes in cell metabolism. Thus, the synthesis of many proteins is stopped and that of new ones is begun. This type of regulation could be called *macroregulation* in contrast to *microregulation*, in which the synthesis of a particular protein, or set of proteins in an operon, is specifically regulated by induction (or derepression) and repression, according to Jacob and Monod (*1*). There is no doubt that the principle of microregulation operates in differentiation. This is best illustrated in the process of lysogenic phage λ induction, where derepression is the key step (*2*).

This article deals with the possibility of tRNA involvement as a critical factor in cell differentiation at the macroregulation level. Various mechanisms for macroregulation can be visualized. The regulation could be at the transcriptional level (synthesis of mRNA), which is not discussed in this article. The regulation could also be at the translational level. The translation of mRNA into protein involves various components, i.e., ribosomes, initiation and termination factors, chain elongation factors, tRNA and aminoacyl-tRNA synthetase (*3*). Among these components for code translation, tRNA is the one most likely to be involved in macroregulation because of the unique feature of being multiple for each amino acid and yet ubiquitous for the synthesis of various proteins of the cell. Controls are likely to be operating at both levels. An excellent treatment on transcriptional and translational controls in cell differentiation is found in a review article by Gross (*4*), and the discussion will not be repeated here.

Degeneracy of the code and degeneracy of corresponding adaptors (tRNA) have been established, although the one codon-one tRNA relationship is generally not found because of "wobbling" (*5*). On the other hand, there is generally a single aminoacyl-tRNA synthetase for each amino acid in bacteria (*6–9*), although some contradictory results have been reported (*10*). In higher organisms, the mitochondrial system seems to have its own synthetases and tRNA's in addition to cytoplasmic synthetases and tRNA's (*11*).

II. Control of Protein Synthesis Mediated by Transfer RNA

The first clue as to the possible role of tRNA in the regulation of protein synthesis may be found in the suppression mechanisms recognized by Yanofsky *et al.* (*12*) and by Benzer and Champe (*13*). The fact that tRNA is indeed the key factor in the suppressor mechanism has now been firmly established (*14–16*). Sueoka and Yamane (*17, 18*) raised various biological questions in which they considered the regulatory function of tRNA, and used methylated albumin-kieselguhr (MAK) columns to fractionate aminoacyl-tRNA's. This made possible small-scale fractionation of tRNA and direct comparison of two tRNA samples, which are suitable for studies of various biological systems. In this section, several ideas on the regulation of protein synthesis with mediation of tRNA are introduced with citations of pertinent examples. It should be noted that the hypotheses are not necessarily mutually exclusive.

A. Modulation Hypothesis

The basic idea of the modulation hypothesis proposed by Ames and Hartman in 1963 (*19*) originates from a genetic consideration by Itano (*24*) of the quantitative differences among different types of hemoglobin controlled by multiple alleles in heterozygotes. Itano stated that if one mutation leads to a codon triplet that corresponds to a less abundant tRNA, the translation of the mutant allele will be slower, thus producing less of the mutant hemoglobin.

Ames and Hartman (*19*) proposed the hypothesis to explain the polarity effect in the biosynthesis of enzymes in the histidine pathway. The hypothesis states that the translation of mRNA is limited by *modulating triplets* which, in turn, correspond to *modulating tRNA*. The following types of models were proposed to fit experimental results. "(a) When the ribosome comes to a modulating triplet, that particular modulator tRNA could cause the ribosome to have a high probability of falling off the messenger. (b) The modulator tRNA could be a rate-limiting one and hold up the ribosome." In the modulator model, the modulation is effected primarily by the modulating tRNA, whose uniqueness is characterized by its scarcity or its low efficiency in the binding to ribosomes. The hypothesis can indeed explain the polarity effect on the rates of synthesis of the enzymes in an operon. In extreme cases, a nonsense triplet leads to a drastic polarity effect (*20, 21*). To remove the rate-limiting step of modulation, a mutation must be introduced either to change the modulating triplet to a nonmodulating triplet or to change the tRNA to nonmodulating tRNA.

Stent (*22*) further elaborated on the model in an effort to explain the induction and repression mechanism. Further studies on the histidine operon led Roth *et al.* (*23*) to suggest the possible role of charged and uncharged histidine tRNA involved in the initiation and repression of the operon as an effector.

B. Adaptor Modification Hypothesis

Another hypothesis (the adaptor modification hypothesis) was proposed in 1964 by Sueoka and Kano-Sueoka (*25*) on the possible role of tRNA in cell differentiation. In contrast to the modulation hypothesis, which deals principally with genetic change in mRNA, the modification hypothesis raises the possibility that codon recognition of a particular adaptor out of a set of degenerate adaptors for an amino acid (isoaccepting tRNA's[1]) is changed by structural modification. In this situation, the mRNA of the genes that accommodate the codon corresponding to the modified adaptor would not be translated properly, while the mRNA of the other genes that do not accommodate the codon would be translated normally. This means that by modifying a specific tRNA molecule, the function of some of the genes can be shut off and the rest of the genes kept functional at the

[1] The term isoaccepting tRNA was coined by Novelli (*26*).

TABLE I
SCHEMATIC REPRESENTATION OF THE ADAPTOR MODIFICATION HYPOTHESIS (*25*)

Codons[a]	Corresponding adaptors[b]	Utilization of codons[c]			
		System 1	System 2	System 3	etc.[d]
C_1	A_1	C_1	—	C_1	
C_2	A_2	C_2	C_2	—	
C_3	A_3	C_3	C_3	C_3	
C_4	A_4	C_4	C_4	C_4	

[a] Synonymous codons for a particular amino acid. Four codons are arbitrarily considered.

[b] For the sake of simplicity, wobble (*5*) is not considered.

[c] Codons in functional mRNA.

[d] Protein-synthesizing systems in which different codons may be used.

translational level. It should be noted that in this hypothesis mutation is not involved. Needless to say, the premise here is that development and differentiation are in general performed not by a series of somatic mutations but by the regulated activation and deactivation of the genes. A simple illustration of this hypothesis can be seen in Table I.

In Table I, C_1, C_2, C_3, and C_4 represent a degenerate set of codons for an amino acid, and A_1, A_2, A_3, and A_4 that of corresponding adaptors. In this scheme, for the sake of simplicity, wobbling of adaptors (*5*) is not considered. System 1 uses all four codons for protein synthesis, system 2 does not use C_1, system 3 does not use C_2, and so on. If adaptor A_1 is functionally removed from the cell, protein synthesis in systems 1 and 3 is shut off without affecting system 2. Inactivation of A_2 then closes systems 1 and 2, but not system 3. If, on the other hand, only A_2, A_3, and A_4 are functional to start with, so that only system 2 is operating, the appearance of A_1 activates systems 1 and 3 without shutting off system 2.

It is noted that the modification itself must be under genetic control and the genetic locus for it should exist. The chemical nature of the modification can be variable, for example, methylation or thiolation of bases, degradation by nuclease, conformational change, etc.

C. Abundancy Hypothesis

The amount of tRNA is variable among different amino acids and also among components of tRNA for a particular amino acid (isoaccepting tRNA's). Yamane (*27*) has reported that in *Escherichia coli* the relative abundance of tRNA reflects the amino acid composition of total protein. This finding needs further investigation as to its generality. A typical example of the variability of isoaccepting tRNA's is seen in the leucine tRNA's of *E. coli*, where one of the five constitutes more than 50% of the total leucine tRNA (*28, 29*). The possibility that this relative abundance of one isoaccepting tRNA reflects the relative abundance of the corresponding codon in mRNA is not necessarily supported, since the abundances of ribosome- or polysome-bound isoaccepting tRNA's in *E. coli* do not mimic those in bulk tRNA for several amino acids (*30, 31*).

Yegian and Stent (*32*) have noted that during leucine starvation of the *E. coli* RC^{str} strain, 10–40% of the total leucine acceptor capacity is protected against periodate oxidation, contrary to the expectation that starvation for a particular amino acid leads to exhaustion of acylated tRNA for that amino acid. About half the protected acceptor capacity comes from its aminoacylation with leucine. In addition, an unidentified substance, which is neither leucine nor any other amino acid, is also attached to the rest of the protected leucine tRNA during leucine starvation. This is found only with the RC^{str} strain in leucine starvation. This protector substance cannot be removed with mild alkali. It can, however, be removed upon incubation with a crude synthetase preparation in the absence of AMP and pyrophosphate. Furthermore, it is found that upon leucine starvation the protector substance attaches itself only to leucine $tRNA_1$, the major species among five different leucine tRNA's. The authors speculate that the protected tRNA complex may cause a functional removal of the tRNA from the systems, and thereby plays some role in the regulation of protein synthesis.

Anderson (*33*), using the S-30 protein synthesizing system from *E. coli*, demonstrated that the concentration of a tRNA can regulate the rate of translation of synthetic polynucleotides. Phenylalanine incorporation was stimulated by increasing the amount of phenylalanine tRNA in the poly U-directed phenylalanine incorporation system up to a $tRNA^{Phe}$:70 S ribosome ratio of seven. When poly (A,G) was used, the rate of polypeptide synthesis could be stimulated in direct proportion to the amount of arginine tRNA corresponding to AGA and AGG in the system, indicating that the *E. coli* extract contained only a rate-limiting amount of the arginine tRNA recognizing AGA and AGG for translation of poly (A,G). Anderson suggested that there are rate-limiting tRNA species in the cells that regulate the rate of protein synthesis at the translational level.

In general, then, any change in the abundance of the individual tRNA species is potentially capable of changing the rate of the translation of mRNA.

D. New Synthesis

The fourth hypothesis involving synthesis of new tRNA was proposed by Subak-Sharpe and Hay in 1965 (*34*). The idea came from the fact that DNA of herpes virus has a markedly higher G + C content than that of the host cell (40–44%). In this situation, the virus might be inefficient in performing its own protein synthesis, if it were to use the host protein-synthesizing system as such, which is presumably balanced for the translation of the host genome. It is expected that the frequency of codons may be quite different between the two systems. Subak-Sharpe and his collaborators found evidence, although not definitive, that there is a new synthesis of viral-specified tRNA after herpes virus infection with at least one new arginine tRNA detected (*35, 36*). This line of thought has been strengthened by recent studies on tRNA of *E. coli* after phage T4 infection (*37–40*). In accordance with the idea of Subak-Sharpe and Hay, the base composition of T4 (34% G + C) is considerably lower than that of *E. coli* (50% G + C).

The new synthesis of tRNA, then, should in principle be able to act as a key-regulating process both in affecting the rate of protein synthesis and in affecting code translation in the manner discussed in Section II, B.

E. Inhibitor Hypothesis

Production of a specific inhibitor that interferes with the function of a particular tRNA is another possibility. So far no clear case is known. The infection of Qβ might fall into this category, although Hung and Overby do not favor this point of view (*41*).

III. Chemical Basis for Adaptor Alteration

This section discusses known mechanisms for tRNA modification pertinent to *in vivo* situations. Alteration of aminoacyl-tRNA synthetase is also included.

A. Chain Cleavage

Transfer RNA is susceptible to various ribonucleases. If ribonuclease attack is specific for particular tRNA's, it could, in principle, be significant in the regulation of protein synthesis. Nishimura and Novelli (*42*) noted that tRNA's of *E. coli* have different susceptibilities to *Bacillus subtilis* ribonuclease or to RNase T1 and that some of the partially digested tRNA's still accept amino acids. Amino acids from the affected aminoacyl-tRNA, however, cannot be transferred into protein (*43*). Several examples have been reported showing that fragments of the tRNA molecule can be reconstituted to form a molecule capable of accepting amino acid in the presence of the synthetase (*44–48*). An *in vivo* example is known where a preexisting leucine tRNA of *E. coli* is "nicked" by a nuclease formed after phage T2 infection without losing leucine acceptor activity (*49*).

B. Conformational Changes

Interconversion between two conformations stable at physiological conditions without involving change in the primary structure has been studied in various tRNA's. The most thoroughly worked-out cases are tryptophan tRNA of *E. coli* (*50–54*) and a leucine tRNA of yeast (*55–57*). In both examples, one form is biologically active (native) and the other inactive (denatured). While the active form is normal in all aspects of tRNA function, the inactive form is devoid of amino acid acceptor activity, normal polyribonucleotide-dependent binding ability to ribosomes, and amino acid incorporation into polypeptides. The change from the active to the inactive form, then, should remove the adaptor from the system. In order to test the possibility that the inactive form may play a role in the regulation of protein synthesis, a method of mild deproteinization that avoids adsorption of tRNA on column material using silicic acid was developed (*58*). The application of this method showed that in both the exponential and stationary phases, more than 90% of the *E. coli* tryptophan tRNA was in the active form (*59*).

Thus, up to the present, there is no evidence that the denaturation of tRNA plays a regulatory role in cell metabolism, although *in vivo* denaturation and renaturation, if they occur, should potentially be capable of cell regulation.

C. Methylation

Transfer RNA contains a number of methylated bases in addition to the four common bases (*60*). The methyl group of the modified bases is derived from the methyl group of methionine through *S*-adenosylmethionine, and the methylation reaction takes place at the polynucleotide level (*61–63*).[2] Six different methylating enzymes specific for tRNA have been purified from *E. coli* and characterized by Hurwitz and his co-workers (*64*, *65*). An enzyme that methylates the adenine residue in rat liver has also been characterized (*66*). Recent nucleotide sequence analyses of tRNA's reveal that a particular methylated nucleotide occupies a definite position along the RNA chain. Moreover, the methylating enzymes appear to be species specific (*67*).

The function of these methylated bases in tRNA has been explored by a number of investigators. The observation made by Mandel and Borek (*61*) that a methionine auxotroph of *E. coli* RC^{rel} synthesizes tRNA deficient in methylated bases when the cells are starved for methionine has made it possible to investigate the role of methylated bases in tRNA molecules.

[2] See article by Srinivasan and Borek in this series, Volume 5, p. 157 (1966).

Peterkofsky and his co-workers (*68, 69*) have investigated the methyl-deficient leucine tRNA of *E. coli* and have found that its coding properties as well as its elution characteristics on the MAK column are different from those of normal tRNA, although the amino acid acceptor capacity of methyl-deficient tRNA is normal in the homologous system. Leucyl-tRNA responds to poly (U,C) and poly (U,G) in the ribosome binding assay. Methyl-deficient leucyl-tRNA manifests a highly significant increase in poly (U,C) response, and a component of methyl-deficient leucyl-tRNA isolated by MAK column chromatography responds exclusively to poly (U,C). *In vitro* methylation of methyl-deficient leucine tRNA changed both the chromatographic properties and the coding properties of the tRNA, suggesting that methylated bases are responsible for the recognition of the right codon (*69*).

Littauer and his co-workers (*70*) studied methyl-deficient phenylalanine tRNA. Methyl-deficient $tRNA^{Phe}$ accepts the amino acid equally as well as normal $tRNA^{Phe}$. However, the methyl-deficient tRNA requires less magnesium than does the normal species for optimal charging. Methyl-deficient phenylalanyl-tRNA can be separated from normal phenylalanyl-tRNA on the MAK column. Examination of the coding properties of normal and methyl-deficient species indicates that the methyl-deficient species seems to have a higher ambiguity for the recognition of the code. Normal and methyl-deficient phenylalanyl-tRNA respond equally well to poly U in the *in vitro* protein-synthesizing system. In contrast, when poly (U,C) is used, the methyl-deficient species is more than twice as active as the normal one. The ribosome binding assay of those two species of tRNA shows that methyl-deficient phenylalanyl-tRNA seems to recognize the codons corresponding to amino acids other than phenylalanine.

There have been several attempts to relate the methylation of tRNA to the regulation of protein synthesis, since the methylases are base- and species-specific and the degree of methylation gives different characteristics to the tRNA molecule. Up to the present, some difference in the methylating activity of tRNA has been noted in T2-infected *E. coli* B (*71*), in *E. coli* K12 after induction of phage λ (*72*), in cancer cells (*73–77*), during morphogenesis of slime mold (*78*), and during insect metamorphosis (*79*). Some findings and interpretation of the data will be summarized below.

Following the infection of *E. coli* B with phage T2, although no change in the overall activity of methylation of tRNA is observed (*80*), there are alterations in the relative levels of base-specific tRNA methylating activities when the *in vitro* assay is made with crude enzyme preparations (*71*). The ratio of methyladenosine or methyluridine to mono- and dimethylguanosine increased twofold when extracts of the infected cells were used. T1 infection did not change the ratio of newly methylated bases. It was also found that, in *E. coli* K12 (λ), there is a reduction of the total tRNA methylase activity for nearly 30 minutes after an inducing dose of UV irradiation (*72*). The reduction seems to come from the appearance of an inhibitor of the methylases, the nature of which has not been characterized.

Work in several laboratories indicates an elevation of methylase activity of two- to tenfold in tumor tissue extracts compared to the appropriate normal controls, as cited in the previous section. Moreover, Tsutsui *et al.* (*73*) have shown that when methyl-deficient *E. coli* tRNA is methylated with extracts of four different tumors *in vitro*, the base ratio of methylated bases is different from that of extracts of normal tissue. Mittelman *et al.* (*75*) have shown that extracts of the SV40-induced tumor are capable of hypermethylating homologous tRNA of normal tissue *in vitro* while extracts of normal tissue are not.

These findings should be interpreted with caution, however, for the following reasons: (a) In order to measure methylase activity, heterologous tRNA's have been used in most of the studies; therefore, the reaction may not reflect the *in vivo* situation. (b) Kaye and Leboy (*77*) have found that not only the rate but also the extent and distribution pattern of methylation is ion-dependent and that at optimal ion concentration most of the extracts of normal organs and tumor tissues of mouse behave similarly. (c) Baguley and Staehelin (*76*) demonstrated that both the protein concentration curve and the time curve for methylation by crude enzyme extracts of rat liver show a plateau, and that this plateau is due to the exhaustion of substrate tRNA molecules by digestion with RNase contamination in crude extracts. By partially purifying one of the methylases from the normal and tumor tissue of rat, it was shown that, although the tumor methylase fraction contained sixfold higher activity, it methylated heterologous tRNA to the same extent that the normal enzyme fraction did. In other words, one of the methylases in the tumor tissue increased in activity but had a substrate specificity identical to that of the normal tissue. Purification of enzymes and a well defined *in vitro* system are essential to a study of the role of methylase in cell differentiation.

D. Oxidation and Reduction of Thiolated Bases

Carbon, Hung, and Jones (*81*) discovered that the exposure of *E. coli* or rabbit liver tRNA to solutions of dilute iodine–KI (pH 7.0) results in a rapid partial inactivation of acceptor ability for several amino acids. Certain components of tRNA specific for a particular amino acid contain an easily oxidizable functional group, and the oxidized, hence inactivated, tRNA's can be reactivated by exposure to reducing agents. There is also evidence of the existence in *E. coli* tRNA of a sulfur-containing base, 2-thiopyrimidine, which reacts with iodine (*81*). The presence in *E. coli* tRNA of another thiolated nucleotide 4-thiouridylic acid, has been shown by Lipsett (*82*). Goehler and Doi (*83*) have described a similar reversible oxidative inactivation of acceptor ability in certain tRNA species of *Bacillus subtilis* and presented evidence that the oxidation of lysyl-tRNA accompanies changes in conformation and binding efficiency to ribosomes. Carbon *et al.* (*81*) suggested that the reversible oxidative inactivation of the acceptor ability would meet the requirement for the adaptor modification hypothesis, and, although an enzymatic oxidative inactivation of specific adaptors has not been discovered, the existence of such an enzyme is possible.

The sulfur of thiolated nucleotides derives from cysteine, and the enzymatic system of thiolation of tRNA catalyzed by a soluble extract of *E. coli* has been described (*84*, *85*).

E. Terminal-CCA Addition

A number of workers have noted that a high proportion of isolated yeast tRNA's lacks 3′-terminal adenosine (*86–88*). In the case of yeast serine tRNA, the molecule without terminal adenosine can be chromatographically separated from the intact molecule (*88*). Purified adenylate pyrophosphorylase can catalyze the addition not only of AMP but also of CMP onto the 3′-end of tRNA molecules (*89*, *90*). Since crude aminoacyl-tRNA synthetase preparations contain the pyrophosphorylase, those tRNA's lacking terminal A or -CCA can be repaired and aminoacylated normally.

During the early developmental stage of the sea urchin embryo (*91*, *92*) and of the *Xenopus* embryo (*93*) there is a rapid incorporation of labeled precursors into 4 S RNA. This incorporation seems to be due to the terminal addition or replacement of the -CCA terminal of tRNA. The possibility exists that the activation of tRNA by the addition of -CCA initiates protein synthesis in the fertilized egg (*91*). However, the tRNA of the unfertilized egg is active in accepting amino acids (*94*), and the above possibility is not proved.

F. Minor Nucleosides

Transfer RNA molecules contain a rather large number of modified nucleosides other than the four major ones (*95*). They are located at nonhydrogen-bonded regions of the molecules when tRNA molecules are arranged according to the "cloverleaf" model (*96*). Much of the functional significance of those modified nucleotides is yet to be elucidated. Some studies on methylated and thiolated nucleosides are summarized in Sections III, C and III, D.

Among the interesting modified bases are N^6-isopentenyladenosine

(*97, 98*) and 2-methylthio-N^6-isopentenyladenosine (*99, 100*).[3] They are located next to the 3′-end of the anticodon site, and therefore presumably play a specific role in the binding of tRNA to ribosomes in the presence of mRNA. Gefter and Russell (*101*) made a detailed survey of the function of the modified adenosine residue in suppressor tyrosine tRNA. Under their experimental conditions three forms of this tRNA exist in *E. coli* cells, all with the same primary nucleotide sequence but differing in the extent of modification of the adenosine residue. These differences do not affect the rate of aminoacylation, but do affect the ability to bind to ribosomes. The one with a fully modified nucleoside (2-methylthio-N^6-isopentenyladenosine) is the most active in ribosomal binding and protein synthesis, the one with the unmodified residue of adenosine shows a marked reduction in these functions, and the one with a partially modified adenosine possesses intermediate abilities. Fittler and Hall made a similar observation with yeast serine tRNA containing N^6-isopentenyladenosine (*102*). Modification of N^6-isopentenyladenosine by iodine in serine tRNA did not affect amino acid acceptor activity but did reduce the ability of seryl-tRNA to bind to ribosomes in the presence of mRNA. It is certain from the above examples that the modification of the base adjacent to the anticodon is necessary for efficient binding to the ribosome-mRNA complex. N^6-Isopentenyladenosine possesses high cytokinin activity (*97*). The relationship between the presence of this material in tRNA and cytokinin activity, however, is not known (*103*).[3]

G. Aminoacyl-tRNA Synthetase

Closely related to tRNA-mediated regulation is the alteration of aminoacyl-tRNA synthetase, which may lead to a change in specificity of tRNA recognition (*104, 105*). Neidhardt and Earhart (*104*) have demonstrated that infection of *E. coli* with T-even phages brings about the appearance of a "new" valyl-tRNA synthetase. A temperature-sensitive mutant having a heat-labile valyl-tRNA synthetase can support the propagation of T4 phage at high temperature. The infection of mutant cells with T4 at 40°C is without effect. If, however, the cells are infected at 30°C and then shifted to 40°C 5 minutes later, phage propagates normally. The appearance of the "new" synthetase can be distinguished by its behavior on a hydroxylapatite column, by its sedimentation behavior in a sucrose gradient, or, as mentioned above, by its temperature stability in a temperature-sensitive mutant having a temperature-sensitive valyl-tRNA synthetase. Two possibilities exist as to its origin, namely, *de novo* synthesis of a new enzyme or a modification of preexistent molecules. The

[3] See article by Hall in this volume.

appearance of the "new" activity is accompanied at all times by a proportional decrease in the original valyl-tRNA synthetase activity. Using a density transfer experiment with D_2O, Chrispeels *et al.* (*106*) showed that the "new" enzyme brought about by phage infection in both the mutant and wild-type cells was due to polypeptide chains synthesized prior to infection. However, this conversion required protein synthesis, since the addition of chloramphenicol at any time blocked the change completely. The role of the altered synthetase in phage-infected cells has yet to be solved.

There are also some indications that there exists a qualitative change in aminoacyl-tRNA synthetase during the development of the sea urchin embryo (*107*). The synthetases isolated from the three different stages of development (unfertilized eggs, blastula, and pluteus) seem to charge yeast tRNA differently with a mixture of various amino acids, which may indicate a difference in the synthetase at different stages of development. More work is necessary before drawing any conclusions concerning this system.

Strehler *et al.* (*105*) investigated the leucyl-tRNA synthetases of various rabbit tissues in order to examine the possibility that individual tissues or organs contain a unique set of aminoacyl-tRNA synthetases. First, they suggested that the heart-muscle enzyme preparation contains three distinct synthetases separable on a DEAE-cellulose column. The enzyme eluted first from the column labeled predominantly those rabbit leucine tRNA species that emerged late from a MAK column; conversely, the enzyme eluted last labeled predominantly the leucine tRNA species that eluted early when a mixture of tRNA from the liver, heart, and reticulocyte was used. Strehler *et al.* further showed that enzyme preparations from reticulocytes scarcely labeled the leucyl-tRNA component that eluted late from the MAK column as compared with liver enzyme preparations. A mixture of tRNA from various organs of the rabbit was fractionated on a reversed phase column, and the differential chargeability of ten amino acids was tested against synthetase preparation from five rabbit tissues. The alanyl-tRNA synthetase of kidney acylated only one of the two components of alanine tRNA, while enzymes from other sources acylated both.

One possibility that could explain these results is the presence of various amounts of mitochondria in different tissues. Mitochondria contain unique synthetases as well as tRNA's (*108*, *109*). A similar finding of multiple synthetases in rat liver (*110*, *111*) should be further examined in this respect.

Recently Kull and Jacobson (*112*) reported three phenylanyl-tRNA synthetases (A,B,C) distinct from mitochondrial phenylalanyl-tRNA synthetase in *Neurospora crassa*. These three enzymes were separable by hydroxylapatite chromatography and had different affinities for different tRNA's of *E. coli* in Tris-Cl buffer. Although a monomer-dimer relationship between the enzyme forms may be regarded as unlikely on the basis of their sedimentation behavior, the possibility of modifications during preparation cannot be eliminated, as the authors pointed out. The functional significance of this multienzyme system in cell metabolism remains to be seen.

IV. Specific Biological Systems

A. Virus–Host

When the cells are infected with viruses, a drastic metabolic shift takes place leading macromolecular synthesis toward the production of virus progeny. In addition, many viruses exhibit a shift in the pattern of protein synthesis from the "early" to the "late" phase during the infection period. The mechanism for these major shifts in the pattern of protein synthesis has not been well understood. Recently factors that endow RNA polymerase with initiation specificity have been identified in various systems. In the case of the "early" to "late" shift in T4 infection, a phage-specific factor seems to be responsible, thus rendering the shift to transcriptional control (*149*).

The involvement of tRNA as a regulating factor at the translation level has been tested as one of the possible mechanisms. The T-even phage system of *E. coli* has been investigated most extensively so far and therefore, will be considered first. Sueoka and Kano-Sueoka (*25*) compared tRNA's isolated from normal and T2-infected *E. coli* cells by MAK column chromatography. Among tRNA's for 17 amino acids examined, only leucyl-tRNA of *E. coli* showed a clear alteration of the elution pattern, giving two major peaks, Leu I and Leu II, with Leu I being the major component. As a result of T2 infection, an alteration of the leucyl-tRNA profile took place in two steps: first, the appearance of a new peak, Leu F (or leucyl-$tRNA_F$) preceding all the normal leucyl-tRNA's and a decrease in the amount of the major peak, Leu I; second, the subsequent disappearance of Leu F and a reduction in the relative amount of Leu I to about half the original. The entire process was completed within 3 minutes after infection at 37°C, and the profile remained unchanged 8 minutes after infection. The alteration is due to a change in the tRNA itself, not in the leucyl-tRNA synthetase. The injection of phage DNA and protein synthesis after infection was found to be necessary for the alteration, since the phage ghost or the addition of chloramphenicol before or at the time of infection did not induce the alteration. All ten strains of *E. coli* and a strain of *Shigella dysenteriae* so far tested have shown the alteration. All T-even phages cause the alteration whereas all T-odd phages and the induction of phage λ do not. Over twenty of the early amber mutants demonstrate normal alteration on the nonpermissive host. Since the alteration is completed within 2–3 minutes after infection, it must be one of the earliest events taking place after infection. The cells infected with UV-irradiated phage, which do not show a normal early-to-late shift, also indicate the presence of normally altered leucyl-tRNA. Therefore, if the alteration of leucine tRNA is involved in the regulation of protein synthesis, it is likely to be related to the cessation of host protein synthesis (*113*). One of the possibilities raised in 1964 (*25*), that the alteration is related to the shift of protein synthesis from the early to the late phase, has therefore been eliminated.

By the use of reversed-phase column chromatography, Waters and Novelli (*114*) confirmed the early changes of leucyl-tRNA after T2 infection. In addition, they extended the study to include late changes of leucyl-tRNA and found that, in addition to the three peaks observed during the early phase of T2 infection, at least one and possibly two new peaks eluted at higher salt concentration than the other three. The new peaks, which appeared 30–40 minutes after infection, may therefore be involved with some late function. Whether the appearance is due to *de novo* synthesis of phage-specific tRNA or the modification of preexisting tRNA is not known.

By reversed-phase column chromatography, leucine tRNA of *E. coli* can be fractionated into five well-resolved components (*28*). Preparations of the leucine tRNA's of uninfected cells and of cells infected with T2 for 8 minutes each gave five components of leucine tRNA (Leu_1 to Leu_5). The relative amount of Leu_1 decreased to about half the original amount after infection, with a corresponding increase in Leu_5 (*29*). The five leucine tRNA components, once separated on the reversed-phase column can be distinguished from each other on MAK columns. The Leu I peak of the MAK column consists of a single major component, Leu_1. Leu_1, Leu_2, and Leu_3 from T2-infected cells gave MAK column profiles identical with those of their counterparts from uninfected cells. On the other hand, Leu_4 and Leu_5 of T2-infected cells gave profiles differing from those of uninfected cells. The Freon reversed-phase column can resolve five components of leucine tRNA of *E. coli* in aminoacylated form (*115*). With this column, it has also been shown that the relative amount of leucyl-$tRNA_1$ from cells infected with T2 for 6 minutes decreases drastically.

Characterization of the new component, leucyl-$tRNA_F$ (Leu_F), which appears 1–2 minutes after infection, has been carried out by Kano-Sueoka and Sueoka (*49*). Leucyl-$tRNA_F$ cannot be deacylated enzymatically. Reacylation of leucyl-$tRNA_F$ also cannot be achieved following deacylation.

It is clear that once $tRNA_F^{Leu}$ is charged with leucine, it cannot be recognized by the synthetase. The results suggest that this loss of synthetase recognition is due to aminoacylation itself. A comparison of hyperchromicity, sedimentation coefficient, and Sephadex G 100 column fractionation of leucyl-$tRNA_F$ shows that it is approximately half the size of normal tRNA. However, the fact that $tRNA_F^{Leu}$ (before aminoacylation) sediments at the same rate as bulk tRNA on the sucrose density gradient, whereas leucyl-$tRNA_F$ (after aminoacylation) sediments much more slowly, indicates that $tRNA_F^{Leu}$ is intact. It is, therefore, an unstable molecule, and conditions that weaken the hydrogen-bonded structure destroy its chargeability. It has been concluded that $tRNA_F^{Leu}$ is derived from $tRNA_I^{Leu}$ by the introduction of a chain break or breaks within the molecule by a phage-induced nuclease.

The codon responses of leucyl-tRNA components thus produced and isolated have been studied by the ribosomal binding technique and by the incorporation of leucine from leucyl-tRNA into protein in the presence of various trinucleotides and polynucleotides (*116, 117*). From the results, the most probable assignments for uninfected *E. coli* leucine tRNA's are: Leu_1, CUG; Leu_2, CUU, CUC; Leu_3, CUA, CUG; Leu_4, UUG?; Leu_5, UUG. The five components of T2-infected *E. coli* leucyl-tRNA showed codon specificities essentially similar to those of their noninfected counterparts. Leucyl-$tRNA_F$ responds to poly (U,G) with high efficiency, but not to any triplets containing U and G so far tested. An unusual feature of the response of leucyl-$tRNA_F$ to poly (U,G) is that, while it binds to the polymer with unusually high efficiency in the absence of ribosomes, its leucine is not incorporated into polypeptide *in vitro*.

Thus, 1–2 minutes after infection, half of Leu_1 undergoes a drastic structural change leading to inactivation of both leucine acceptor activity and codon response. However, further alteration of Leu_1 is prevented somehow after the third minute, when 40% of Leu_1 has been modified. Analyses of polysome-bound tRNA from *E. coli* before and after the phage infection reveal that the amount of Leu_1 bound to polysomes decreases drastically after infection, indicating a greatly reduced usage of Leu_1, and consequently a reduced usage of the CUG codon, after T2 infection (*31*). This situation is established 1.5 minutes after infection and remains unchanged for 10 minutes after infection. Studies by the *in vitro* protein-synthesizing system also have shown that the mRNA of T2 rarely contains the CUG codon. From these two facts, namely, (a) the breakdown of Leu_1 to 60% of the original level within 2–3 minutes after infection, and (b) the rare occurrence of the CUG codon corresponding to Leu_1 in T2 mRNA, the following model was proposed to explain the cessation of host protein synthesis without inhibiting T2 protein synthesis: Phage T2 infection induces a specific ribonuclease that cleaves leucyl-tRNA$_1$ at the site of the translation of *E. coli* mRNA on ribosomes; this, in turn, leads to the cessation of host protein synthesis. This process eliminates about half the Leu$_1$ molecules in the cells. Phage mRNA carries little or no CUG codon corresponding to Leu$_1$, and thus its potential for protein synthesis is unimpaired (*31*). The model fits the phenomena so far observed on the alteration of leucine tRNA during the early period of T2 infection. Host protein synthesis and the degradation of Leu$_1$ stop at about the same time. The translation of phage M12 RNA does not occur when M12 and T4 doubly infect *E. coli* (*118*). Of course, rigorous proof of the model remains to be obtained. The above phenomenon is the event that takes place at an extremely early period of infection and is due to the modification of preexisting host tRNA molecules.

Recently several laboratories have reported that new phage-specified tRNA's have been synthesized after T-even phage infection. The presence of thiolated bases in tRNA led Weiss and his collaborators to examine possible changes in sulfur-containing RNA's after phage infection (*37*, *38*). They discovered that infection of *E. coli* with T4 phage induces changes in the MAK column chromatographic profiles of ^{35}S-labeled tRNA and that these changes are phage-induced and involve protein synthesis, since chloramphenicol blocks them. When T4 infected or uninfected cells were exposed to radioactive sulfate and the labeled tRNA's were subjected to MAK column chromatography, a striking difference was observed in the ^{35}S elution pattern. The first observable change in the labeled RNA profile occurred between 4.5 and 7.5 minutes after infection, and was due to *de novo* addition of sulfur. Further, some of the tRNA labeled with ^{35}S after T4 infection is T4-specific tRNA synthesized after infection. Of the three partially resolved ^{35}S-labeled peaks of *E. coli* tRNA seen on the MAK column, the amount of the third peak increased strikingly after infection. This third peak hybridized quite well with T2 or T4 DNA, but not with *E. coli* or T7 DNA. Also, T4 [^{35}S]tRNA differs from T4 mRNA, since the ^{35}S was not found in the T4 mRNA fraction and since T4 mRNA did not compete with the T4 [^{35}S]tRNA during hybridization. Their observations suggest a synthesis of new tRNA transcribed from the phage genome. Final proof of this came from an experiment on hybridization between T4 DNA and aminoacylated tRNA (*38*). A suitable condition for the hybridization was established, and leucyl-tRNA from T4-infected cells was shown to hybridize specifically with T4 DNA. Moreover, since only leucyl-tRNA taken from a certain area of the MAK column elution profile was hybridized, it appeared that a new species of leucyl-tRNA is specified by the T4 genome. They also found a new prolyl-tRNA after T4 infection (*38*). However, there is no direct evidence that leucyl-tRNA or prolyl-tRNA corresponds to new sulfur-containing tRNA. Daniel *et al.* (*39*, *40*) have also shown that ^{32}P-labeled "soluble" RNA isolated from T4 infected cells specifically hybridizes with T4 DNA and that the hybridized RNA contains ψMP. Since ψMP apparently occurs only in tRNA, the hybridized RNA is most likely tRNA. Later they discovered that the hybridized RNA can accept a significant amount of amino acid if a mixture of [^{14}C]amino acids is used for the charging reaction (*40*). There have been reports showing that labeled "soluble" RNA after T2 infection has elution patterns different from those in the uninfected cells (*119*) and that there exists a low molecular weight RNA hybridizable to T4 DNA (*120*). The nature of the RNA is, however, not known.

There is also a report on the alteration in function of prolyl-tRNA of *E. coli* after infection with phage Qβ (*41*). It has been shown that prolyl-tRNA from infected cells responded to polycytidylic acid two to three times less efficiently when the tRNA was tested for its ability to bind to ribosomes in the presence of polynucleotides. In a tRNA-dependent, *in vitro* protein-synthesizing system, tRNA from infected cells showed a markedly decreased ability to incorporate proline in the presence of poly C. However, a close examination of the report suggests that the difference might be due to a component other than $tRNA^{Pro}$ existing in Qβ-infected tRNA preparations that may inhibit the incorporation of proline. This point should be clarified further.

Alteration of the translation mechanism caused by animal virus infection has been reported by Subak-Sharpe *et al.* (*35*). Their rationale for searching for a new viral tRNA in cells infected with herpes virus is based on the following reasoning. The base composition of herpes virus DNA is 68% G + C, which is 24–28% higher than that of host DNA. A population of transfer RNA's of the host cell should be optimally adapted to translate mRNA of a G + C content of 40–44%. Therefore, the herpes virus, having a higher G + C content, must encounter inefficiency in translating its own genome by host translation machinery. It is highly probable, then, that the virus introduces its own tRNA and that this responds to high G + C codons. Labeled tRNA was isolated from herpes-infected and noninfected cells, and it was shown that (a) herpes DNA hybridized only with tRNA from infected cells, not from noninfected cells; (b) the MAK column elution profile of arginyl-tRNA of infected cells was different from that of noninfected cells; and (c) arginyl-tRNA synthesized after infection contained one or two species of new tRNA with different numbers of nucleotides up to the first G from the 3′-CCA terminal end when examined by RNase T1 digestion of arginyl-tRNA and DEAE-cellulose column chromatography. These above observations indicate that most likely the new arginyl-tRNA is specified by the virus.

B. Growth Conditions

Bacillus subtilis has a dormant stage, spore, in its life cycle. A set of genes related to sporulation must be activated to form spores, which are morphologically distinct from vegetative cells. Germination of the spore is also a process of changing a completely dormant state to a metabolically active vegetative cell. Whether or not control at the translational level is involved in such major metabolic shifts can be tested in this relatively simple system of cellular differentiation.

Alteration of the elution profile of aminoacyl-tRNA's of *B. subtilis* in different physiological states has been reported for lysine, serine, tyrosine, and valine. Doi and his co-workers (*121*, *122*) investigated changes in valine tRNA during sporulation. Valyl-tRNA from exponentially growing cells of strain 23 or 168 was eluted from the MAK column as two major peaks, the first larger than the second. However, a reversal in the ratio of these took place during early sporulation, returning to the exponential type in late sporulation. The ratio of the two valyl-tRNA's also changed significantly during step-up or step-down growth conditions and during the stationary phase of one asporogenous mutant. The RNase T1 digest of the two valyl-tRNA peaks indicated that the two have different nucleotide sequences at the 3′-hydroxyl end, suggesting the presence of two distinct valine tRNA species. The change in the relative amounts of the two components is, therefore, due to the increased valine-accepting activity of one (peak I) of the two valine tRNA's (*122*). However, this should be interpreted with caution. Careful analysis of charging kinetics by Heyman *et al.* (*123*) indicates that the two components have different charging kinetics, the first component on the MAK column acylating more slowly than the second. It was pointed out that, in various other experiments reported (*116*, *117*), the two components were not charged to a fully saturated level. Heyman *et al.* (*123*) concluded that peak I of $tRNA^{Val}$ always exists in larger quantity than peak II, although the amount of I increases during the exponential phase while II remains the same throughout the growth cycle. In any case, the relative amounts of the two components change at different physiological states of the cells.

Lazzarini (*124*) has shown that when *B. subtilis* are grown in an enriched (TY) medium containing tryptone and yeast extract, the tRNA of vegetative cells contains one major component of lysyl-tRNA as revealed by MAK column chromatography. On the other hand, spore tRNA contains twice as much lysine-accepting activity as that of vegetative tRNA relative to other amino acids, and comparison of vegetative and spore lysyl-tRNA on MAK columns indicates that spore tRNA contains an extra component that cannot be found in vegetative tRNA in appreciable amounts. Consequently, the formation of a new species of lysine tRNA specific to the spores is observed during sporulation. The work was extended, showing that the presence of the new component was medium-dependent and not obligatory to spore formation (*125*). Comparison of the lysyl-tRNA profile of vegetative cells and spores obtained on another enriched sporulation medium (NSM) containing nutrient broth revealed no significant difference. The growth rates on these media are the same; however, the levels of terminal growth attainable on the two media differ considerably. When twice-concentrated NSM was used, in which the stationary cell concentration was equivalent to that of the TY medium, the appearance of a new component was noted. In a similar way, by reducing the concentration of the nutrient in the TY medium, hence, reducing the final growth level, no new components could be detected. Therefore, the concentration of complex nutrients does affect the accumulation of a new component. However, a high terminal growth level alone is not sufficient for the accumulation of a new component since the tRNA of spores formed in a synthetic medium allowing a high terminal growth level does not contain this component. Lazzarini and Santangelo (*125*) also tested four asporogenous mutants blocked at the early stage of sporulation and found that three out of four exhibited a new component. Since cells grown in a non-sporulating medium do not contain this new component, this component may be important to sporulation. However, from the case of spores grown in a synthetic medium, the new component is not essential to spore formation. The biological function of this new component, therefore, is not known nor is its origin. The question remains whether it has the same primary sequence as that of the vegetative species or whether it is transcribed from a new cistron.

Comparison on MAK columns of aminoacyl-tRNA from cells grown in a rich (Penassay) and a poor (SCM) medium revealed that among the three peaks obtained for seryl-tRNA from cells grown in the Penassay medium, only the first two were observed in tRNA of cells grown in the minimum medium (*126*). Here again, there is no evidence as to whether the third peak, characteristic of Penassay grown cells, is due to a new synthesis or to a modification of preexisting tRNA.

In *B. subtilis* there are two $tRNA^{Tyr}$ species, Tyr I (exponential) and Tyr II (stationary). Unlike lysine tRNA's, the tRNA from exponentially growing cells contains predominantly the exponential form, Tyr I, and that from stationary cells almost exclusively the stationary form, Tyr II, when the cells are grown in a Penassay medium (*127*, *128*). The biological and biochemical properties of these two species of tyrosine tRNA, including tyrosine acceptor activity, *in vitro* assay for codon response, and incorporation into polypeptides, have been examined. No functional difference between the two could be found, the only difference being the elution patterns on either MAK column or reversed-phase chromatography. Again, whether or not the two are transcribed from the same gene is not known since no primary sequence study has been done. The nucleotide sequence from the 3′-end to the first G is at least the same. Alteration of the elution profile of tyrosyl-tRNA is medium-dependent. If a minimum medium or a minimum medium plus casein hydrolyzate (SCM) is used, both forms can be recovered at both the exponential and stationary phases, although the proportions of the two change.

The relative amount of the major component, $tRNA_1^{Leu}$, of *E. coli* tRNA isolated from stationary cells is somewhat less than that of exponentially growing cells (*127*). The correlation between this and T-even phage infection is not known, but in both cases, the synthesis of protein, DNA and RNA of the cells is repressed.

Kwan *et al.* (*129*) reported that anaerobiosis of *E. coli* induces a change in isoleucine tRNA as well as in the 50 S ribosome subunits. When they studied miscoding of isoleucine with polyuridylic acid in the presence of streptomycin, they found that the optimal magnesium concentration was lower in the extract of anaerobically grown cells and that the elution pattern on a DEAE-Sephadex column of the isoleucyl-tRNA from anaerobically grown cells differs from that of aerobically grown cells. The differences are rather small, however, and nothing is known concerning the nature and significance of the change.

The alteration of the phenylalanine tRNA of *E. coli* under somewhat artificial conditions has been studied by Wettstein and Stent (*130*). MAK column chromatography of the phenylalanyl-tRNA from cells grown under optimum conditions revealed a single peak. However, the tRNA from bacteria grown in a medium with less than 0.1 μM iron contained one or two additional peaks of abnormal phenylalanyl-tRNA. The abnormal phenylalanyl-tRNA accumulated during growth in the iron-free medium. However, enough of the normal component may still be present to carry out protein synthesis at a more or less normal rate. The abnormal phenylalanyl-tRNA can be converted to a normal species *in vivo* in the absence of RNA synthesis if protein synthesis is allowed. This suggests that the abnormal species is an undermodified molecule of normal phenylalanine tRNA, and that limited iron concentration in the growth medium inhibits the synthesis of some enzyme that modifies the molecule. In this case, thiolation, or methylation, is shown to be unlikely. The biological significance of the abnormal modification of tRNA is not known.

Shearn and Horowitz (*131*, *132*) extensively studied the tRNA of *Neurospora crassa* obtained from vegetative and ethionine-treated cultures. Ethionine derepresses the synthesis of tyrosinase. Although no qualitative alterations of tRNA have been found, some quantitative changes of specific acceptor activity were observed for a few amino acids. Shearn and Horowitz concluded, however, that the observed changes most likely result from partial RNase digestion of the tRNA components for the following reasons (*132*). (a) The relative changes in specific acceptor activity observed are small compared with the large net loss by ethionine treatment (40% loss of tRNA within 5 hours and 75% by 20 hours). (b) The alterations observed do not generally affect the relative proportion of synonymous tRNA's. (c) Any theory in which tRNA plays a regulatory role must postulate that the availability of tRNA is limiting the rate of protein synthesis. The last requirement is apparently not met, since after ethionine treatment of mycelium for 5 hours, a time when 40% of the general tRNA content of the mycelium has been lost, there is only a 13% decline in the rate of protein synthesis. They have shown that considerable ribonuclease activity is apparent in the ethionine-derepressed cultures (*131*). The authors raise the possible significance of a net breakdown of RNA and soluble protein in this system, which contributes to the synthesis of other kinds of macromolecules (insoluble proteins, polysaccharides, and/or lipids) using the catabolic products.

C. Different Tissues and Organs

A wide variety of results are available on the chromatographic pattern of various aminoacyl-tRNA's from different organs of the same species as well as of different species. Before each case is analyzed, it should be mentioned that the relative proportions of tRNA's for various amino acids are quite similar among most mammalian species and among tissues or organs of the same animal (*133*). Accordingly, we do not expect drastic changes in many tRNA species among the different organs or species.

Holland and his co-workers (*133–135*) made a wide survey of aminoacyl-tRNA's of several organs of several species. The tRNA's isolated from the brain, liver, kidney, and skeletal muscles of ox, rabbit, mouse, and chick were studied for alanine, glycine, leucine, lysine, serine, phenylalanine, threonine, tyrosine, and valine using MAK column chromatography. The elution profiles of most of the aminoacyl-tRNA's tested were similar, irrespective of tissue, organ, or species. The reproducibility was checked by *in vitro* vs. *in vivo* charging, by charging with heterologous enzymes, and by reversing radioactive isotopes. The following are a few cases that showed some differences. Seryl-tRNA from the liver of mouse and rabbit has an extra minor peak that does not exist in kidney or muscle tRNA. This is due to changes in both the synthetase and tRNA. The elution profiles of tyrosyl-tRNA and leucyl-tRNA of avian and mammalian cells are different although different organs of chick have the same profile for tyrosyl-tRNA.

TABLE II
CHROMATOGRAPHIC CHARACTERISTICS AND SOURCE OF AMINOACYLATED tRNA[a]

Aminoacyl-tRNA	Species	Organ	Characteristic
Alanine	Rabbit	Muscle	No detectable differences[b] (*133*)
		Liver	
	Mouse	Muscle	
		Embryo	
Arginine	Rabbit and ox	Brain	Difference from liver and kidney noted in brain[c] (*136*)
		Liver	
		Kidney	
Glycine	Rabbit	Kidney	No detectable differences[b] (*133*)
		Liver	
Leucine	Rabbit	Brain	No detectable differences[b] (*133*)
		Liver	
		Muscle	
	Mouse	Liver	Differences between avian and mammalian sources[b] (*133*)
		Brain	
		Fibroblasts	
Lysine	Rabbit	Liver	No detectable differences[b] (*133*)
		Muscle	
		Kidney	
	Mouse	Muscle	
		Embryo	
	Human	Liver	
Methionine	Rabbit and ox	Brain	Difference from liver and kidney noted in brain[c] (*136*)
		Liver	
		Kidney	
Phenylalanine	Rabbit	Liver	No detectable differences[b] (*133*)
		Muscle	
		Kidney	
	Mouse	Liver	
		Embryo	
	Chick	Liver	
	Ox	Muscle	
Serine	Rabbit	Liver	Difference noted between kidney, muscle, and liver[b] (*134*)
		Kidney	
		Muscle	
	Mouse	Liver	
		Kidney	
		Muscle	
	Rabbit and ox	Brain	Difference from liver and kidney noted in brain[c] (*136*)
		Liver	
		Kidney	
Threonine	Rabbit	Brain	No detectable differences[b] (*133*)
		Liver	
		Muscle	
		Kidney	

TABLE II (*Continued*)

Aminoacyl-tRNA	Species	Organ	Characteristic
	Mouse	Liver	
	Chick	Liver	
Tyrosine	Rabbit	Liver	No detectable differences between mammalian organs; difference between avian and mammalian sources[b] (*133*)
	Mouse	Liver	
		Kidney	
		Embryo	
	Human	Liver	
	Hamster	Liver	
	Chick	Liver	
		Brain	
Valine	Rabbit	Muscle	No detectable differences[b] (*133*)
		Liver	
	Mouse	Liver	

[a] An extension of the Table arranged by Taylor *et al.* (*133*).
[b] Analyzed by MAK column chromatography.
[c] Analyzed by reversed-phase column chromatography.

Hatfield and Caicuts (*136*) examined several aminoacyl-tRNA's from the brain, liver, and kidney of ox and rabbit by reversed-phase column chromatography, and found that the elution profiles of methionyl-, arginyl-, and seryl-tRNA's of the brain differed from those of other organs tested. A summary of the elution characteristics and source of tRNA is given in Table II.

D. Differentiation and Development

Aminoacyl-tRNA's from immature red cells of a 4-day-old avian embryo and from adult reticulocytes have been compared (*137*). Of five aminoacyl-tRNA's (arginine, leucine, methionine, serine, and tyrosine) examined by MAK and Freon reversed-phase columns, only methionyl-tRNA showed differences in the relative amounts of the two peaks. Since aminoacylation was carried out *in vivo*, the alteration of the methionine tRNA pattern during the development of red blood cells could have been due to the change either in tRNA itself or in the synthetase.

Changes in the distribution of lysyl-tRNA's during early embryogenesis of the sea urchin have also been observed (*138*). The lysine tRNA of the unfertilized egg and the two-celled stage were compared in the particulate fraction, containing ribosomes and mitochondria, and in the soluble fraction. Lysyl-tRNA of the sea urchin has two components on a MAK column. The distribution of the two were different in both the particulate fraction and the soluble fraction in the two stages of development.

The cell-free protein synthesizing system containing endogenous mRNA provides a useful tool for the analysis of the mechanism of cell differentiation. The ratio of incorporation of tyrosine to leucine in cell-free extracts of pupae of *Tenebrio molitor* rises as much as tenfold during the course of adult development (*139*). The adult cuticular protein, produced toward the end of pupation, contains tyrosine at high concentration. The ratio of incorporation of tyrosine to leucine is always high, if the supernatant fraction (both enzyme fraction and tRNA) isolated from the last stages of adult development is used, even with first-day pupal ribosomes and the mRNA complex. In addition, the tryptic digest of the *in vitro* product is similar to that of the adult cuticular protein. The result can be interpreted to mean that the first-day pupae contain mRNA for cuticular protein. However, for translation of the message, both tRNA and enzyme from the late stage are necessary. In *Tenebrio*, the juvenile hormone controls the translation of a specific mRNA in a manner similar to pupation (*140*). The hormone induces the second pupal molt. Thus, when enzymes and tRNA from hormone-treated animals were used, a low ratio of Tyr:Leu incorporation was always obtained irrespective of the source of the ribosome mRNA complex. Recently Ilan (*141*) has found a new species of leucyl-tRNA and a new leucyl-tRNA synthetase activity before the adult molt. On the other hand, hormone-treated animals, which go through an additional pupal stage, do not show the new tRNA or synthetase activity.

Modification of the translation machinery has been considered as one of the possible mechanisms for amino acid substitutions of variable regions of immunoglobulin. To test this idea, Mach *et al.* (*142*) analyzed aminoacyl-tRNA patterns of plasmocytoma tumors, which produce different but homogeneous immunoglobulins. They found a difference in leucine tRNA and threonine tRNA when the tRNA's were fractionated on MAK columns. Leucyl-tRNA gave three peaks. The relative amounts of the first two peaks were different in tumor IV and 5563, and the second peak was absent in tumor 41. In the case of threonyl-tRNA, one of the peaks was missing in tumor 41. These differences were the result of differences in tRNA preparations, since the same liver synthetase preparation was used for charging all the tumor tRNA's. Yang and Novelli (*143*) compared more tRNA's from plasma cell tumors, MPC62 and MPC47, by reversed phase column chromatography. Among the seven specific tRNA's tested, they found a drastic difference in the seryl-tRNA pattern of two tumors. By Freon reversed-phase column chromatography, Mushinski and Potter (*144*) compared the leucyl-tRNA profiles of five different mouse plasma cell tumors, all of which produce κ-type light chains, with each other and with that of normal liver. They found differences in the profiles of the different tumors, which were also different from the profile of normal liver. These differences consisted of the nearly total loss of certain peaks or an increase in the relative amounts of other peaks compared with similar chromatographic patterns of normal liver tRNA. They also showed that the change in pattern was due to tRNA but not to the preparation of the synthetase. The fact that these quantitative variations took place within a very closely related group of similarly differentiated tissues led them to suggest that genes for the different leucine tRNA's are under independent control.

Even if those differences of the elution patterns really reflect the differences in the spectra of tRNA's used in these tumor cells, it is not yet certain whether or not the specific changes in tRNA are related to the production of different immunoglobulins or related to the nature of the malignancy of the cells.

Anderson and Cherry (*145*) analyzed the leucine tRNA's of soybean cotyledons and hypocotyls. Leucine tRNA of cotyledon can be fractionated into six peaks, Leu_1 to Leu_6 by reversed phase chromatography, while only the first four of these are found in the hypocotyl. When a heterologous enzyme preparation was used, the hypocotyl synthetase preparation only slightly acylated the last two peaks of the cotyledon tRNA, those not found in the homologous hypocotyl system, and the charging of hypocotyl tRNA with cotyledon enzyme revealed a small amount in the last two components. That is, hypocotyl contained a small amount of Leu_5 and Leu_6 but the hypocotyl enzyme could not recognize them. Cotyledons seem to have either a different enzyme from hypocotyl or two enzymes, one similar to hypocotyl, the other capable of acylating Leu_5 and Leu_6. In any case, these results seem to indicate that only four leucyl-tRNA's are engaged in protein synthesis in hypocotyl or that Leu_5 and Leu_6 are utilized at the rate-limiting level. It was shown that Leu_5 and Leu_6 were not derived from mitochondria.

Vold and Sypherd (*146*) searched for differences in the MAK chromatographic pattern of aminoacyl-tRNA's from wheat embryo and seedlings. Among eleven aminoacyl-tRNA's studied, the authors found that the profiles for lysine, proline, and serine showed slight but significant differences. Their tRNA preparations included tRNA from mitochondria or chloroplasts. Therefore, the possibility remains that the differences they observed may be contributed by those subcellular organelles.

E. Neoplasia

Several cell lines of normal and tumor origin have been examined for tRNA patterns (*134, 135*). The MAK column elution patterns of tRNA's isolated from the HeLa cells (human cervical carcinoma), MBK (bovine kidney epithelial cells), hamster cells transformed by adenovirus 3, and chick embryo fibroblasts have been observed for alanine, glycine, leucine, lysine, phenylalanine, threonine, tyrosine, and valine tRNA's. Most of these showed similar elution profiles for a given amino acid irrespective of the origin or the kind of cell lines. The relative amounts of the two components of glycyl-tRNA of the HeLa cells were different from those of other cells. Holland *et al.* (*135*) found the elution characteristics of tyrosyl-tRNA specific to fibroblastic and epithelial cell types. Thus tyrosyl-tRNA of MBK, monkey kidney cells (CU-1), and human amnion epithelial cells eluted on the MAK column showed single peaks with similar patterns, while all fibroblast tyrosyl-tRNA's showed single peaks but eluted much later than the epithelial or organ type. This was shown to be true with either continuously cultured lines or primary culture cells when mouse L cells, BHK21, and continuously cultured avian cells were compared with their primary culture counterparts. An exception is found in continuous human fibroblast cells (line WI-38), which contain epithelial or organ-type profiles. It has also been demonstrated that a fibroblast type of tyrosine tRNA is present in fibroblasts of body tissues of chick embryo. Only tumor cell lines provided exceptions to the general observation that fibroblast cells have a fibroblastic type of tyrosine tRNA and epithelial cells contain a predominantly epithelial type or organ type of tyrosine tRNA. Thus HeLa cells, Ehrlich ascites tumor cells (mouse carcinoma) and adeno-7-transformed hamster cells contain both types, which can be aminoacylated with enzyme from any source; SV40-transformed hamster cells and Rous virus-transformed hamster cells contain only the fibroblastic type; and human lymphatic leukemia cells (white blood cells) contain only the epithelial-organ type. The growth conditions of the cells do not influence their characteristic profiles. The origin of the two tyrosine tRNA's is not known. There could be two different genes for tyrosine tRNA in the cell, and either of the genes could be expressed, depending on the epithelial or fibroblastic type. Or there could be one gene, with one tRNA the modified product of the other. In any case, again the question to be answered is whether the difference between the two types occurs as a trivial consequence of differentiation or whether the difference itself plays a role in differentiation. The presence of two types of tyrosine tRNA seems to be specific for certain tumor cells, although this is not always true. Here again, the biochemical characteristics of these two tyrosine tRNA's is not known, and whether or not the characteristic pattern is related to tumorogenesis or to cell transformation remains to be shown.

Taylor *et al.* (*134*) compared the MAK column chromatographic patterns of tRNA from Ehrlich ascites tumor and mouse sarcoma-1 cells and from normal rabbit or mouse tissues. The elution profiles of alanyl-, leucyl-, lysyl-, and threonyl-tRNA's of tumors and normal tissues were nearly identical. However, the glycyl-, phenylalanyl-, seryl-, and tyrosyl-tRNA's of the Ehrlich ascites tumor and of normal tissues gave different profiles, and the phenylalanyl-tRNA pattern was different between mouse sarcoma-1 and normal cells. In all cases, a definite shift in the eluting position was observed, implying that normal tRNA had undergone some modification. However, the relationship between chromatographic change and a generation of tumor is not known.

Axel, Weinstein, and Farber (*147*) examined patterns of tRNA during carcinogenesis by comparing tRNA of normal rat liver and tRNA from the liver of the animal during the feeding of a hepatic carcinogen, ethionine. Ethionine is known to ethylate rat liver tRNA, and the fact that radioactivity of the labeled ethyl-group coincides fairly well with the OD_{260} profile of bulk tRNA indicates that many species of tRNA are ethylated. Ethylated tRNA's have amino acid acceptance capacity equivalent to that of normal tRNA. The leucyl-tRNA of an animal on an ethionine diet for one month, during which time there was no appearance of hepatoma, had a different elution profile on the MAK column. Whereas the normal tRNA had two major peaks, ethylated tRNA had only one of the two peaks. The coding property of the two kinds of leucyl-tRNA revealed that ethylated leucyl-tRNA responded only to poly (U,C) while normal leucyl-tRNA responded to both poly (U,C) and poly (U,G). The same workers also examined leucine tRNA from a hepatoma induced one year before by ethionine feeding but maintained by transplantation and a normal diet. The tumor gave the normal leucyl-tRNA profile. Therefore, the change observed during ethionine feeding was not maintained in the tumor itself, presumably because ethionine was no longer present. Thus the possibility here is that alterations in tRNA may play a role in the initiation of hepatoma.

Baliga *et al.* (*148*) compared MAK column elution profiles of aminoacyl-tRNA's for 18 amino acids from Novikoff hepatoma with those of normal liver and found that histidyl-, tyrosyl-, and asparaginyl-tRNA showed the occurrence in a hepatoma of a new species in addition to the normal component. Small shifts in the elution profile of six other tRNA's were also noted by eluting at a NaCl concentration higher than the normal counterpart.

V. Conclusion

The main theme of the present review is to consider the possible role of tRNA in cell differentiation, the most fundamental level of which is the organized series of turning-on and turning-off processes of numerous genes involved in cell metabolism. We have discussed various aspects of translational control through tRNA. There are numerous examples of alterations (modifications and/or new syntheses) of tRNA's and of the chemical mechanisms by which such can be achieved. The examples cover virus infection, embryo development, different tissues, different growth conditions, and neoplasia. In each instance, the question remains whether or not the alteration in tRNA plays a critical role in the control. Needless to say, the trivial alternatives are: (a) the secondary effect of some other primary process, about which we know nothing; and (b) various kinds of artifacts that alter tRNA during the isolation procedure. The artifact could be quite subtle in the sense that different cells may have different cell components responsible for changing tRNA during its preparation. Upon discovery of an alteration, every effort must be exerted to eliminate the possibility of artifacts. Factors that could lead to erroneous conclusions from the chromatographic profiles are: (a) nuclease action during tRNA isolation or during aminoacylation reaction; (b) aggregation of tRNA; (c) conformational change of tRNA; and (d) incomplete charging.

For conclusive evidence that tRNA plays a key role in cell differentiation, more information should eventually be obtained for the following criteria:

1. Chemical (structural) nature of the alteration.
2. Elimination of artifacts.
3. New synthesis or modification of preexisting tRNA.
4. Functional change by alteration—amino acid acceptance, ribosome binding, and codon response.
5. Frequency of actual usage of tRNA—polysome-bound frequency.
6. *In vitro* demonstration that the tRNA alteration is the cause of the differentiation.
7. Genetic basis of the control—identification of the gene locus ultimately responsible for the alteration.
8. Interrelation of the control by tRNA alteration with other related metabolic processes—transcription, DNA replication.

In conclusion, the study of the regulatory role of tRNA in cell differentiation is in its infancy and still remains a hypothesis, although its future seems promising, particularly in the animal systems where mRNA is stable and exists without immediate translation. It is evident from this review that this subject includes a wide variety of phenomena and principles, each of which should be subjected to rigorous detailed study before any conclusion is given on the biological significance of the phenomenon.

Acknowledgments

This review was supported in part by Grant GM10923 from the National Institutes of Health and Grant E398B from the American Cancer Society, New Jersey Division.

We would like to thank Mrs. K. Dannenhauer, Dr. J. Kan, and Dr. J. L. Arceneaux for their critical reading of the manuscript.

References

1. F. Jacob and J. Monod, *J. Mol. Biol.* **3,** 318 (1961).
2. W. F. Dove, *Ann. Rev. Genet.* **2,** 305 (1968).
3. See *Cold Spring Harbor Symp. Quant. Biol.* **34,** (1969).
4. P. R. Gross, *Ann. Rev. Biochem.* **37,** 631 (1968).
5. F. H. C. Crick, *J. Mol. Biol.* **19,** 548 (1966).
6. E. B. Keller and R. S. Anthony, *Federation Proc.* **22,** 231 (1963).
7. T. Yamane and N. Sueoka, *Proc. Natl. Acad. Sci. U.S.* **51,** 1178 (1964).
8. A. N. Baldwin and P. Berg, *J. Biol. Chem.* **241,** 831 (1966).
9. R. Calender and P. Berg, *Biochemistry* **5,** 1681 (1966).
10. C. T. Yu, *Cold Spring Harbor Symp. Quant. Biol.* **31,** 565 (1966).
11. See review article: M. M. K. Nass, *Science* **165,** 25 (1969). See also Barnett and Brown (*108*) and Barnett *et al.* (*109*).
12. C. Yanofsky, D. R. Helinski, and B. D. Maling, *Cold Spring Harbor Symp. Quant. Biol.* **26,** 11 (1961).
13. S. Benzer and S. P. Champe, *Proc. Natl. Acad. Sci. U.S.* **47,** 1025 (1961).
14. M. R. Capecchi and G. N. Gussin, *Science* **149,** 417 (1965).
15. D. L. Engelhardt, R. E. Webster, R. C. Wilhelm, and N. D. Zinder, *Proc. Natl. Acad. Sci. U.S.* **54,** 1791 (1965).
16. J. D. Smith, J. N. Abelson, B. F. C. Clark, H. M. Goodman, and S. Brenner, *Cold Spring Harbor Symp. Quant. Biol.* **31,** 479 (1966).
17. N. Sueoka and T. Yamane, *Proc. Natl. Acad. Sci. U.S.* **48,** 1454 (1962).
18. N. Sueoka and T. Yamane, *in* "Symposium on Informational Macromolecules" (H. J. Vogel, V. Bryson, and J. O. Lampen, eds.), p. 205. Academic Press, New York, 1963.
19. B. N. Ames and P. E. Hartman, *Cold Spring Harbor Symp. Quant. Biol.* **28,** 349 (1963).
20. A. Newton, J. R. Beckwith, D. Zipser, and S. Brenner, *J. Mol. Biol.* **14,** 290 (1965).
21. A. Newton, *Cold Spring Harbor Symp. Quant. Biol.* **31,** 181 (1966).
22. G. S. Stent, *Science* **144,** 816 (1964).
23. J. R. Roth, D. F. Silbert, G. R. Fink, M. J. Voll, D. Antón, P. E. Hartman, and B. N. Ames, *Cold Spring Harbor Symp. Quant. Biol.* **31,** 383 (1966).
24. H. A. Itano, *in* "Abnormal Hemoglobins in Africa" (J. H. P. Jonxis, ed.), p. 3. Oxford Univ. Press, London and New York, Blackwell, Oxford, 1963.
25. N. Sueoka and T. Kano-Sueoka, *Proc. Natl. Acad. Sci. U.S.* **52,** 1535 (1964).
26. G. D. Novelli, *Ann. Rev. Biochem.* **36,** 449 (1967).
27. T. Yamane, *J. Mol. Biol.* **14,** 618 (1965).
28. A. D. Kelmers, G. D. Novelli, and M. P. Stulberg, *J. Biol. Chem.* **240,** 3979 (1965).
29. J. Kan, T. Kano-Sueoka, and N. Sueoka, *J. Biol. Chem.* **243,** 5584 (1968).
30. F. O. Wettstein, *Cold Spring Harbor Symp. Quant. Biol.* **31,** 595 (1966).
31. T. Kano-Sueoka and N. Sueoka, *Proc. Natl. Acad. Sci. U.S.* **62,** 1229 (1969).
32. C. D. Yegian and G. S. Stent, *J. Mol. Biol.* **39,** 45 (1969).
33. W. F. Anderson, *Proc. Natl. Acad. Sci. U.S.* **62,** 566 (1969).
34. H. Subak-Sharpe and J. Hay, *J. Mol. Biol.* **12,** 924 (1965).
35. H. Subak-Sharpe, W. M. Shepherd, and J. Hay, *Cold Spring Harbor Symp. Quant. Biol.* **31,** 583 (1966).

36. H. Subak-Sharpe, *Symp. Soc. Gen. Microbiol.* **18,** 47 (1968).
37. W. T. Hsu, J. W. Foft, and S. B. Weiss, *Proc. Natl. Acad. Sci. U.S.* **58,** 2028 (1967).
38. S. B. Weiss, W. T. Hsu, J. W. Foft, and N. H. Scherberg, *Proc. Natl. Acad. Sci. U.S.* **61,** 114 (1968).
39. V. Daniel, S. Sarid, and U. Z. Littauer, *Proc. Natl. Acad. Sci. U.S.* **60,** 1403 (1968).
40. V. Daniel, S. Sarid, and U. Z. Littauer, *FEBS Letters* **2,** 39 (1968).
41. P. P. Hung and L. R. Overby, *J. Biol. Chem.* **243,** 5525 (1968).
42. S. Nishimura and G. D. Novelli, *Biochim. Biophys. Acta* **80,** 574 (1964).
43. S. Nishimura and G. D. Novelli, *Proc. Natl. Acad. Sci. U.S.* **53,** 178 (1965).
44. A. A. Bayev, I. Fodov, A. D. Mirzabekov, V. D. Axelrod, and L. Y. Kazarinova, *Molekularnaya Biol.* **1,** 859 (1967).
45. K. Oda, F. Kimura, F. Harada, and S. Nishimura, *Biochim. Biophys. Acta* **179,** 97 (1969).
46. T. Seno, M. Kobayashi, and S. Nishimura, *Biochim. Biophys. Acta* **174,** 408 (1969).
47. T. Seno, M. Kobayashi, and S. Nishimura, *Biochim. Biophys. Acta* **182,** 280 (1969).
48. N. Imura, G. B. Weiss, and R. W. Chambers, *Nature* **222,** 1147 (1969).
49. T. Kano-Sueoka and N. Sueoka, *J. Mol. Biol.* **37,** 475 (1968).
50. W. J. Gartland and N. Sueoka, *Proc. Natl. Acad. Sci. U.S.* **55,** 948 (1966).
51. K. H. Muench, *Cold Spring Harbor Symp. Quant. Biol.* **31,** 539 (1966).
52. T. Ishida and N. Sueoka, *Proc. Natl. Acad. Sci. U.S.* **58,** 1080 (1967).
53. T. Ishida and N. Sueoka, *J. Biol. Chem.* **243,** 5329 (1968).
54. W. J. Gartland, T. Ishida, M. Nirenberg, and N. Sueoka, *J. Mol. Biol.* **44,** 403 (1969).
55. T. Lindahl, A. Adams, and J. R. Fresco, *Proc. Natl. Acad. Sci. U.S.* **55,** 941 (1966).
56. T. Lindahl, A. Adams, M. Geroch, and J. R. Fresco, *Proc. Natl. Acad. Sci. U.S.* **57,** 178 (1967).
57. A. Adams, T. Lindahl, and J. R. Fresco, *Proc. Natl. Acad. Sci. U.S.* **57,** 1684 (1967).
58. N. Sueoka and J. Hardy, *Arch. Biochem. Biophys.* **125,** 558 (1968).
59. T. Ishida and N. Sueoka, unpublished results.
60. E. Borek and P. R. Srinivasan, *Ann. Rev. Biochem.* **35,** 275 (1966).
61. L. R. Mandel and E. Borek, *Biochem. Biophys. Res. Commun.* **4,** 14 (1961).
62. M. Gold, J. Hurwitz, and M. Anders, *Biochem. Biophys. Res. Commun.* **11,** 107 (1963).
63. E. Fleissner and E. Borek, *Biochemistry* **2,** 1093 (1963).
64. J. Hurwitz, M. Gold, and M. Anders, *J. Biol. Chem.* **239,** 3462 (1964).
65. J. Hurwitz, M. Gold, and M. Anders, *J. Biol. Chem.* **239,** 3474 (1964).
66. B. C. Baguley and M. Staehelin, *Biochemistry* **7,** 45 (1968).
67. P. R. Srinivasan and E. Borek, *Biochemistry* **3,** 616 (1964).
68. A. Peterkofsky, C. Jesensky, and J. D. Capra, *Cold Spring Harbor Symp. Quant. Biol.* **31,** 515 (1966).
69. J. D. Capra and A. Peterkofsky, *J. Mol. Biol.* **33,** 591 (1968).
70. U. Z. Littauer, M. Revel, and R. Stern, *Cold Spring Harbor Symp. Quant. Biol.* **31,** 501 (1966).
71. E. Wainfan, P. R. Srinivasan, and E. Borek, *Biochemistry* **4,** 2845 (1965).
72. E. Wainfan, P. R. Srinivasan, and E. Borek, *Cold Spring Harbor Symp. Quant. Biol.* **31,** 525 (1966).
73. E. Tsutsui, P. R. Srinivasan, and E. Borek, *Proc. Natl. Acad. Sci. U.S.* **56,** 1003 (1966).

74. R. L. Hancock, *Cancer Res.* **27,** 646 (1967).
75. A. Mittelman, R. H. Hall, D. S. Yohn, and J. T. Grace, Jr., *Cancer Res.* **27,** 1409 (1967).
76. B. C. Baguley and M. Staehelin, *European J. Biochem.* **6,** 1 (1968).
77. A. M. Kaye and P. S. Leboy, *Biochim. Biophys. Acta* **157,** 289 (1968).
78. D. Pillinger and E. Borek, *Proc. Natl. Acad. Sci. U.S.* **62,** 1145 (1969).
79. B. S. Baliga, P. R. Srinivasan, and E. Borek, *Nature* **208,** 555 (1965).
80. M. Gold, R. Hausmann, U. Maitra, and J. Hurwitz, *Proc. Natl. Acad. Sci. U.S.* **52,** 292 (1964).
81. J. A. Carbon, L. Hung, and D. S. Jones, *Proc. Natl. Acad. Sci. U.S.* **53,** 979 (1965).
82. M. N. Lipsett, *J. Biol. Chem.* **240,** 3975 (1965).
83. B. Goehler and R. H. Doi, *Proc. Natl. Acad. Sci. U.S.* **56,** 1047 (1966).
84. R. S. Hayward and S. B. Weiss, *Proc. Natl. Acad. Sci. U.S.* **55,** 1161 (1966).
85. M. N. Lipsett and A. Peterkofsky, *Proc. Natl. Acad. Sci. U.S.* **55,** 1169 (1966).
86. V. M. Ingram and J. A. Sjöquist, *Cold Spring Harbor Symp. Quant. Biol.* **28,** 133 (1963).
87. J. Apgar, G. A. Everett, and R. W. Holley, *J. Biol. Chem.* **241,** 1206 (1966).
88. M. H. Makman and G. L. Cantoni, *Biochemistry* **5,** 2246 (1966).
89. J. J. Furth, J. Hurwitz, R. Krug, and M. Alexander, *J. Biol. Chem.* **236,** 3317 (1961).
90. D. D. Anthony, J. L. Starr, D. S. Kerr, and D. A. Goldthwait, *J. Biol. Chem.* **238,** 690 (1963).
91. V. R. Glišin and M. V. Glišin, *Proc. Natl. Acad. Sci. U.S.* **52,** 1548 (1964).
92. P. R. Gross, K. Kraemer, and L. I. Malkin, *Biochem. Biophys. Res. Commun.* **18,** 569 (1965).
93. R. Bachvarova, E. H. Davidson, V. Allfrey and A. E. Mirsky, *Proc. Natl. Acad. Sci. U.S.* **55,** 358 (1966).
94. L. Stavy and P. R. Gross, *Proc. Natl. Acad. Sci. U.S.* **57,** 735 (1967).
95. H. A. Sober, ed., "Handbook of Biochemistry," selected data for Molecular Biology, pH-62. Chem. Rubber Publ. Co. Cleveland, Ohio, 1968.
96. R. W. Holley, J. Apgar, G. A. Everett, J. T. Madison, M. Marquisee, S. H. Merrill, J. R. Penswick, and A. Zamir, *Science* **147,** 1462 (1965).
97. R. H. Hall, M. J. Robins, L. Stasiuk, and R. Thedford, *J. Am. Chem. Soc.* **88,** 2614 (1966).
98. K. Biemann, S. Tsunakawa, T. Sonnenbichler, H. Feldmann, D. Dutting, and H. G. Zachau, *Angew. Chem.* **78,** 600 (1966).
99. W. J. Burrows, D. J. Armstrong, F. Skoog, S. M. Hecht, J. T. A. Boyle, N. J. Leonard, and J. Occolowitz, *Science* **161,** 691 (1968).
100. F. Harada, H. J. Gross, F. Kimura, S. H. Chang, S. Nishimura, and U. L. RajBhandary, *Biochem. Biophys. Res. Commun.* **33,** 299 (1968).
101. M. L. Gefter and R. L. Russell, *J. Mol. Biol.* **39,** 145 (1969).
102. F. Fittler and R. H. Hall, *Biochem. Biophys. Res. Commun.* **25,** 441 (1966).
103. R. H. Hall, *in* "Biochemistry and Physiology of Plant Growth Substances" (F. Wightman and G. Setterfield, eds.), p. 47. Runge Press, Ottawa, Canada, 1968.
104. F. C. Neidhardt and C. F. Earhart, *Cold Spring Harbor Symp. Quant. Biol.* **31,** 557 (1966).
105. B. L. Strehler, D. D. Hendley, and G. P. Hirsch, *Proc. Natl. Acad. Sci. U.S.* **57,** 1751 (1967).

106. M. J. Chrispeels, R. F. Boyd, L. S. Williams, and F. C. Neidhardt, *J. Mol. Biol.* **31,** 463 (1968).
107. C. Ceccarini, R. Maggio, and G. Barbata, *Proc. Natl. Acad. Sci. U.S.* **58,** 2235 (1967).
108. W. E. Barnett and D. H. Brown, *Proc. Natl. Acad. Sci. U.S.* **57,** 452 (1967).
109. W. E. Barnett, D. H. Brown, and J. L. Epler, *Proc. Natl. Acad. Sci. U.S.* **57,** 1775 (1967).
110. A. Vescia, *Biochem. Biophys. Res. Commun.* **29,** 496 (1967).
111. O. O. Favorova, T. N. Spasokukotskaya, and L. L. Kiselev, *Molekulyarnaya Biol.* **2,** 69 (1968).
112. F. J. Kull and K. B. Jacobson, *Proc. Natl. Acad. Sci. U.S.* **62,** 1137 (1969).
113. T. Kano-Sueoka and N. Sueoka, *J. Mol. Biol.* **20,** 183 (1966).
114. L. C. Waters and G. D. Novelli, *Proc. Natl. Acad. Sci. U.S.* **57,** 979 (1967).
115. L. C. Waters and G. D. Novelli, *Biochem. Biophys. Res. Commun.* **32,** 971 (1968).
116. T. Kano-Sueoka, M. W. Nirenberg, and N. Sueoka, *J. Mol. Biol.* **35,** 1 (1968).
117. J. Kan, M. Nirenberg, and N. Sueoka, in preparation.
118. S. Hattman and P. H. Hofschneider, *J. Mol. Biol.* **35,** 513 (1968).
119. T. W. Tillack and D. W. E. Smith, *Virology* **36,** 212 (1968).
120. B. C. Baguley, P. L. Bergquist, and R. K. Ralph, *Biochim. Biophys. Acta* **138,** 51 (1967).
121. I. Kaneko and R. H. Doi, *Proc. Natl. Acad. Sci. U.S.* **55,** 564 (1966).
122. R. H. Doi, I. Kaneko, and R. T. Igarashi, *J. Biol. Chem.* **243,** 945 (1968).
123. T. Heyman, S. Serov, B. Desseaux, and J. Legault-Demare, *Biochim. Biophys. Acta* **145,** 596 (1967).
124. R. A. Lazzarini, *Proc. Natl. Acad. Sci. U.S.* **56,** 185 (1966).
125. R. A. Lazzarini and E. Santangelo, *J. Bacteriol.* **94,** 125 (1967).
126. R. H. Doi, I. Kaneko, and B. Goehler, *Proc. Natl. Acad. Sci. U.S.* **56,** 1548 (1966).
127. N. Sueoka, T. Kano-Sueoka, and W. J. Gartland, *Cold Spring Harbor Symp. Quant. Biol.* **31,** 571 (1966).
128. J. L. Arceneaux and N. Sueoka, *J. Biol. Chem.* **244,** 5959 (1969).
129. C. N. Kwan, D. Apirion, and D. Schlessinger, *Biochemistry* **7,** 427 (1968).
130. F. O. Wettstein and G. S. Stent, *J. Mol. Biol.* **38,** 25 (1968).
131. A. Shearn and N. H. Horowitz, *Biochemistry* **8,** 295 (1969).
132. A. Shearn and N. H. Horowitz, *Biochemistry* **8,** 304 (1969).
133. M. W. Taylor, C. A. Buck, G. A. Granger, and J. J. Holland, *J. Mol. Biol.* **33,** 809 (1968).
134. M. W. Taylor, G. A. Granger, C. A. Buck, and J. J. Holland, *Proc. Natl. Acad. Sci. U.S.* **57,** 1712 (1967).
135. J. J. Holland, M. W. Taylor, and C. A. Buck, *Proc. Natl. Acad. Sci. U.S.* **58,** 2437 (1967).
136. D. Hatfield and M. Caicuts, *Federation Proc.* **28,** 349 (1969).
137. J. C. Lee and V. M. Ingram, *Science* **158,** 1330 (1967).
138. S. S. Yang and D. G. Comb, *J. Mol. Biol.* **31,** 139 (1968).
139. J. Ilan, *J. Biol. Chem.* **243,** 5859 (1968).
140. J. Ilan, J. Ilan, and N. Patel, *J. Biol. Chem.* in press.
141. J. Ilan, *Cold Spring Harbor Symp. Quant. Biol.* **34,** (1969). In press.
142. B. Mach, H. Koblet, and D. Gros, *Cold Spring Harbor Symp. Quant. Biol.* **32,** 269 (1967).
143. W. K. Yang and G. D. Novelli, *Biochem. Biophys. Res. Commun.* **31,** 534 (1968).
144. J. F. Mushinski and M. Potter, *Biochemistry* **8,** 1684 (1969).

145. M. B. Anderson and J. H. Cherry, *Proc. Natl. Acad. Sci. U.S.* **62,** 202 (1969).
146. B. S. Vold and P. S. Sypherd, *Proc. Natl. Acad. Sci. U.S.* **59,** 453 (1968).
147. R. Axel, I. B. Weinstein, and E. Farber, *Proc. Natl. Acad. Sci. U.S.* **58,** 1255 (1967).
148. B. S. Baliga, E. Borek, I. B. Weinstein, and P. R. Srinivasan, *Proc. Natl. Acad. Sci. U.S.* **62,** 899 (1969).
149. A. A. Travers, *Nature* **223,** 1107 (1969).

N^6-(Δ^2-Isopentenyl)adenosine: Chemical Reactions, Biosynthesis, Metabolism, and Significance to the Structure and Function of tRNA

Ross H. Hall

Department of Biochemistry, McMaster University, Hamilton, Ontario, Canada

I. Introduction 57
II. Occurrence of N^6-(Δ^2-Isopentenyl)adenosine and Its Derivatives in Nature 60
III. N^6-(Δ^2-Isopentenyl)adenosine and Related Compounds in the Free State 61
IV. Synthesis of N^6-(Δ^2-Isopentenyl)adenosine 62
V. Synthesis of N^6-(Δ^2-Isopentenyl)-2-methylthioadenosine . . . 64
VI. Chemical Properties of N^6-(Δ^2-Isopentenyl)adenosine and Derivatives 64
VII. Chemical Reactions of N^6-(Δ^2-Isopentenyl)adenosine . . . 64
VIII. Role of N^6-(Δ^2-Isopentenyl)adenosine in the Function of tRNA . 67
A. Immobilization of the Δ^2-Isopentenyl Group with Iodine . . 67
B. Effect of the Deletion of the Δ^2-Isopentenyl Group from $tRNA^{Tyr}$ on the Codon–Anticodon Interaction 68
C. Effect of Deletion of Base "Y" from $tRNA^{Phe}$ 70
IX. Biosynthesis of N^6-(Δ^2-Isopentenyl)adenosine 70
X. Enzymatic Degradation of N^6-(Δ^2-Isopentenyl)adenosine . . 75
XI. Biological Activity of N^6-(Δ^2-Isopentenyl)adenosine 76
XII. Discussion of Possible Significance of N^6-(Δ^2-Isopentenyl)-adenosine to the Control of tRNA Function 80
Appendix 82
References 84

I. Introduction

Until recently, each nucleic acid was thought to be composed of four nucleoside structures. This view has changed markedly in the last fifteen years, and at present 9 deoxyribonucleosides (5 are modified nucleosides) and 39 ribonucleosides (35 are modified nucleosides), most of which are known to occur in DNA or in RNA, have been identified.[1] In fact, DNA

[1] A list of nucleotides and their properties has been compiled by Dunn and Hall (*1*). A more complete review of all the modified nucleosides has been prepared by Hall (*2*).

and RNA may be visualized as consisting of four major types of nucleosides, each additional component being a structural modification of one of the eight types of nucleosides. It is generally considered that these modified components confer unique properties critical to the biological functioning of nucleic acids; this concept adds a new dimension to the study of nucleic acid structure and function.

Modified nucleosides occur in DNA, rRNA, and tRNA; however, with respect to assessing their biological significance, most progress has been made with studies on tRNA. This progress arises mainly from the facts that the primary sequences of several tRNA molecules are now known (more are in the process of being determined) and that three biochemical functions may be tested. One of the principal generalizations from this work is that the modified nucleosides in tRNA occur in definite locations in the primary sequence; there is no evidence of random modification of tRNA structure. The modified nucleosides presumably are placed in specific locations for definite reasons, but nothing is yet known about their specific functions.

Because of their structural variety, it is impossible to treat the modified nucleosides as a single class of nucleic acid components. Therefore, in attempting to assess their specific significance to the function of nucleic acid molecules, one must consider each structural type separately. The majority of the known modified nucleosides constitute relatively simple modifications of the basic nucleoside structure, for example, replacement of OH with S (4-thiouridine), NH_2 with OH (inosine), or attachment of a methyl group. Although such nucleosides represent the simplest form of structural modification, they can give rise to changes in hydrogen-bonding, base-stacking characteristics, or covalent bonding capability. In other words, one such alteration at a strategic location in the nucleic acid molecule could cause a significant change in three-dimensional structure with a corresponding change in function.

In addition to this relatively simple type of structural modification, we now recognize the existence of hypermodified nucleosides that result from a more complex alteration. The hypermodified nucleoside we define by three distinct characteristics: (a) a relatively large side chain; (b) a functional group (organic chemistry definition) such as hydroxyl, carboxyl, and the allylic double bond in the side chain; (c) location adjacent to the 3′ end of the anticodon, with considerable potential bearing on the reactivity of the tRNA molecule.

The known anticodon sequences and the nucleoside adjacent to their 3′ ends are shown in Table I. Not all tRNA molecules have modified nucleosides in this position, although such molecules appear to be in a minority. In some cases, the nucleoside in this position has not yet been identified, presumably because its structure is sufficiently unusual to pre-

TABLE I
NUCLEOSIDE ADJACENT TO THE 3′ END OF THE ANTICODON

tRNA	Anticodon (5′ → 3′)	Adjacent nucleoside
Yeast		
Ala	I G C	1-Methylinosine (*76*)
Ile	I A U	*N*-(Nebularin-6-ylcarbamoyl)threonine (*3*, *77*)
Phe	Gm A A	Unidentified, "Y" (*41*)
Ser I	I G A	N^6-(Δ^2-Isopentenyl)adenosine (*8*)
Ser II	I G A	N^6-(Δ^2-Isopentenyl)adenosine (*8*)
Tyr	G Ψ A	N^6-(Δ^2-Isopentenyl)adenosine (*9*)
Val	I A C	Adenosine (*78*, *79*)
Escherichia coli		
fMet	C A U	Adenosine (*80*)
Met	C* A U	Unidentified (*81*)
Su^+_{iii} Tyr	C U A	N^6-(Δ^2-Isopentenyl)-2-methylthioadenosine (*6*, *35*, *39*)
Rat liver		
Ser	I G A	N^6-(Δ^2-Isopentenyl)adenosine (*10*)

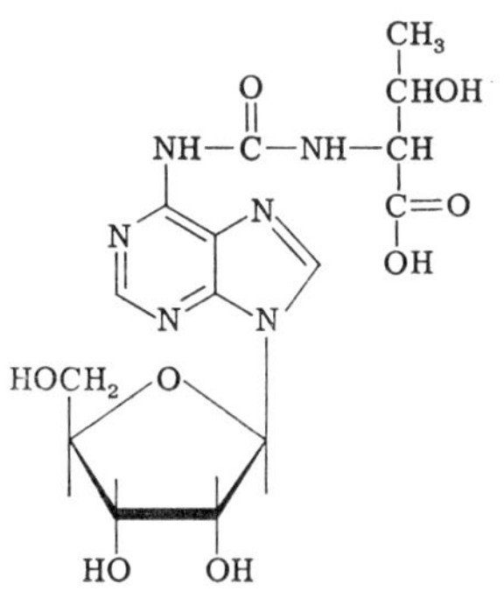

(I)

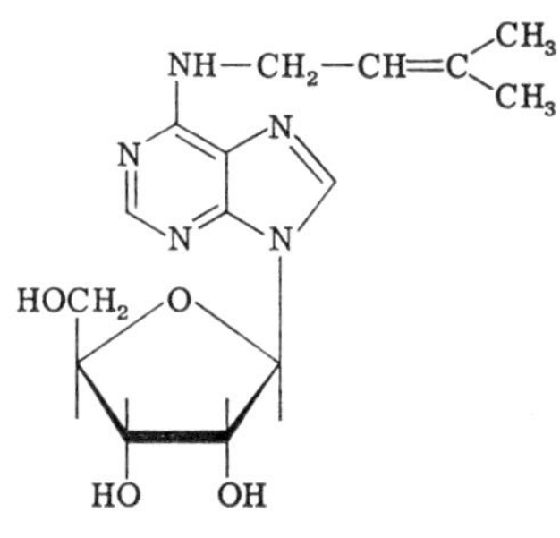

(II)

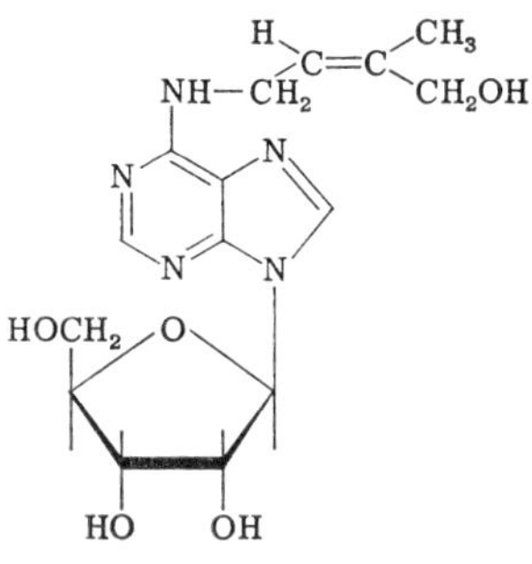

(III)

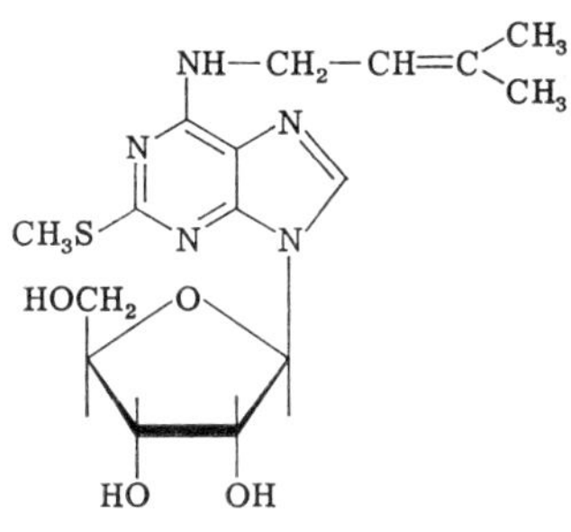

(IV)

vent easy recognition. Two nucleosides in this list satisfy the definition of a hyper-modified nucleoside. *N*-(Nebularin-6-ylcarbamoyl)threonine (I) has only recently been identified (*3*). The other, N^6-(Δ^2-isopentenyl)-adenosine (II), was first characterized in 1966 (*4, 5*) and has received much attention. Its significance to tRNA function, its mode of biosynthesis, and its chemical and biological properties in relation to its occurrence in tRNA are the subject of this article.

II. Occurrence of N^6-(Δ^2-Isopentenyl)adenosine and Its Derivatives in Nature

Transfer RNA's containing the Δ^2-isopentenyl group occur in most, if not all, organisms. Three such nucleosides have been identified in tRNA's: N^6-(Δ^2-isopentenyl)adenosine (II); *N*-(*cis*-4-hydroxy-3-methylbut-2-enyl)-adenosine (III); and N^6-(Δ^2-isopentenyl)-2-methylthioadenosine (IV). The natural distribution of these compounds is given in Table II.

The parent nucleoside (II) has been found in the tRNA's of all classes of organisms: bacteria, yeast, plant, and animal. Compound (III) has been found only in the tRNA's of plants, and compound (IV) only in those

TABLE II
OCCURRENCE OF N^6-(Δ^2-ISOPENTENYL)ADENOSINE AND DERIVATIVES IN UNFRACTIONATED tRNA

Source of tRNA	Moles/100 moles of total nucleosides[a] N^6-(Δ^2-Isopentenyl)-adenosine	N^6-(*cis*-4-Hydroxy-3-methylbut-2-enyl)-adenosine	N^6-(Δ^2-Isopentenyl)-2-methylthio-adenosine	Reference
Baker's yeast	0.06	—	—	(*33*)
Calf liver	0.05	—	—	(*33*)
Chick embryo	0.03	—	—	(*33*)
Human liver	0.05	—	—	(*33*)
Immature peas	0.003	0.005	—	(*82*)
Spinach leaves	0.02	0.01	—	(*82*)
Immature corn kernels	—	0.01	—	(*82*)
Mature corn kernels (seed)	—	Present	—	[b]
Escherichia coli B	—	—	Present	(*35, 39*)
Lactobacillus acidophilus	Present	—	—	(*43*)
Lactobacillus plantarum	Present	—	—	(*43*)
Tobacco pith cells grown in culture	Present	—	—	(*45*)

[a] These values are minimal.
[b] McLennan and Hall, unpublished data.

of *Escherichia coli*. In studies in my laboratory, neither (III) nor (IV) has been detected in the tRNA of mammalian tissues. It remains to be seen whether the phylogenetic segregation of (III) and (IV) will be upheld when more analytical data become available.

There appears to be a definite relationship among tRNA's that contain the Δ^2-isopentenyl group. N^6-(Δ^2-Isopentenyl)-2-methylthioadenosine (IV) has been detected in tyrosine tRNA (*6*), serine-I and -II tRNA's (*7*), and phenylalanine tRNA (*7, 7a*). The parent nucleoside, N^6-(Δ^2-isopentenyl)-adenosine (II) occurs in the serine tRNA (*8*) and tyrosine tRNA (*9*) of yeast and in rat liver serine tRNA (*10*). All these tRNA's respond to codons whose first letter is U. Further, Nishimura *et al.* (*7*) have shown that the serine-III tRNA of *E. coli*, which responds to codons AGC and AGU, does not contain (IV) but rather an analog of (I).

III. N^6-(Δ^2-Isopentenyl)adenosine and Related Compounds in the Free State

A number of compounds with structures related to (II) have been isolated from natural sources. The free base, N^6-(Δ^2-isopentenyl)adenine (V), has been isolated from liquid cultures of *Corynebacterium fascians* (*11, 12*) and *Agrobacterium tumefaciens* (*13*), pathogens known to cause abnormal development in plants. The *trans* analog of (II), N^6-(*trans*-4-hydroxy-3-methylbut-2-enyl)adenine (VI), its ribonucleoside and ribonucleotide have been isolated from extracts of *Zea mays* kernels (milk stage) (*14, 15*). The base (VI), termed zeatin, was identified by Letham *et al.* (*16*) and synthesized by Shaw *et al.* (*17*). Letham (*18*) isolated zeatin from coconut milk and has shown that it is probably the principal plant growth factor in coconut milk. Zeatin and its ribosyl derivative have been isolated from cultures of the puffball fungus, *Rhizopogon roseolus* (*19*). Koshimizu *et al.* (*20*) identified dihydrozeatin, N^6-(4-hydroxy-3-methylbutyl)adenine (VII), in extracts of the immature seeds of *Lupinus luteus*. The isolated compound is optically active with a negative rotation.

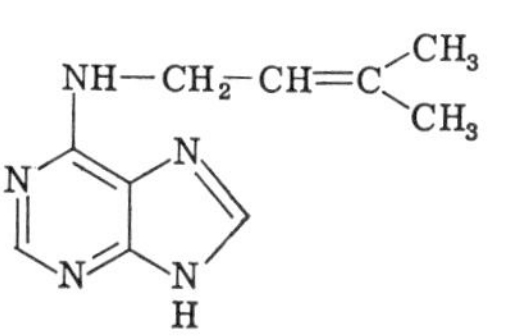

N^6-(Δ^2-Isopentenyl)adenine

(V)

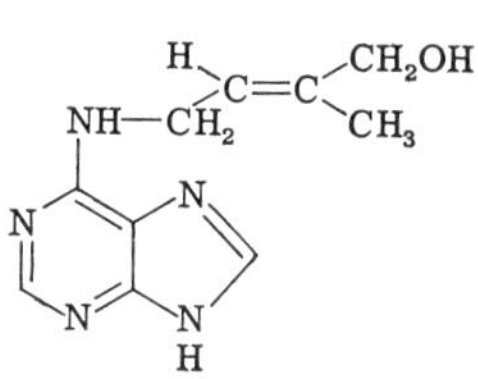

Zeatin

(VI)

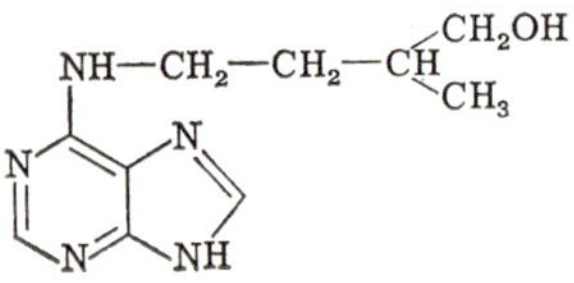

Dihydrozeatin
(VII)

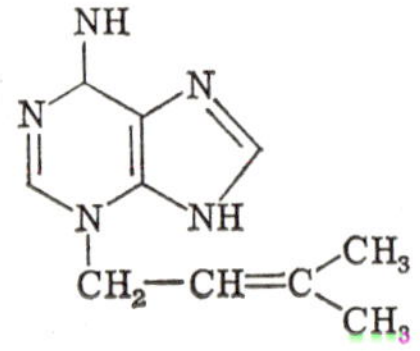

Triacanthine
(VIII)

The leaves of the honey locust tree, *Gleditsia triacanthos* L., contain an alkaloid material isolated and identified as 3-(Δ^2-isopentenyl)adenine (VIII) (*21*) and called triacanthine. The amount of VIII varies with the age of the leaf, reaching a maximum in young leaves and decreasing as the leaf ages (*22*). Triacanthine does not act directly as a plant growth factor (cytokinin activity) (*23*); however, when autoclaved, it is converted into an active substance. It has been suggested that rearrangement of the side chain from the N-3 position to the N-6 position may occur; on this basis, Rogozinska (*22*) suggested that triacanthine represents an inert storage form for the active cell division factor. This concept is attractive, in view of the fact that enzymatic degradation of (II) occurs in plant tissues (*24*). In order to maintain the required nucleoside level in the tissue, a continuous renewal of the supply would be necessary (see Section XI for a discussion of the biological activity of N^6-(Δ^2-isopentenyl)adenosine). Janot *et al.* (*25*) and Monseur and Adriaens (*26*) isolated triacanthine from the leaves of *Holarrhena floribunda* and *Chidlowia sanguinea*, although these two groups of workers did not identify their isolated compounds. Leonard and Deyrup (*21*) later showed that these natural products are identical to triacanthine.

Precedents exist for the occurrence of the side chain of compound (III) in nature. The plant *Galega officinalis* L. contains two guanidine derivatives, compounds (IX) and (X). Olomucki *et al.* (*27*) have demonstrated that the hydroxylated form (X) occurs in the *cis* configuration.

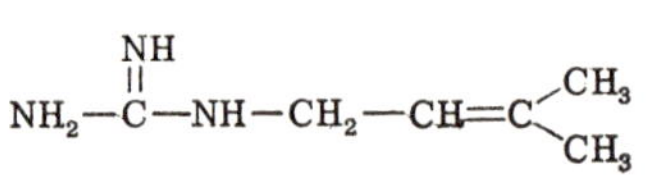

(IX)

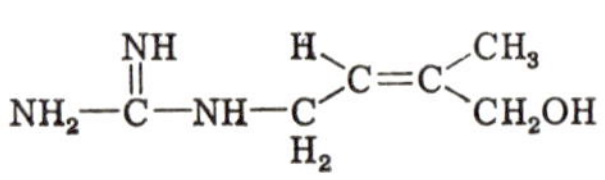

(X)

IV. Synthesis of N^6-(Δ^2-Isopentenyl)adenosine

N^6-(Δ^2-isopentenyl)adenosine (II) is readily synthesized by two different routes. In one method, the corresponding alkyl side chain in the form of the amine is condensed with 6-chloro-9-(β-D-ribofuranosyl)purine (XI) (*4*, *28*). In the second method, the corresponding alkyl side chain, in the form

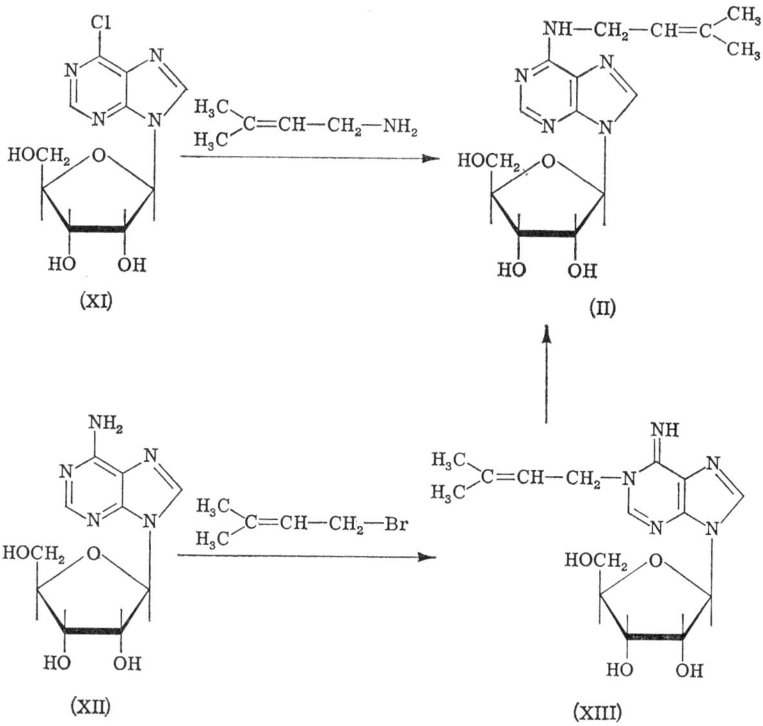

of the bromide, is used to alkylate adenosine (XII) directly (*29*, *30*, *31*). The initial compound formed is 1-(Δ^{2}-isopentenyl)adenosine (XIII), which rearranges to (II) when the N-1-substituted intermediate is heated in aqueous solution at pH 7.5.

The facile rearrangement of the N-1 isomer to the N-6 isomer raises the question as to which isomer exists in the native tRNA, even though the isolated nucleoside has been rigorously identified as the N-6 isomer. This question has a parallel in tRNA structure since both N-1 and N-6 methyladenosine occur in the nucleic acids of various organisms (*32*). The influence of the isopentenyl side chain on the oligonucleotide structure and chemical reactivities of the surrounding oligonucleotide segment of the tRNA molecule would vary greatly, depending on whether the side chain is attached at the N-1 or N-6 position of the adenine. In order to answer this question, we hydrolyzed tRNA with acid under conditions that would release the purine base without permitting the N-1 → N-6 rearrangement and found only the N-6 isomer (*33*). Grimm and Leonard (*29*) also demonstrated that under the conditions used for the isolation of (II) from tRNA (*4*, *8*, *34*), the N-1 → N-6 rearrangement could not occur.

V. Synthesis of N^6-(Δ^2-Isopentenyl)-2-methylthioadenosine (IV) (35)

When 2,6-*bis*-(methylthio)purine (XIV) is refluxed with Δ^2-isopentenylamine, it yields N^6-(Δ^2-isopentenyl)-2-methylthiopurine (XV). A mercuri salt of XV is prepared and condensed with 1-bromo-2,3,5-*O*-tribenzoylribofuranose. After deblocking, the desired nucleoside (IV) is obtained. Compound (XV) can also be prepared by condensing 2-methylthio-6-chloropurine with the corresponding alkylamine.

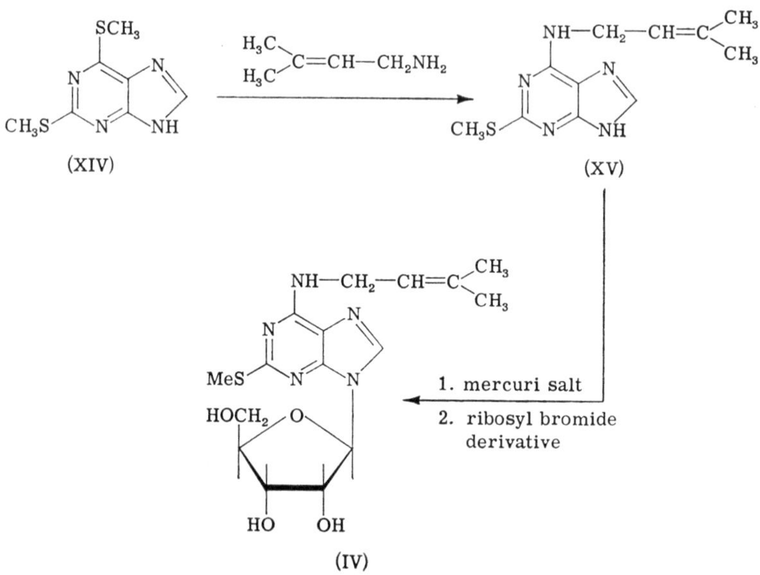

VI. Chemical Properties of N^6-(Δ^2-Isopentenyl)adenosine and Derivatives

The basic properties, together with the ultraviolet absorption spectra, are given in the Appendix. N^6-(Δ^2-isopentenyl)adenosine is a relatively unstable compound, and on storage it slowly decomposes to give two or three unidentified products. The rate of decomposition is not appreciably slowed even when the compound is stored in the dry state at $-70°$. The lability of this nucleotide can be attributed to its very reactive allylic double bond (see Section VII).

VII. Chemical Reactions of N^6-(Δ^2-Isopentenyl)adenosine

When N^6-(Δ^2-isopentenyl)adenosine (II) is hydrolyzed in dilute acid under conditions that sever the glycosyl bond, the free base is not obtained;

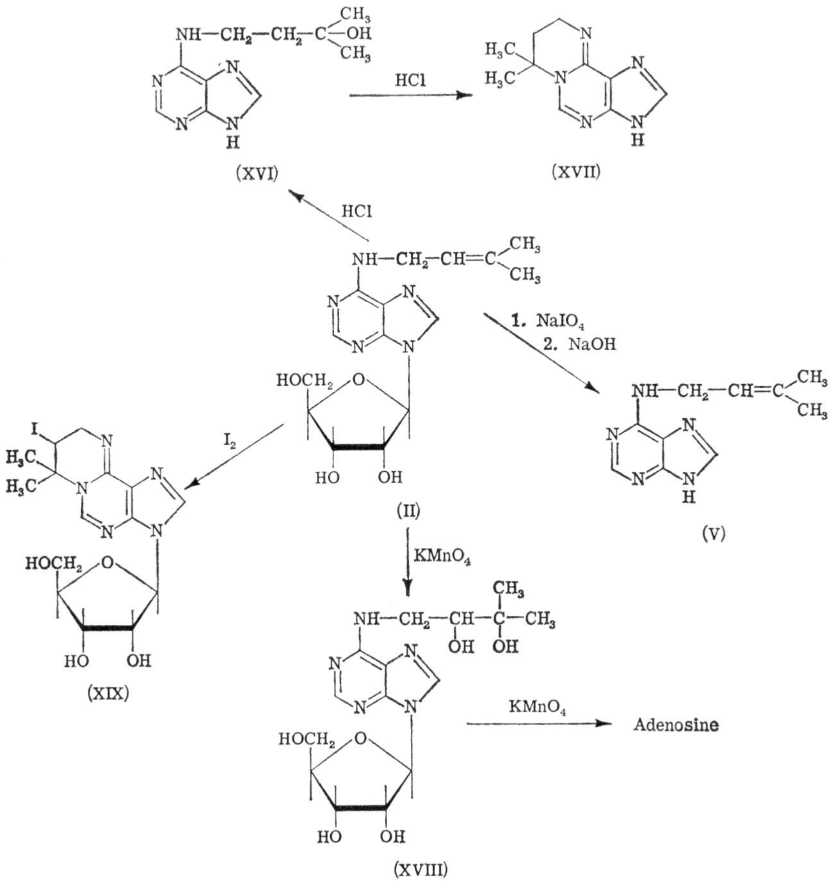

rather, the hydrated product, *N*^{6}-(3-hydroxy-3-methylbutyl)adenine (XVI), is formed. On continued acid treatment, the hydroxyl group is expelled to form what is presumably a carbonium intermediate that undergoes ring closure to form 3*H*-7,7-dimethyl-7,8,9-trihydropyrimido-[2,1-*i*]-purine[2] (XVII) (*4*). Compound (XVII), although stable to acid, is degraded by alkali solution to form a product that appears to be an imidazole derivative. The ready hydration of the allylic double bond under relatively mild chemical conditions suggests that the N-1 of the purine residue actively

[2] The numbering of this ring system is as follows:

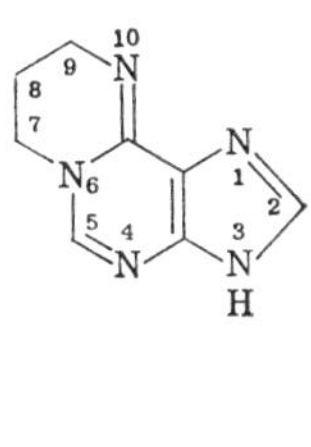

promotes reaction of the side chain. The spatial arrangement of the allylic double bond of the side chain and the N-1 position, therefore, may well play an important role in the biological function of N^6-(Δ^2-isopentenyl)-adenosine (Fig. 1). The hydration of the double bond of (I) by an enzyme present in many tissues to form N^6-(3-hydroxy-3-methylbutyl)adenosine (XX) underscores this point (see Section X).

The mechanism of formation of (XVI) and (XVII) in acid solution has been studied by Martin and Reese (*36*). Their results suggest that the glycosyl bond is broken, the free base (V) forming a carbonium ion that can either accept a hydroxyl group from the aqueous solution to form (XVI) or accept the proton to form (XVII). The relative rates of formation of the products are shown in Fig. 2.

The free base of compound (II), N^6-(Δ^2-isopentenyl)adenine (V), can be produced by treating (II) with periodate, followed by treatment with sodium hydroxide (*4*).

Treatment of compound (II) with dilute aqueous permanganate (pH 7.0) for 10 minutes at 25° gives the dihydroxylated product (XVIII) in approximately 50% yield. Adenosine in 30% yield is also formed. The cleavage mechanism of the side chain to form adenosine has not been investigated. The mildness of the cleavage reaction of the side chain has been exploited to remove the Δ^2-isopentenyl side chain from tRNA without disruption of the rest of the tRNA molecule (see Section IX).

Iodine reacts rapidly with compound (II) in aqueous solution under mild conditions (25°, pH 7.0) to form 7,7-dimethyl-8-iodo-7,8,9-trihydro-pyrimido-3-(β-D-ribofuranosyl)[2,1-*i*]purine (XIX) (33). The nuclear magnetic resonance (NMR) spectrum of (XIX) shows two single peaks at δ 1.68 and 1.84 (ratio 60:40); the two peaks together account for six protons. Therefore, these peaks represent the methyl groups, and their existence suggests the presence of two isomers. Further, the unsplit nature of each peak indicates the absence of a proton at C-7, which supports the

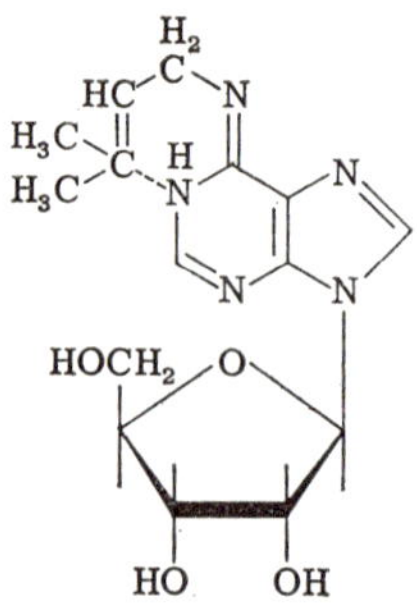

FIG. 1. Spatial arrangement of N^6-(Δ^2-isopentenyl)adenosine.

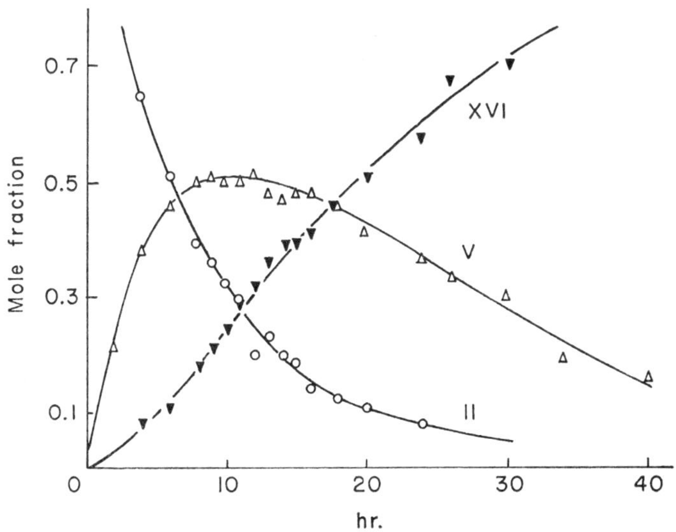

FIG. 2. Hydrolysis of N^6-(Δ^2-isopentenyl)adenosine (II) in N-hydrochloric acid at 41°C; ○ = mole fraction of nucleoside (II); △ = mole fraction of aglycon (V); ▼ = mole fraction of N^6-(3-hydroxy-3-methylbutyl)adenine (XVI). Data taken from Martin and Reese (*36*).

assigned structure (XIX). The ultraviolet absorption spectra of (XIX) are similar to those of N^1,N^6-dimethyladenosine (*37*), and this similarity lends additional support to the assignment of the structure.

Under these conditions iodine reacts quantitatively with the N^6-(Δ^2-isopentenyl)adenosine residues of tRNA. Furthermore, the reaction is selective; iodine does not react with other known components of tRNA except with the thiolated nucleosides. In an experiment conducted in my laboratory, Dr. F. Fittler showed that compound (II) cannot be detected in yeast tRNA after the iodine treatment (unpublished data). Kline *et al.* (*38*) showed that the iodinated nucleoside product in the tRNA molecule is identical to that obtained from compound (II) alone.

VIII. Role of N^6-(Δ^2-Isopentenyl)adenosine in the Function of tRNA

Three experiments provide data relevant to the question of the significance of this component to tRNA function.

A. Immobilization of the Δ^2-Isopentenyl Groups with Iodine

The selective reaction of iodine with the N^6-(Δ^2-isopentenyl)adenosine residues enables an investigation of the amino acid acceptance and binding

TABLE III
BIOLOGICAL ACTIVITY OF IODINE-TREATED tRNA

Amino Acid Accepting Activity (in cpm)[a]		
Sample	Serine	Incorporated phenylalanine
Iodine-treated tRNA	918	2280
Untreated tRNA	822	2325
No tRNA	43	81

Binding of tRNA[b]

	Aminoacyl-tRNA bound to ribosomes: Untreated		Aminoacyl-tRNA bound to ribosomes: I_2-treated	
	Cpm	Pmole	Cpm	Pmole
$tRNA^{Phe}$ with poly U	1730	3.6	1780	3.7
$tRNA^{Phe}$ without poly U	142	0.3	156	0.3
$tRNA^{Ser}$ with poly (U,C)	676	3.4	238	1.15
$tRNA^{Ser}$ without poly (U,C)	132	0.8	7	0.4

[a] The assay was carried out according to the procedure of Hoskinson and Khorana (*83*). In each assay 1.7 A^{260} units of unfractionated yeast tRNA was incubated with 0.68 pmole of [^{14}C]amino acid; serine, 120 μCi/μmole phenylalanine 360 μCi/μmole.

[b] The assay was carried out according to Nirenberg and Leder (*84*). Data were taken from Fittler and Hall (*85*).

capacity of the iodine-treated tRNA. The data presented in Table III show that iodine-treated tRNA does not lose its ability to accept serine residues; on the other hand, the ability of the $tRNA^{Ser}$ to bind with the appropriate messenger–ribosome complex is reduced to approximately one-third of the original capacity. If a close fit of the Δ^2-isopentenyl side chain into the tertiary structure of the anticodon loop is assumed, it seems reasonable that a modification of the side chain, particularly one that fixes the side chain into a rigid configuration as well as adding a bulky iodine atom, must affect the conformation of the anticodon loop.

B. Effect of the Deletion of the Δ^2-Isopentenyl Group from $tRNA^{Tyr}$ on the Codon–Anticodon Interaction

Suppressor tyrosine tRNA (Su^+_{III}) contains the modified nucleoside N^6-(Δ^2-isopentenyl)-2-methylthioadenosine (IV) (*6, 35, 39*) adjacent to the 3′ end of the anticodon (see Table I). Gefter and Russell (*40*) isolated three molecular species of this tRNA. All three species have an identical sequence and differ only with respect to the extent of the modification of the adenosine residue adjacent to the 3′ end of the anticodon. Species 3 is completely modified [N^6-(Δ^2-isopentenyl)-2-methylthioadenosine], species 2 lacks the 2-methylthio substituent and contains N^6-(Δ^2-isopentenyl)adenosine, and species 1 has only adenosine in this position. The availability of these three molecular species enabled Gefter and Russell to compare their abilities to support protein synthesis.

All three forms can be charged with tyrosine and, in an *in vitro* protein synthesizing system requiring the suppressor tyrosine tRNA, species 3 and 2 support protein synthesis, whereas species 1 (containing only adenosine) does not significantly support protein synthesis (Fig. 3). Furthermore, the species that does not support protein synthesis was defective in its ability to bind to ribosomes. These experiments demonstrate convincingly that loss of the Δ^2-isopentenyl group prevents $tRNA^{Tyr}$ from carrying out its basic transfer function.

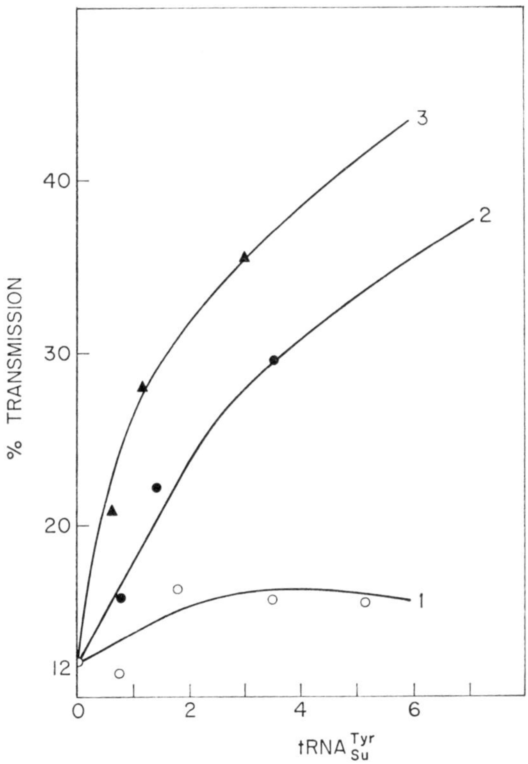

FIG. 3. Suppressor activities of $tRNA^{Tyr}_{Su}$ containing modified nucleosides adjacent to the anticodon. Curve 1: adenosine; curve 2: N^6-(Δ^2-isopentenyl)adenosine; curve 3: N^6-(Δ^2-isopentenyl)-2-methylthioadenosine. Data taken from Gefter and Russell (*40*).

C. Effect of Deletion of Base "Y" from $tRNA^{Phe}$

The nucleoside originally called Y that occurs in $tRNA^{Phe}$ (*41*) has not been identified and may not be related structurally to compound (II). Nevertheless, its location in the primary sequence of $tRNA^{Phe}$ adjacent to the 3′ end of the anticodon (Table I) makes information about its properties pertinent to the present discussion.

Thiebe and Zachau (*42*) found that when an aqueous solution of yeast $tRNA^{Phe}$ is incubated for about 2 hours at pH 2.9 and 37°, the base of "Y" is selectively eliminated. The sugar-phosphate backbone of the molecule is not disrupted, and there appears to be little nonspecific damage to the molecule. The acid-treated $tRNA^{Phe}$ can be charged to the same extent with phenylalanine, although the rate of charging is about one-third of that of untreated $tRNA^{Phe}$. The acid-treated tRNA, however, has completely lost its codon recognition, does not bind with ribosomes in the presence of poly U or other polynucleotides, and does not transfer its amino acid into growing peptide chains.

These results show that the anticodon region of the tRNA molecule is not an essential part of that region within the molecule responsible for recognition of the homologous amino acyl synthetase. On the other hand, the results suggest that the structural integrity of the anticodon loop is essential for the binding in the codon · tRNA · ribosome complex. The disruption in the three-dimensional structure of the anticodon loop illustrated by the reaction of N^6-(Δ^2-isopentenyl)adenosine with iodine or deletion of base Y is probably considerable, and this in itself is sufficient to prevent binding with the anticodon. However, deletion of the isopentenyl side chain, leaving the anticodon structure intact, results in loss of activity.

Apart from the biological significance of the results, the study with "Y" and the iodination reaction of compound (II) demonstrate that a knowledge of the chemical reactivities of these unique components of tRNA enables one to design chemical reactions that selectively alter the structure of the tRNA at their locations. The availability of such highly selective reactions provides a powerful experimental tool.

IX. Biosynthesis of N^6-(Δ^2-Isopentenyl)adenosine

The three questions of immediate concern are: (a) the source of the Δ^2-isopentenyl group; (b) the mode of attachment of the Δ^2-isopentenyl group (in particular whether it is attached to a monomeric precursor or to the preformed tRNA molecule); and (c) the nature of the mechanism that controls the biosynthesis.

The Δ^2-isopentenyl group is derived from mevalonic acid, the source of the five-carbon building blocks of isoprenoid compounds. Utilization of mevalonic acid as the precursor of the Δ^2-isopentenyl groups of tRNA can be readily demonstrated using the acetate-requiring bacteria, *Lactobacillus acidophilus*. Since part of the acetate requirement for this microorganism can be replaced by mevalonic acid, mevalonic acid probably enters the cell intact. When the bacteria are grown in the presence of labeled mevalonic acid, the tRNA becomes radioactive (*43, 44*). All this radioactivity can be accounted for in the N^6-(Δ^2-isopentenyl)adenosine residues of tRNA. Chen and Hall (*45*) carried out an analogous experiment using tobacco tissue grown in culture on media containing [2-^{14}C]DL-mevalonic acid. A significant amount of radioactivity was incorporated into the N^6-(Δ^2-isopentenyl)-adenosine residues of the tRNA. No other component of the tRNA became radioactive.

It would be desirable to demonstrate the incorporation of mevalonate *in vivo* in other organisms, but when we repeated the experiments using yeast, mammalian cell cultures, and *Escherichia coli*, no labeling of the tRNA occurred. The mevalonic acid does not appear to be taken up in these cells. Bloch and Goodwin (*46*) noted that mevalonic acid penetrates whole cells with difficulty. With respect to yeast Fittler *et al.* (*43*) found that [^{14}C]acetate is incorporated into the N^6-(Δ^2-isopentenyl)adenosine residues of the tRNA. The label also enters other components of the tRNA, but the specific activity of the N^6-(Δ^2-isopentenyl)adenosine isolated is 50 times that calculated on the basis of random incorporation. Acetate is a precursor of mevalonate; thus the data suggest that mevalonic acid is also the precursor of the Δ^2-isopentenyl side chains of tRNA of yeast. Although it has not been possible to investigate a greater variety of organisms, we can reasonably conclude that for most organisms mevalonate is the source of the Δ^2-isopentenyl group.

The biosynthetic mechanism of N^6-(Δ^2-isopentenyl)adenosine parallels that of the methylated components of tRNA; the isopentenyl side chain is attached to an adenylic acid residue in the preformed tRNA molecule. In order to prove this point experimentally by studying the mechanism *in vitro* it is necessary to obtain a tRNA substrate that lacks Δ^2-isopentenyl groups. Since tRNA as isolated would be expected to contain its full complement of Δ^2-isopentenyl groups, a key technical factor in this experimental approach is the ability to remove the Δ^2-isopentenyl group. Treatment of N^6-(Δ^2-isopentenyl)adenosine with permanganate under mild conditions removes the side chain and leaves adenosine (Section VII). The reaction conditions are mild enough to apply to tRNA without damaging the molecule. Kline *et al.* (*38*) applied this reaction to samples of the tRNA of *L. acidophilus* in which the Δ^2-isopentenyl groups were labeled with ^{3}H or ^{14}C, and showed that about 25% of the Δ^2-isopentenyl side chain could be removed under these conditions. This chemical reaction, therefore, becomes a useful technique for producing the desired tRNA substrate. The reaction does not appear to cause a significant amount of nonspecific damage to the tRNA molecule, since yeast tRNA treated under these conditions retains about 90% of its amino acid acceptor capability. The data discussed in Section VIII show that alteration of the structure of N^6-(Δ^2-isopentenyl)adenosine residues in tRNA does not interfere with the amino acid-accepting activity.

The mode of formation of N^6-(Δ^2-isopentenyl)adenosine has been studied *in vitro* using a partially purified enzyme from yeast (*38*) and a crude enzyme system from rat liver. The enzyme catalyzes the transfer of the Δ^2-isopentenyl group from Δ^2-isopentenyl pyrophosphate to permanganate-treated homologous tRNA (Table IV). The yeast enzyme system does not utilize permanganate-treated rat liver tRNA as a substrate. The enzyme requires Δ^2-isopentenyl pyrophosphate; the Δ^3 isomer does not serve as a substrate.

One important question concerning the biosynthetic pathway is the specificity of the reaction. The fact that the enzyme system catalyzes the

TABLE IV

INCORPORATION OF [4-^{14}C]Δ^2-ISOPENTENYL GROUPS INTO THE N^6-(Δ^2-ISOPENTENYL)-ADENOSINE RESIDUES OF tRNA

tRNA	Enzyme system	Δ^2-isopentenyl pyrophosphate incorporated into 25 A_{260} units tRNA (pmoles)
Baker's yeast, untreated	Yeast	17
Baker's yeast + $KMnO_4$	Yeast	129
Baker's yeast + I_2 then $KMnO_4$	Yeast	20
Baker's yeast + $KMnO_4$ then I_2	Yeast	121
Rat liver, untreated	Yeast	15
Rat liver + $KMnO_4$	Yeast	9
Escherichia coli B, untreated	Yeast	54
Escherichia coli B + $KMnO_4$	Yeast	45

The unfractionated tRNA was incubated with a partially purified enzyme system from yeast and [4-^{14}C]Δ^2-isopentenyl pyrophosphate for 30 minutes at 37°. The tRNA was hydrolyzed, and the radioactivity of the N^6-(Δ^2-isopentenyl)adenosine was measured. (All the radioactivity is associated with this nucleoside.) These data are taken from Kline *et al.* (*38*).

In addition, similar results have been obtained using a crude enzyme system and homologous tRNA from cultured tobacco pith tissue (*45*) and from rat liver (*47*). The rat liver system also catalyzed the incorporation of labeled Δ^2-isopentenyl groups into untreated *E. coli* tRNA.

incorporation of Δ^2-isopentenyl groups into the permanganate-treated tRNA, and not into untreated tRNA, is presumptive evidence that the enzyme system catalyzes the attachment of the isopentenyl group to those adenine residues from which the Δ^2-isopentenyl side chain was deleted. Furthermore, the nucleoside product of the reaction has been identified as N^6-(Δ^2-isopentenyl)adenosine (*47*). Confirmatory evidence for the specific nature of the reaction stems from an experiment in which the N^6-(Δ^2-isopentenyl)adenosine residues of the tRNA were rendered insensitive to permanganate oxidation.

Treatment of tRNA with aqueous iodine solution converts the N^6-(Δ^2-isopentenyl)adenosine residues to (XIX) *in situ,* as descrided in Section VII. Since (XIX) no longer contains the allylic double bond, it is insensitive to oxidation under the mild conditions used in these experiments. The rationale behind the experiment, therefore, is to immobilize the isopentenyl groups of the N^6-(Δ^2-isopentenyl)adenosine residues by treatment with iodine, followed by treatment with permanganate. Yeast tRNA treated first with iodine and subsequently with permanganate does not accept the isopentenyl side chain (Table IV). The two treatments per se are not harmful to the tRNA since reversal of the two procedures yields a tRNA capable of accepting the isopentenyl group. Because of the selectivity of both the iodine and permanganate reactions for the N^6-(Δ^2-isopentenyl)-adenosine residues, this experiment shows that the enzyme system is specific for those adenosine residues in tRNA that normally contain a Δ^2-isopentenyl group.

With respect to the nature of the recognition sites in the tRNA molecule, it is significant that the two molecular species of yeast tRNA containing N^6-(Δ^2-isopentenyl)adenosine, $tRNA^{Tyr}$ (*48*) and $tRNA^{Ser}$ (*8*), possess the common sequence A-i^{6}A-A-Ψ-C-U-U in the anticodon loop and that the first four base-pairs of the supporting arm are identical in both primary sequences. The corresponding seven-base sequencc for rat liver $tRNA^{Ser}$ (*10*) and *E. coli* Su^+_{iii} $tRNA^{Tyr}$ (*6*) are A-i^{6}A-A-Ψm-C-C-A and A-i^{6}A-A-Ψ-C-U-G, respectively. The striking feature of all these sequences is the occurrence of three consecutive adenosine residues; the middle adenosine contains the Δ^2-isopentenyl side chain. These results all point toward the presence of a specific receptor site common to tRNA molecules containing N^6-(Δ^2-isopentenyl)adenosine.

In the experiments with yeast and rat liver tRNA, the small amount of incorporation of radioactive label into untreated tRNA is not considered experimentally significant; on the other hand, untreated tRNA obtained from *E. coli* B has the capacity to accept a significant number of the Δ^2-isopentenyl groups catalyzed by both mammalian and yeast crude enzyme systems (Table IV). This fact is interpreted as meaning that at the time the cells are harvested the reactions for modifying tRNA are lagging behind the rate of transcription of tRNA molecules; consequently, a significant number of molecules are structurally incomplete. The fact that the modifying reactions proceed somewhat independently of the transcription of tRNA molecules has been well documented in the case of methylation of nucleic acids [see review by Srinivasan and Borek (*49*)]. Goodman *et al.* (*6*) report that in $tRNA^{Tyr}_{Su^{+}\,III}$ synthesized in starved *E. coli* before infection with phage ϕ80, the adenosine residue adjacent to the 3′ end of the anticodon remains largely unmodified.

Labeled N^6-(Δ^2-isopentenyl)adenosine was identified in the *E. coli* B tRNA after exposure to the enzyme system and labeled mevalonic acid, and this product accounted for all the radioactivity present (*38*). The fact that N^6-(Δ^2-isopentenyl)adenosine and not N^6-(Δ^2-isopentenyl)-2-methylthioadenosine is obtained indicates that the incompletely modified residue in the tRNA is adenosine, not 2-methylthioadenosine. This result, together with the fact that Gefter and Russell (Section VIII) find N^6-(Δ^2-isopentenyl)adenosine rather than 2-methylthioadenosine in species 2 of the incompletely modified $tRNA^{Tyr}_{Su\,III}$, suggests that the order of modification of the *E. coli* tRNA is attachment first of the isopentenyl group, followed by attachment of the methylthio group.

The pivotal role of mevalonic acid as the precursor of isoprenoid compounds as well as an essential component of tRNA adds a new facet to the intermediary metabolism of isoprenoids and in particular to the control of this metabolism. Some studies are under way concerning the effect of isoprenoid products on the synthesis and metabolism of mevalonic acid. There is evidence, for example, that cholesterol fed to rats suppresses the conversion of β-hydroxy-β-methylglutarate to mevalonate in the liver (*50*). Dorsey and Porter (*51*) demonstrated that geranyl and farnesyl pyrophosphate inhibit the conversion of mevalonic acid to 5-phosphomevalonic acid in an *in vitro* liver system. The cell therefore appears to have feedback mechanisms for controlling the availability of mevalonic acid, either by limiting its production or by blocking its conversion to the next intermediate on the pathway.

It is apparent that any control mechanism that affects the isoprenoid biosynthetic pathway prior to the synthesis of Δ^2-isopentenylpyrophosphate could have a bearing on the rate of attachment of the isopentenyl group to tRNA. Perhaps the cell has a regulatory mechanism that ensures a supply of isopentenyl groups for tRNA modification independent of the supply for isoprenoid biosynthesis, and perhaps there is some kind of an interlocking control mechanism. These questions have particular relevance to the biological function of those molecular species of tRNA containing N^6-(Δ^2-isopentenyl)adenosine, since the data cited above indicate that these molecules will not function properly without their isopentenyl group.

X. Enzymatic Degradation of N^6-(Δ^2-Isopentenyl)adenosine

During the normal course of catabolism of tRNA, the components are released as mononucleotides, nucleosides, or possibly as free bases. The major nucleotide components of tRNA are metabolized to form reusable products. On the other hand, many of the modified constituents of tRNA do not appear to be metabolized in mammals. For example, most of the known modified components, including the hypermodified nucleoside *N*-(nebularin-6-ylcarbamoyl)threonine, have been isolated from urine in the form of the nucleoside or free base [see Hall (*2*) for a list of the modified nucleosides found in mammalian urine]. N^6-(Δ^2-Isopentenyl)-adenosine, on the other hand, has not been detected in urine (G. B. Chheda, private communication).

Many tissues contain enzymes that metabolize N^6-(Δ^2-isopentenyl)-adenosine (II) (*24*, *52*). Investigations of the metabolism of compound (II) in chicken bone marrow and tobacco pith tissue grown in culture show that crude extracts from these tissues metabolize (II) *in vitro* to give a number of products that are further metabolized to give inosine (XXI) or hypoxanthine (XXII). Of the several products formed in the course of the metabolism, three have been identified. They suggest the accompanying degradation scheme.

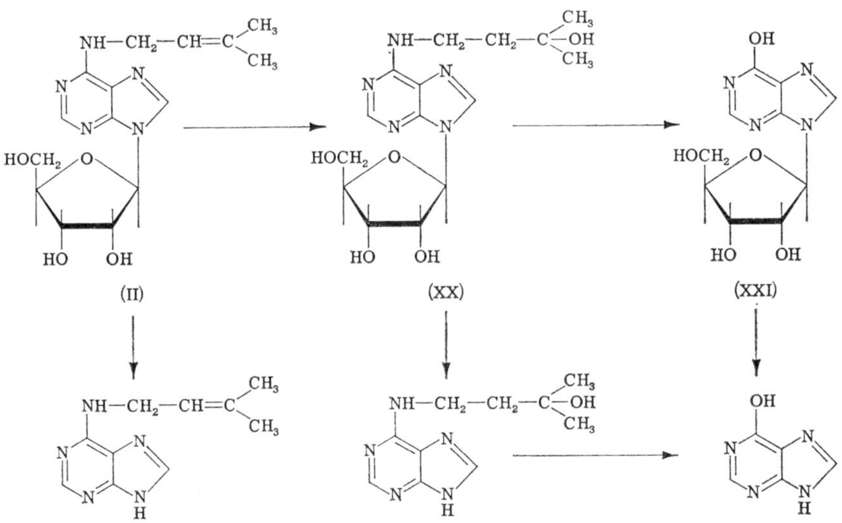

It appears that (II) is subjected to at least two competing enzymatic pathways. In one, hydrolysis occurs to form the free base, and in the second, the allylic double bond of the side chain is hydrated to form

N^6-(3-methyl-3-hydroxybutyl)adenosine (XX). Compound XX is subsequently converted to the free base (XVI). Other products have been found in the enzymatic reaction mixture but have not yet been identified. It is instructive to note that the free base (V), as well as compounds (XVI) and (XIX), exhibit potent cytokinin activity (*53*). It is not certain whether any of these enzymes are unique or whether they are general enzymes present in the tissues.

The pathway for metabolism of N^6-(Δ^2-isopentenyl)adenosine may vary in different organisms, possibly because of different physiological requirements. Miura and Miller (*54*) cultured the fungus *Rhizopogon roseolus* in the presence of [8-^{14}C]N^6-(Δ^2-isopentenyl)adenine and observed that a portion of the compound was hydroxylated to form N^6-(*trans*-4-hydroxy-3-methylbut-2-enyl)adenine (VI) (zeatin).

XI. Biological Activity of N^6-(Δ^2-Isopentenyl)adenosine

The occurrence of N^6-(Δ^2-isopentenyl)adenosine and/or its derivatives in the tRNA of all species, the fact that it is located adjacent to the 3′ end of the anticodon of some species, and the presence of degradative enzymes in plant and animal tissue that metabolize it suggest that this component possesses a degree of significance to the proper functioning of the cell out of proportion to its relatively small amount. This significance is further enhanced by the fact that N^6-(Δ^2-isopentenyl)adenosine per se exhibits biological activity in several experimental systems. Although no direct evidence exists as yet for a relationship between the occurrence of this nucleoside in tRNA and its biological activity, it might be helpful to consider briefly the nature of the observed biological activity.

This activity is most readily demonstrated in plant systems, since N^6-(Δ^2-isopentenyl)adenosine is a member of a class of compounds known as cytokinins. These compounds stimulate growth when added exogenously to experimental plant systems. Compounds with cytokinin activity have attracted the attention of plant physiologists for several years, and a great deal has been learned about the nature of the stimulatory activity at the biological level, although little has been uncovered about the mechanism of action.

Cytokinin activity is conveniently assayed using tobacco pith tissue grown in culture, and in this system an active compound promotes cell division and cell differentiation (*55, 56*). Cytokinins also evoke physiological responses in a number of other systems; for example, stimulation of budding (*57, 58*), the retardation of senescence of excised leaves (*59, 60*), stimulation of the rate of germination of lettuce seeds (*33, 61*), and the potentiation of antibiotic activity against microorganisms (*62*). Whether the primary locus of action of the cytokinins is the same in all biological systems remains a speculative question. Each of the several known compounds that exhibit cytokinin activity does not have an identical spectrum of activity in all these biological test systems.

Compounds that exhibit cytokinin activity possess in common an N^6-substituted adenine, and the most active members of this group are the furfuryl, benzyl, and Δ^2-isopentenyl derivatives [see review by Skoog *et al.* (*63*)].

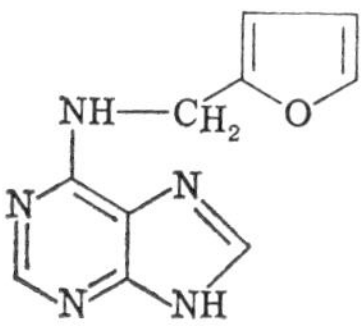

N^6-Furfuryladenine

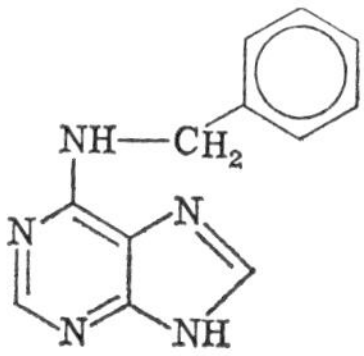

N^6-Benzyladenine

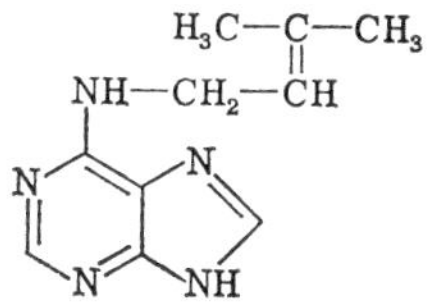

N^6-(Δ^2-Isopentenyl)-adenine

(V)

Most of the earlier studies in this field have been carried out with synthetic compounds such as N^6-furfuryladenine (kinetin) and N^6-benzyladenine. The intense nature of the biological activity of these derivatives led plant physiologists to the conclusion that the activity is an indication of regulatory hormone activity in plants [see, for example, Mothes (*64*)]. Therefore, they reasoned that there should be naturally occurring cytokinins in plants. Zeatin (VI), found in the immature fruit of several species, and N^6-(Δ^2-isopentenyl)adenosine (II) and its hydroxylated derivative (III), detected in tRNA, are the first naturally occurring purine derivatives with cytokinin activity identified in plant tissue. These compounds exhibit tenfold or greater activity in the standard tobacco-callus tissue-culture assay system than N^6-furfuryladenine (*63*, *65*). The fact that these compounds exhibit a hormonelike activity when added exogenously to experimental systems does not necessarily imply, however, that these particular compounds exert hormonelike activity in growing plants.

The mechanism of action of cytokinins in plant tissue is still obscure, and perhaps the most informative results have been obtained in metabolic studies using excised leaves. The antisenescence effect of the cytokinins appears to be related to maintenance of protein and nucleic acid synthesis. Osborne (*66*) reported that addition of puromycin, which inhibits polypeptide formation, to the culture medium containing detached *Xanthium* leaves reverses the stimulatory effect of N^6-furfuryladenine. On the other hand, addition of actinomycin D, which inhibits DNA-dependent RNA synthesis, does not interfere with the stimulatory action of N^6-furfuryladenine. In another study on excised tobacco leaves, N^6-benzyladenine was found to prevent destruction of chlorophylls *a* and *b*; in fact it maintains photosynthetic activity at a high level compared to controls (*67*). These latter workers also showed that N^6-benzyladenine causes an increase in rate of nucleic acid and protein synthesis in chloroplasts excised from tobacco leaves.

The effect of N^6-substituted adenine derivatives has also been studied in several animal systems. Buckley *et al.* (*68*) reported that N^6-furfuryladenine induces mitotic divisions throughout the digestive tract of the salamander *Triturus viridescens*. Tetraploid divisions were also reported. This effect follows after the animal has been exposed to the compound for 9 days and is most evident in the duodenum. After the animals are withdrawn from exposure, mitotic activity gradually decreases, but the number of tetraploid divisions increases (*69*). These workers also noted that exposure to N^6-furfuryladenine causes the absence of the central spindle in the treated tissue and that the spindle reappears in the tissue during the recovery period (in the absence of N^6-furfuryladenine). Grillo and Polsky (*70*) also investigated the effect of N^6-furfuryladenine on *Triturus viridescens* and concluded that this derivative does not stimulate mitosis but interferes with karyokinesis, probably by prolonging the actual mitotic time. This effect appears in routine cytological preparation as a higher number of cells in mitosis than normal. In other words, the effect of N^6-furfuryladenine in this system is strikingly similar morphologically to the effect of colchicine.

These experiments and many others conducted in animal systems have been carried out with the thought that the cytokinin effect in plant systems might be duplicated in animal systems. Almost all the work has been done with the synthetic compounds, such as N^6-furfuryladenine, and the results have generally been negative. These derivatives may not be the appropriate test compounds. The animal system may require administration of the cytokinin in another form, such as the nucleoside, and, perhaps more important, the animal systems may be more discriminating than the plant systems and hence able to respond only to the correct chemical structure. If there is a process in animal tissue comparable to the cytokinin process in plant tissue, it will be necessary, in order to study it, to characterize the natural components involved in growth and differentiation in animal cells and to work with these compounds.

The biological activity of N^6-(Δ^2-isopentenyl)adenosine in plant systems and its natural occurrence in both plant and animal cells raises the question as to whether it might have a demonstrable effect in animal systems analogous to the cytokinin effect in plants. Some preliminary probing of the biological activity of N^6-(Δ^2-isopentenyl)adenosine in animals cells has been carried out. Grace *et al.* (*71*) reported that it is a

TABLE V
CELLS DERIVED FROM HUMAN MYELOGENOUS LEUKEMIA (*86*) CULTURED IN THE PRESENCE OF N^6-(Δ^2-ISOPENTENYL)ADENOSINE FOR 48 HOURS[a]

Concentration (μg/ml)	Total cell count[b] ($\times 10^5$)	Viability (%)
0.1	3.9	70
1	2.3	40
2	2.9	16
3	2.5	10
4	2.3	10
Control (medium above)	3.2	70

[a] Data were taken from Grace *et al.* (*71*).
[b] Initial input: 3.4×10^5, 70% viable.

potent inhibitor of a line of cells derived from human myelogenous leukemia cells (leukemic myeloblasts) grown in culture (see Table V). There is no significant inhibition in cultured human lymphoblastic leukemic cells (LKID) or Burkitt lymphoma (P-3HR-1) cells at comparable concentrations. This nucleoside is also a potent inhibitor of Sarcoma-180 cells. The free base (V) does not inhibit the human leukemic myeloblast at concentrations up to 50 μg/ml, and this result emphasized the need to apply the test compound in a suitable form. The inhibitory activity of N^6-(Δ^2-isopentenyl)adenosine to cultured mammalian cells may be common to other N^6-substituted adenosine derivatives. Hampton *et al.* (*72*) reported that N^6-furfuryladenosine inhibits the growth of a strain of adult human fibroblasts grown in culture. Under similar conditions, this compound has no effect on the strains of cultured cells (HeLa, Sarcoma-180) or on most fibroblast cells. These results are difficult to interpret. Many adenosine derivatives are toxic to mammalian cells, plant cells and bacteria [see, for example, a discussion of the nucleoside antibiotics in Fox *et al.* (*73*)].

Cultured human and rat lymphocytes stimulated by phytohemagglutinin are completely inhibited by N^6-(Δ^2-isopentenyl)adenosine (10^{-5} *M*) (*74, 75*). At a lower concentration (10^{-7} *M*), the effect is reversed and the nucleoside stimulates DNA synthesis and mitosis (*74*). It remains to be seen whether the underlying molecular events induced by N^6-(Δ^2-isopentenyl)adenosine in animal cells are related in any way to the molecular events initiated in plant tissue.

The results obtained in plant systems prove neither that the cytokinin effect represents a natural process in plants nor that N^6-(Δ^2-isopentenyl)-adenosine (II) or a derivative is a natural regulatory hormone. Nevertheless, the observations are suggestive, and if (II) per se is a regulatory hormone, its biosynthesis and metabolism in the cell assume additional significance. The relevant observations are correlated in a scheme shown

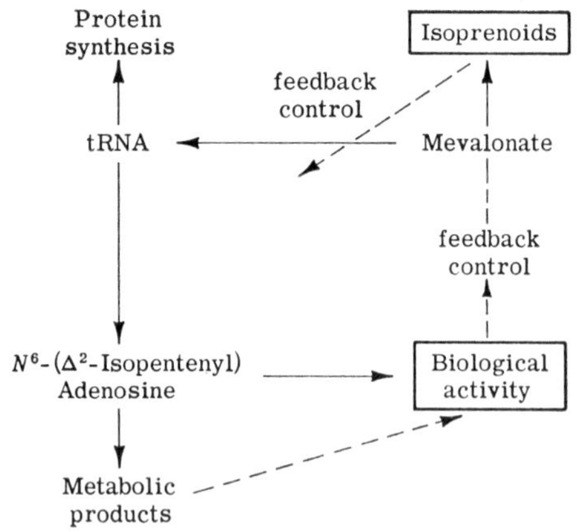

FIG. 4. Scheme showing proposed interrelation of the biosynthesis of N^6-(Δ^2-isopentenyl)adenosine, its metabolism, and its biological activity.

in Fig. 4. The central concept is that because of the physiological activity of (II), its cellular concentration in the free state is critical and, further, the concentration is governed by a balance between the rate of release from tRNA and the activity of the metabolic enzyme systems. A subsidiary concept to this scheme is that the total regulation mechanism would include mevalonic acid; this introduces the possibility of the existence of interlocking control pathways involving isoprenoids, tRNA, and N^6-(Δ^2-isopentenyl)adenosine.

XII. Discussion of Possible Significance of N^6-(Δ^2-Isopentenyl)-adenosine to the Control of tRNA Function

In considering the possible role of N^6-(Δ^2-isopentenyl)adenosine in a tRNA, two facts stand out: the strategic location in the anticodon loop, and the highly reactive allylic double bond in the side chain. The biochemical evidence cited in Section VIII suggests that perhaps the Δ^2-isopentenyl group with its reactive double bond in some way facilitates the association-disassociation process during the amino acid transfer and translocation steps in peptide bond formation. The Δ^2-isopentenyl group or some similar group might not seem to be absolutely essential for this process because a few molecular species of tRNA appear not to contain a modified nucleoside adjacent to the 3′ end of the anticodon. These molecules (for example, $tRNA^{Val}$, $tRNA^{fMet}$) at least have an adenosine residue in this position, and the possibility exists that a very labile substituent not yet recognized is attached to this particular residue.

With respect to the interaction of tRNA, messenger RNA, and ribosomes, not only is there the problem of bringing together all the combining sites of the reacting molecules, but there is also the problem of ensuring the fidelity of the interaction of anticodon and codon. Perhaps the Δ^2-isopentenyl group helps maintain the integrity of the complementary interaction of the codon · anticodon pair. If this is the case, the anticodon might be visualized as a tetranucleotide sequence, the fourth nucleotide being N^6-(Δ^2-isopentenyl)adenosine. The possible mechanisms by which the Δ^2-isopentenyl group might participate in this macromolecular interaction is an intriguing subject for speculation, but the presence of the reactive allylic double bond suggests that the mechanism may depend on something other than the sheer bulk of the five-carbon chain.

In attempting to divine the possible significance of the N^6-(Δ^2-isopentenyl)adenosine residue to tRNA, it would be helpful to consider broader aspects of the role of tRNA in protein synthesis. Growing evidence suggests that, in addition to regulation of protein synthesis by selective transcription of messenger RNA molecules, regulation may involve a translational process, and such a mechanism would place the tRNA in a central position [see Hall (*2*) for a summary of the evidence].

It is possible to visualize two basic mechanisms by which the tRNA could participate in such a regulatory process. The problem of maintaining fidelity of the codon-anticodon interaction has already been commented on; it is conceivable, however, that the hypermodified nucleoside could distort the codon-anticodon interaction in such a way as to permit abnormal codon recognition. Second, certain tRNA's may be specific to certain protein molecules. Therefore, the availability of such tRNA's would determine whether synthesis of a particular protein would take place. It might be convenient, therefore, if the cell had a rapid mechanism for activating and deactivating these tRNA molecules. In this respect, a chemical group like the Δ^2-isopentenyl side chain has certain advantages: it is readily accessible to molecular interactions; it is relatively unique to a few tRNA molecules; it occurs in only one position in these particular tRNA molecules; it is reactive and lends itself to a facile one-step enzymatic alteration in structure that could prevent the anticodon-codon interaction; and, most important, the structural alteration in the anticodon region would still permit the loading of the tRNA molecule with its corresponding amino acid, thereby allowing the tRNA molecule to remain charged but inactive in the transfer step.

Finally, although these considerations are directed at N^6-(Δ^2-isopentenyl)adenosine, they could apply in a broader context to any modified nucleoside lying adjacent to the 3′ end of the anticodon. N^6-(Δ^2-isopentenyl)adenosine represents the first-described member of a group of hypermodified nucleosides that possesses unique chemical and physical properties that may be essential to the proper functioning of tRNA molecules.

APPENDIX

N^6-(Δ^2-Isopentenyl) adenosine

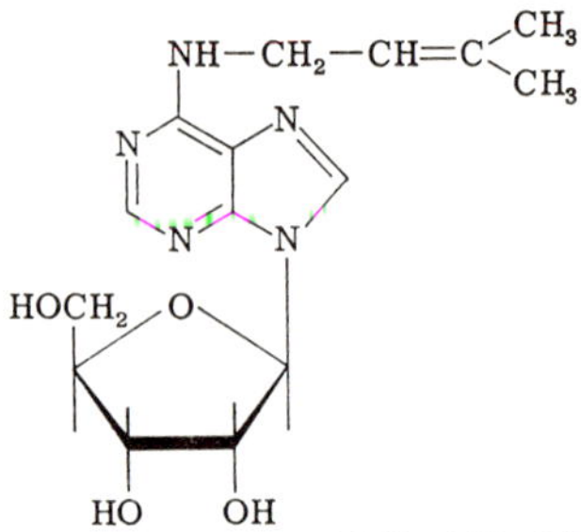

$C_{15}H_{21}N_5O_4$

	M. W. 335.36	
M. P. 145 – 147°		(*4*)
$[\alpha]^{25}_{546}$ – 97° (0.07, ethanol)		(*4*)
p*K* (basic) 3.76		(*36*)

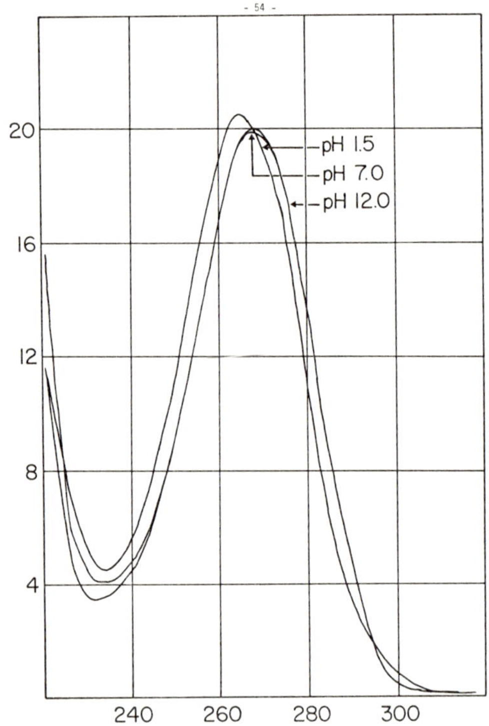

FIG. 5. Ultraviolet absorption spectra of N^6-(Δ^2-isopentenyl)adenosine.[3]

[3] The ultraviolet absorption spectra were obtained on a Cary Model 14 spectrophotometer. We are indebted to N. J. Leonard for supplying a sample of N^6-(Δ^2-isopentenyl)-2-methylthioadenosine.

*N*⁶-(Δ²-Isopentenyl)-2-methylthioadenosine

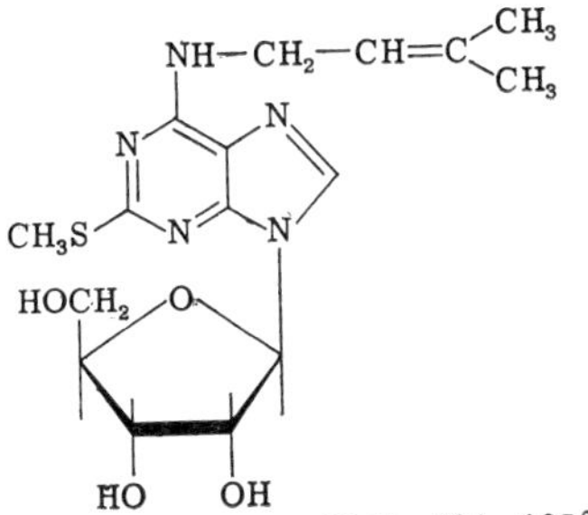

$C_{16}H_{23}N_5O_4S$

M. W. 381.147

M. P. 194–195°

(*35*)

- 55 -

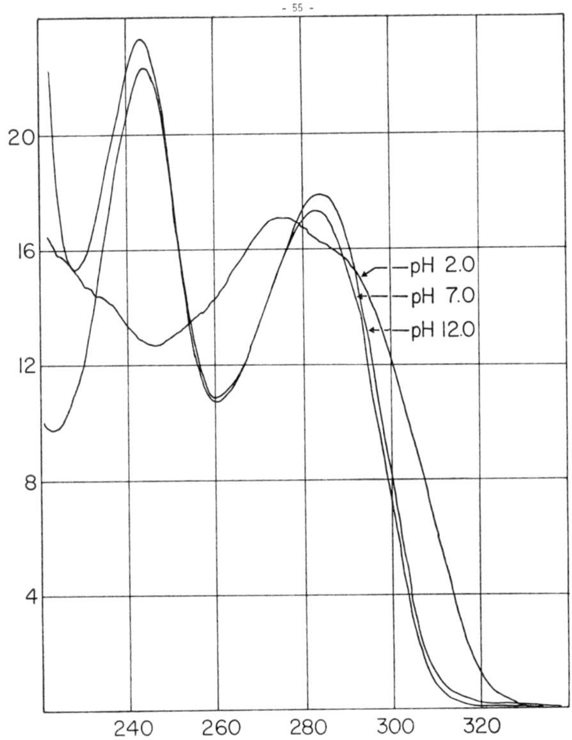

FIG. 6. Ultraviolet absorption spectra of *N*⁶-(Δ²-isopentenyl)-2-methylthioadenosine.[3]

References

1. D. B. Dunn and R. H. Hall, *in* "Handbook of Biochemistry" (H. A. Sober, ed.), p. G-3. Chemical Rubber Co., Cleveland, Ohio, 1968.
2. R. H. Hall, "The Modified Nucleosides in Nucleic Acids." Columbia Univ. Press, New York, in press.
3. M. P. Schweizer, G. B. Chheda, L. Baczynskyj, and R. H. Hall, *Biochemistry* **8,** 3283 (1969).
4. R. H. Hall, M. J. Robins, L. Stasiuk, and R. Thedford, *J. Am. Chem. Soc.* **88,** 2614 (1966).
5. K. Biemann, S. Tsunakawa, K. Sonnenbichler, H. Feldmann, D. Dütting, and H. G. Zachau, *Angew. Chem.* **78,** 600 (1966).
6. H. M. Goodman, J. Abelson, A. Landy, S. Brenner, and J. D. Smith, *Nature* **217,** 1019 (1968).
7. S. Nishimura, Y. Yamada, and H. Ishikura, *Biochim. Biophys. Acta* **179,** 517 (1969).
7a. M Uziel and H. G. Gassen, *Biochemistry* **8,** 1643 (1969).
8. H. G. Zachau, D. Dütting, and H. Feldmann, *Z. Physiol. Chem.* **347,** 212 (1966).
9. J. T. Madison and H. Kung, *J. Biol. Chem.* **242,** 1324 (1967).
10. M. Staehelin, H. Rogg, B. C. Baguley, T. Ginsberg, and W. Wehrli, *Nature* **219,** 1363 (1968).
11. J. P. Helgeson and N. J. Leonard, *Proc. Natl. Acad. Sci. U.S.* **56,** 60 (1966).
12. D. Klämbt, G. Thies, and F. Skoog, *Proc. Natl. Acad. Sci. U.S.* **56,** 52 (1966).
13. D. Klämbt, *Wissen. Z. Univ. Rostock,* (*Math.-Naturwissen. Reihe*) **16,** 623 (1967).
14. D. S. Letham, *Life Sci.* **5,** 551 (1966).
15. C. O. Miller, *Proc. Natl. Acad. Sci. U.S.* **54,** 1052 (1965).
16. D. S. Letham, J. S. Shannon, and I. R. C. McDonald, *Tetrahedron* **23,** 479 (1967).
17. G. Shaw, B. M. Smallwood, and D. V. Wilson, *J. Chem. Soc.* 921 (1966).
18. D. S. Letham, *in* "Biochemistry and Physiology of Plant Growth Substances" (F. Wightman and G. Setterfield, eds.), p. 19. Runge Press, Ottawa, 1968.
19. C. O. Miller, *Science* **157,** 1055 (1967).
20. K. Koshimizu, T. Kusaki, T. Mitsui, and S. Matsubara, *Tetrahedron Letters* **14,** 1317 (1967).
21. N. J. Leonard and J. A. Deyrup, *J. Am. Chem. Soc.* **84,** 2148 (1962).
22. J. H. Rogozinska, *Bull. Acad. Polon. Sci.* **15,** 313 (1967).
23. J. H. Rogozinska, J. P. Helgeson, and F. Skoog, *Physiol. Plantarum* **17,** 165 (1964).
24. C.-M. Chen, D. M. Logan, B. D. McLennan, and R. H. Hall, *Plant Physiol.* **43,** S-18 (1968).
25. M. Janot, A. Cavé, and R. Goutaul, *Bull. Soc. Chem. France* 896 (1959).
26. X. G. Monseur and E. L. Adriaens, *J. Pharm. Belg.* 279 (1960).
27. M. Olomucki, G. Desvages, N-V. Thoai, and J. Roche, *Compt. Rend. Acad. Sci.* **260,** 4519 (1965).
28. R. H. Hall and M. J. Robins, *in* "Synthetic Procedures in Nucleic Acid Chemistry" (W. W. Zorbach and R. S. Tipson, eds.), p. 210. Wiley (Interscience), New York, 1968.
29. W. A. H. Grimm and N. J. Leonard, *Biochemistry* **6,** 3625 (1967).
30. W. A. H. Grimm, T. Fujii, and N. J. Leonard, *in* "Synthetic Procedures in Nucleic Acid Chemistry" (W. W. Zorbach and R. S. Tipson, eds.), p. 212. Wiley, New York, 1968.
31. N. J. Leonard, S. Achmatowicz, R. N. Loeppky, K. L. Carraway, W. A. H. Grimm,

A. Szweykowska, H. Q. Hamzi, and F. Skoog, *Proc. Natl. Acad. Sci. U.S.* **56,** 709 (1966).
32. D. B. Dunn, *Biochim. Biophys. Acta* **46,** 198 (1961).
33. M. J. Robins, R. H. Hall, and R. Thedford, *Biochemistry* **6,** 1837 (1967).
34. H. G. Zachau, D. Dütting, and H. Feldmann, *Angew. Chem.* **78,** 392 (1966).
35. W. J. Burrows, D. J. Armstrong, F. Skoog, S. M. Hecht, J. T. A. Boyle, N. J. Leonard, and J. Occolowitz, *Science* **161,** 691 (1968).
36. D. M. G. Martin and C. B. Reese, *J. Chem. Soc.* (C) 1731 (1968).
37. A. D. Broom, L. B. Townsend, J. W. Jones, and R. K. Robins, *Biochemistry* **3,** 494 (1964).
38. L. Kline, F. Fittler, and R. H. Hall, *Biochemistry* **8,** 4361 (1969).
39. F. Harada, H. J. Gross, F. Kimura, S. H. Chang, S. Nishimura, and U. L. Raj-Bhandary, *Biochem. Biophys. Res. Comm.* **33,** 299 (1968).
40. M. L. Gefter and R. L. Russell, *J. Mol. Biol.* **39,** 145 (1969).
41. U. L. RajBhandary, S. H. Chang, A. Stuart, R. D. Faulkner, R. M. Hoskinson, and H. G. Khorana, *Proc. Natl. Acad. Sci. U.S.* **57,** 751 (1967).
42. R. Thiebe and H. G. Zachau, *European J. Biochem.* **5,** 546 (1968).
43. F. Fittler, L. K. Kline, and R. H. Hall, *Biochemistry* **7,** 940 (1968).
44. A. Peterkofsky, *Biochemistry* **7,** 472 (1968).
45. C-M. Chen and R. H. Hall, *Phytochemistry* **8,** 1687 (1969).
46. K. Bloch and T. W. Goodwin, cited in "Ciba Foundation Symposium on Biosynthesis of Terpenes and Sterols," p. 45. Little, Brown, Boston, Massachusetts, 1959.
47. F. Fittler, L. K. Kline, and R. H. Hall, *Biochem. Biophys. Res. Commun.* **31,** 571 (1968).
48. J. T. Madison and H. Kung, *J. Biol. Chem.* **242,** 1324 (1967).
49. P. R. Srinivasan and E. Borek, This series, **5,** 157 (1966).
50. M. D. Siperstein and V. M. Fagan, *J. Biol. Chem.* **241,** 602 (1966).
51. J. K. Dorsey and J. W. Porter, *J. Biol. Chem.* **243,** 4667 (1968).
52. B. D. McLennan, D. M. Logan, and R. H. Hall, *Proc. Am. Assoc. Cancer Res.* **9,** 47 (1968).
53. R. H. Hall and B. I. S. Srivastava, *Life Sci.* **7,** 7 (1968).
54. G. A. Miura and C. O. Miller, *Plant Physiol.* **44,** 372 (1969).
55. E. Linsmaier and F. Skoog, *Physiol. Plantarum* **18,** 100 (1965).
56. T. Murashige and F. Skoog, *Physiol. Plantarum* **15,** 473 (1962).
57. J. Benes, K. Veres, L. Chvojka, and A. Friedrich, *Nature* **206,** 830 (1965).
58. T. Engelbrecht, *Wissen. Z. Univ. Rostock (Math.-Naturwissen. Reihe)* **16,** 647 (1967).
59. D. J. Osborne and D. R. McCalla, *Plant Physiol.* **36,** 219 (1961).
60. B. I. S. Srivastava and G. Ware, *Plant Physiol.* **40,** 62 (1965).
61. C. G. Skinner, J. R. Claybrook, F. D. Talbert, and W. Shive, *Arch. Biochem. Biophys.* **65,** 567 (1956).
62. R. H. Hall and G. O. Gale, *Proc. Soc. Exptl. Biol. Med.* **103,** 234 (1960).
63. F. Skoog, H. Q. Hamzi, A. M. Szweykowska, N. J. Leonard, K. L. Carraway, T. Fujii, J. P. Helgeson, and R. N. Loeppky, *Phytochemistry* **6,** 1169 (1967).
64. K. Mothes, *Wissen. Z. Univ. Rostock (Math.-Naturwissen. Reihe)* **16,** 619 (1967).
65. N. J. Leonard, S. M. Hecht, F. Skoog, and R. Y. Schmitz, *Proc. Natl. Acad. Sci. U.S.* **59,** 15 (1968).
66. D. J. Osborne, *J. Sci. Food Agr.* **16,** 1 (1965).
67. E. G. Romanko, Kh. Ya. Khein, and O. N. Kulayeva, *Biokhimiya* **33,** 547 (1968).
68. W. B. Buckley, E. R. Witkus, and C. A. Berger, *Nature* **194,** 1200 (1962).
69. S. P. Kevin, E. R. Witkus, and C. A. Berger, *Expt. Cell. Res.* **41,** 259 (1966).

70. R. S. Grillo and R. Polsky, *Exptl. Cell. Res.* **44,** 375 (1966).
71. J. T. Grace, M. T. Hakala, R. H. Hall, and J. Blakeslee, *Proc. Am. Assoc. Cancer Res.* **8,** 23 (1967).
72. A. Hampton, J. J. Biesele, A. E. Moore, and G. B. Brown, *J. Am. Chem. Soc.* **78,** 5695 (1956).
73. J. J. Fox, K. A. Watanabe, and A. Bloch, This series. **5,** 251 (1966).
74. R. Gallo, J. Whang-Peng, and S. Perry, *Science* **165,** 400 (1969).
75. B. Hacker and T. Feldbush, *Proc. Am. Assoc. Can. Res.* **10,** 34 (1969).
76. R. W. Holley, J. Apgar, G. A. Everett, J. T. Madison, M. Marquisee, S. H. Merrill, J. R. Penswick, and A. Zamir, *Science* **147,** 1462 (1965).
77. S. Takemura, M. Murakami, and M. Miyazaki, *J. Biochem.* (*Japan*), **65,** 553 (1969).
78. A. A. Bayev, T. V. Venkstern, A. D. Mirzabekov, A. I. Krutilina, L. Li, and V. D. Axelrod, *Mol. Biol.* (*U.S.S.R.*) **1,** 754 (1967).
79. S. Takemura, T. Mizutani, and M. Miyazaki, *J. Biochem.* (*Japan*) **64,** 827 (1968).
80. S. K. Dube, K. A. Marcker, B. F. C. Clark, and S. Cory, *Nature* **218,** 232 (1968).
81. S. Cory, K. A. Marcker, S. K. Dube, and B. F. C. Clark, *Nature* **220,** 1039 (1968).
82. R. H. Hall, L. Csonka, H. David, and B. D. McLennan, *Science* **156,** 69 (1967).
83. R. M. Hoskinson and H. G. Khorana, *J. Biol. Chem.* **240,** 2129 (1965).
84. M. W. Nirenberg and P. Leder, *Science* **145,** 1399 (1964).
85. F. Fittler and R. H. Hall, *Biochem. Biophys. Res. Commun.* **25,** 441 (1966).
86. S. Iwakata and J. T. Grace, Jr., *N.Y. J. Med.* **64,** 2279 (1964).
87. R. H. Hall, *in* "Biochemistry and Physiology of Plant Growth Substances" (F. Wightman and G. Setterfield, eds.), p. 47. Runge Press, Ottawa, 1968.

Nucleotide Biosynthesis from Preformed Purines in Mammalian Cells: Regulatory Mechanisms and Biological Significance

A. W. Murray,
Daphne C. Elliott, and
M. R. Atkinson

*School of Biological Sciences,
Flinders University of South Australia,
Bedford Park, South Australia*

I. Introduction 87
II. Catalytic and Regulatory Properties of Enzymes Involved in Formation and Utilization of Purine Nucleosides and Bases . . 90
A. 5′-Nucleotidases 90
B. Aminohydrolases 91
C. Purine Nucleoside Phosphorylases 94
D. Purine Phosphoribosyltransferases 96
E. Purine Nucleoside Kinases 98
III. Supply of Purine Nucleosides and Bases 99
IV. Chemotherapy and Control of Growth 103
A. Conversion of Drugs into Active Metabolites 103
B. Control of Purine Nucleotide Biosynthesis by Drugs and Their Metabolites 105
C. Defects in Purine Utilization and Nucleotide Formation by *de Novo* Synthesis: Gout, the Lesch-Nyhan Syndrome, and Drug Resistance 107
V. General Conclusions 111
References 113

I. Introduction

All cells need a balanced supply of nucleotides as precursors of DNA, RNA, and the nucleotide coenzymes involved in group transfer and bioenergetic processes. It is becoming increasingly clear that the most fundamental control mechanisms of living cells operate at the level of enzymatic formation of polymers and coenzymes from their precursors, and that adequate supplies of these precursors (nucleoside triphosphates and amino acids) are ensured, during normal cell growth and function, by self-balancing networks of enzymatic reactions. Control of growth and function by "starving" the fundamental polymerases of their precursors would be a very unselective and inefficient process. It is likely that the study of purine nucleotide biosynthesis will not reveal the ultimate control points of the normal cell's economy, but will clarify the picture already emerging of a highly efficient production line using regulation of enzyme activity by metabolites to ensure adequate provision of essential building blocks, with minimum waste of energy or of nutrients. However, artificial interference with the supply of purine nucleotides can override the natural control systems that regulate formation of nucleic acids and protein, and this is the probable mechanism of action of many drugs and antimetabolites that result in artificial control of growth and cell division. It is not yet possible to say whether "starvation" of specialized cells in multicellular organisms by restriction of biosynthesis of purine nucleotides plays any part in differentiation or morphogenesis, but the rapid response of microorganisms to changes in concentrations of purine bases and nucleosides indicates this possibility.

If all cells made their purine nucleotides by a single sequence of enzymatic reactions, selective chemotherapy involving inhibition of purine nucleotide biosynthesis would be very limited; it is now known that this is not the case and that two types of pathways lead to the formation of nucleotides. The existence of these dual pathways was emphasized by Kornberg (*1*), who pointed out that nucleotides can be formed by a *de novo* or "do it yourself" pathway from simple metabolites or, alternatively, by utilization of preformed bases or nucleosides. It is one of the main purposes of this review to show that differences in the dependence of cells on *de novo* synthesis of purine nucleotides or utilization of preformed purines is a major factor in selective chemotherapy. While many types of cells that have enzymes for the utilization of purine bases and nucleosides also carry out *de novo* purine nucleotide biosynthesis, the existence of these enzymes permits conversion of antimetabolites that are base or nucleoside analogs into nucleotide analogs that are often powerful regulators of enzyme activity.

As pointed out in the first discussions of purine-utilization pathways (*1–3*), the terminal enzymes involved are nucleoside kinases that phosphorylate nucleosides directly to nucleotides:

$$\text{Nucleoside} + \text{ATP} \rightleftharpoons \text{nucleoside 5'-phosphate} + \text{ADP}$$

and purine phosphoribosyltransferases ("nucleotide pyrophosphorylases") that convert purine bases such as hypoxanthine or adenine into their 5′-phosphoribosyl derivatives:

$$\text{Purine} + \text{PRPP}^{1} \rightleftharpoons \text{nucleotide} + \text{PP}_i$$

[1] Abbreviations not listed at the beginning of this volume are: PRPP, 5-phos-

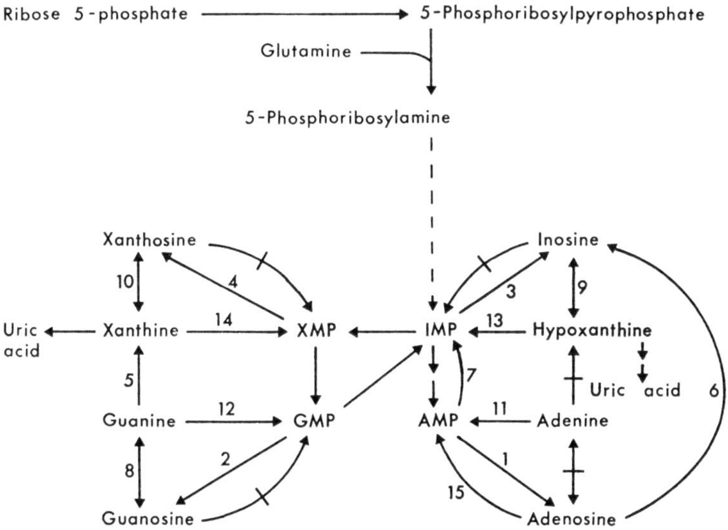

FIG. 1. Pathways of purine biosynthesis, interconversion and degradation. The numbered reactions are specifically discussed in the text and refer to the following enzymes: 1, 2, 3, and 4: 5′-nucleotidase; 5: guanine deaminase; 6: adenosine deaminase; 7: AMP deaminase; 8, 9, and 10: purine nucleoside phosphorylase; 11: adenine phosphoribosyltransferase; 12, 13, and 14: hypoxanthine-guanine phosphoribosyltransferase; 15: adenosine kinase. Reactions marked with a cross-bar (⇸) are not thought to be significant in animal tissues.

Section II of this review deals with the catalytic and regulatory properties of purine-utilization enzymes and includes a comparison of natural and artificial regulators. Except in the special case of nucleotidases, nucleoside phosphorylases and aminopurine hydrolases (see Fig. 1), the enzymes that make purine fragments available by degradation of DNA, RNA, and nucleotide coenzymes are not discussed here, and the reader is referred to recent reviews on nucleases (*4–7*). In Section III the availability of purines and purine nucleosides is discussed, with particular reference to the availability of these compounds in mammalian plasma for utilization by specialized cells. In Section IV the function of purine-utilization enzymes in the conversion of purine and purine nucleoside antimetabolites into active nucleotide derivatives is discussed, as is the significance of this activation in immunosuppressive and cancer chemotherapy. Recent studies on defects in function or regulation of purine utilization and *de novo* path-

---

phoribosyl pyrophosphate; K_i, inhibitor constant; K_m, Michaelis constant; V_{max}, maximal velocity extrapolated to saturating substrate levels; 6-thio IMP, 6-thioinosine 5′-phosphate; 6-thioXMP, 6-thioxanthosine 5′-phosphate; 6-thioGMP, 6-thioguanosine 5′-phosphate; 6-methylthioIMP, 6-methylthioinosine 5′-phosphate.

ways are discussed in Section IV, C, with special reference to gout and the Lesch-Nyhan syndrome (lack of hypoxanthine-guanine phosphoribosyltransferase), and in Section V there is a general discussion of the significance of purine utilization.

Several reviews have dealt with regulation of enzyme activity (*8*, *9*); with pathways utilizing pyridines (*10*) and pyrimidines (*11*) for nucleotide biosynthesis; and with *de novo* pathways leading to deoxyribonucleotides (*12*) and to purine nucleotides (*13*). In a general review of regulation of nucleotide biosynthesis, Blakley and Vitols (*14*) have dealt with many aspects of regulation of enzymes involved in purine utilization.

II. Catalytic and Regulatory Properties of Enzymes Involved in Formation and Utilization of Purine Nucleosides and Bases

A. 5′-Nucleotidases

5′-Nucleotidases (Fig. 1, reactions 1–4) catalyze the hydrolytic cleavage of phosphate from the 5′-position of nucleoside monophosphates and are widely distributed in bacterial, plant and animal tissues. Earlier literature concerning these enzymes has been reviewed (*15*).

Although there is no evidence that more than one enzyme from a particular source is involved in hydrolyzing a wide range of 5′-monophosphates, the 5′-nucleotidases from different sources differ considerably in their properties. Partially purified 5′-nucleotidase from rat liver acetone powder (*16*) had an absolute requirement for a bivalent cation. With this enzyme and the 5′-nucleotidase from chicken liver (*17*), the rate of hydrolysis of IMP was over five times that with AMP, which may be relevant in the release of hypoxanthine by liver (cf. Section III). De Lamirande *et al.* (*18*, *19*) examined the intracellular distribution of 5′-nucleotidase in rat liver and showed that it was mainly in the nuclear and microsomal fractions. However, Fritzson (*20*) has reported 5′-nucleotidase activity in a soluble fraction of rat liver; the activity decreased over the first few days after partial hepatectomy and it was suggested that 5′-nucleotidase may be involved in slowing down growth by depleting the cells of intermediates essential for nucleic acid biosynthesis. 5′-Nucleotidase has recently been isolated in an essentially homogeneous form from rat liver microsomal fraction and plasma membranes (*21*). The enzyme had a low K_m for AMP (12 μM) and was shown to be a lipoprotein; the phospholipid portion was identified as sphingomyelin. AMP hydrolysis was competitively inhibited by UMP, GMP, IMP, and CMP, suggesting a single active site for all nucleotides.

Other animal tissue 5′-nucleotidases that have been studied include those from *Bothrops atrox* venom (*22*), sheep brain (*23, 24*), calf intestine (*25*), human aorta wall (*26*), and bovine pituitary gland (*27*). The pituitary enzyme was essentially homogeneous in the ultracentrifuge and had a molecular weight of about 237,000. Low-molecular-weight fragments with 5′-nucleotidase activity were obtained after treatment with pronase (*27*). It is interesting that the nucleotide specificity of these fragments was altered; modification during isolation procedures may partially explain the differences in substrate specificity obtained by different workers with enzyme from the same tissue.

The 5′-nucleotidases from animal tissues are highly specific for 5′-monophosphate derivatives; 5′-deoxyribonucleotides are also hydrolyzed, but at a lower rate (*20, 22, 25, 28, 29*).

There is little information available concerning regulation of 5′-nucleotidase activity; strict control would be expected, to prevent catabolism of nucleotides required for nucleic acid and coenzyme synthesis. Ipata (*23, 24*) has shown that hydrolysis of AMP by a soluble 5′-nucleotidase from sheep brain is strongly inhibited by ATP, UTP, and CTP and that the inhibitions are allosteric in nature. Strong inhibition of 5′-nucleotidase by triphosphates has also been reported with enzyme solubilized from a particulate fraction of Ehrlich ascites-tumor cells (*30*) and with enzyme in extracts from heart and lung (*31*). Hydrolysis of AMP by both enzymes was more strongly inhibited by ATP than was hydrolysis of IMP; again this may be relevant to the observation that hypoxanthine is the major purine released from some tissues (see Section III). Strong competitive inhibition by ATP of AMP hydrolysis by 5′-nucleotidase from rat heart has been reported (*32*). Endogenous heat-stable inhibitors of 5′-nucleotidase activity have been reported in crude homogenates of rat liver microsomes (*28*) and Ehrlich ascites-tumor cells (*29*); the possibility that inhibition may be due to triphosphates present in the homogenates should be considered.

B. Aminohydrolases

1. Adenine Deaminase

Adenine deaminase has not been detected in animal tissues and is not considered in this review. The reader is referred to papers on adenine deaminase from microorganisms (*33–36*).

2. Guanine Deaminase

Guanine deaminase (Fig. 1, reaction 5) was originally purified from rat liver (*37*) and shown to react with guanine and 8-azaguanine, but not with guanosine or GMP (*38*). The enzyme has a high affinity for guanine and 8-azaguanine with K_m values of 5 and 70 μM, respectively (*38*). An enzyme that catalyzes the deamination of 6-thioguanine, 1-methylguanine, and 1-methyl-6-thioguanine has been purified from rabbit liver (*39*), and a detailed study of the structural requirements for binding has been carried out (*40*). Recent studies indicate that guanine deaminase from rat brain and other tissues has a complicated pattern of regulation and distribution (*41–43*), and it has been shown that several tissues contain an inhibitor of its activity associated with a particulate fraction; the inhibitor appears to be a protein and shows noncompetitive inhibition (*42*). Josan and Krishnan (*43*) have recently shown that rat liver guanine deaminase activity is regulated by GTP. The addition of 0.19 mM GTP converted a sigmoidal response to increasing concentrations of guanine (Hill plot slope 2.4) to a hyperbolic response (Hill plot slope 1.1), decreased the concentration of guanine required for half-maximum velocity from 24 μM to 14 μM and increased the maximum velocity by 70%.

3. Adenosine Deaminase

In this section only the adenosine deaminases (Fig. 1, reaction 6) from animal tissues are considered; enzymes from *Aspergillus oryzae* that are less specific, deaminating adenine, adenosine, and adenine nucleotides, have been described (*44–46*).

Adenosine deaminase activity has been found in a number of animal tissues but is frequently very low [for references see Brady and O'Connell (*47*)]. The enzyme has been extensively purified from calf intestinal mucosa (*47*), ox heart (*48*), and calf spleen (*49*), and partially purified from rat heart and erythrocytes (*32*). Evidence has been obtained for an essential sulfhydryl group in the enzyme molecule (*50*), and molecular weights of between 32,000 and 35,000 have been reported (*51, 52*). The reaction has no metal ion requirements, and there is no evidence that adenosine deaminase is subject to specialized regulatory effects. The mammalian enzyme does not deaminate adenine or adenine nucleotides, but catalyzes deamination of a range of adenosine analogs including 2′-deoxyadenosine, 3′-deoxyadenosine, 2,6-diaminopurine ribonucleoside, 6-hydroxylaminopurine ribonucleoside, and isopropylidene adenosine (*48, 52–55*). These workers have also shown adenosine deaminase to catalyze the dechlorination of 6-chloropurine ribonucleoside. N^6-methyladenosine and a number of 2-substituted adenosines are not deaminated but are strong inhibitors of adenosine binding, often with higher affinity for the enzyme than adenosine (*48, 56*).

4. Adenylate Deaminase

The activity of adenylate deaminase (Fig. 1, reaction 7) from several tissues is markedly influenced by monovalent cations and nucleoside triphosphates; however, the responses to these two parameters varies with enzyme from different sources.

AMP deaminase from human erythrocytes has an absolute requirement for K^+ or NH_4^+; no activity was observed with Li^+, Na^+, Rb^+, or Cs^+ (*57*). The deaminase is stimulated by ATP in the presence of Li^+, Na^+, Rb^+, or Cs^+, but not by ATP alone, and the presence of ATP also decreases the concentration of K^+ required for half-maximal activation. Some activation of the erythrocyte enzyme also has been found with ITP, UTP (*57*), and CTP (*57*, *58*); the enzyme from rabbit erythrocytes is strongly inhibited by P_i (*58*), and that from human erythrocytes by 2,3-diphosphoglycerate (*59*).

Early studies showed that AMP deaminase from brain is stimulated by ATP (*60–62*), but the interactions of the triphosphate and cation responses are different from those with the erythrocyte enzyme.

Deaminase purified from calf brain is strongly stimulated by ATP in the *absence* of monovalent metal cation (*63–66*), and the stimulation is inhibited by GTP (*64*, *66*). In the absence of ATP, brain deaminase is stimulated by monovalent cations but the specificity differs from that of the erythrocyte enzyme. Li^+ is the most efficient activator, followed by Na^+ and K^+; almost no activation is obtained with NH_4^+. In the absence of ATP or cation, the response curves of both the erythrocyte (*58*, *59*) and brain (*63*, *65*, *66*) enzymes to AMP concentration are highly sigmoidal.

AMP deaminase has been purified from rabbit muscle (*67*, *68*), and some of its catalytic and physical properties have been studied (*68–72*). Responses to cation were not reported by Lee (*67*, *69*), who carried out all assays in the presence of excess Na^+ or K^+. However, in the absence of ATP, muscle AMP deaminase is stimulated by monovalent cations (*68*), and in the absence of the activating cations by ATP and ADP (*71*); this stimulation is strongly inhibited by P_i. Smiley and Suelter (*71*) also reported inhibition of the ATP and ADP stimulation by GTP, and it has been reported that desensitization to the GTP inhibition can be obtained by treatment with mercurials (*72*). Monovalent cation and ATP stimulation and GTP inhibition have also been reported for AMP deaminase from Ehrlich ascites-tumor cells (*73*); here it was shown that *O*-adenylyl methylenediphosphonate can mimic the stimulatory effect of ATP. The enzyme from rat liver was also stimulated by ATP (*74–76*) and inhibited by GTP (*74*) and P_i (*76*).

AMP deaminase from several sources is highly specific for 5′-AMP and does not deaminate adenine, 2,6-diaminopurine, adenosine, 2′- or 3′-AMP or 3′,5′-cyclic AMP (*57*, *66*, *70*). The crystalline enzyme from rabbit muscle slowly deaminates dAMP (*70*). Lee (*70*) also reported no deamination of ADP by the muscle enzyme, but it has recently been shown (*77*) that ADP is slowly but quantitatively converted into IDP by enzyme purified from muscle as described by Smiley *et al.* (*68*); this result is of interest as the natural occurrence of ITP has been reported (*78*) and IDP is a substrate for nucleoside diphosphokinase (*79*).

The function of AMP deaminase in animal tissues may be to ensure the proper balance between adenine and guanine nucleotides (hence stimulation by ATP and inhibition by GTP) (*64, 66*) or to act as a "buffer" system to regulate the AMP concentration (*76*). The relative concentrations of ATP and GTP may act to balance the activity of AMP deaminase at a particular level, but it is difficult to envisage alteration of the concentrations of these two nucleotides as a sensitive control mechanism in view of the high ATP levels and low GTP levels in most tissues (*80*). It seems feasible for sensitive control to be related to the strong regulatory effects of AMP on metabolism, particularly on reactions leading to the production of ATP [for references see Lee and Wang (*76*)]. AMP deaminase from liver (*76*), brain (*63, 65, 66*), and muscle (*68, 71*) require high concentrations (0.4–4 m*M*) of AMP for half-maximal activity even in the presence of optimal ATP and cation concentrations, and it is likely that the low AMP concentration in cells would always be a major factor limiting deaminase activity, particularly where response of activity to AMP concentration is highly sigmoidal. As discussed by Krebs (*81*), a small percentage decrease in ATP concentration can lead to a large percentage increase in AMP concentration. However, Setlow and Lowenstein's (*82*) observations on the effects of Na^+, K^+, and Mg^{2+} on the ATP stimulation and GTP inhibition may have physiological significance. Na^+ was effective in overcoming the GTP inhibition whereas K^+ did not reverse the inhibition at concentrations below 200 m*M*; it was suggested that an influx of Na^+ into the cell accompanied by an efflux of K^+ would lead to reversal of the GTP inhibition. In the absence of monovalent cation and presence of 1 m*M* ATP at pH 7.2, 5 m*M* Mg^{2+} decreased the concentration of AMP required for half-maximal activity from 3.3 to 0.9 m*M*.

C. Purine Nucleoside Phosphorylases

The early literature on the properties of the purine nucleoside phosphorylases (Fig. 1, reactions 8–10) has been reviewed (*83*), and in this section only the more recent work on enzymes from animal tissues is dealt with.

Although purine nucleoside phosphorylases have been isolated from brain (*84*), liver (*85*), Ehrlich ascites-tumor cells (*86–90*), spleen (*91–93*), fish muscle (*94*), and rabbit bone marrow (*95*), the enzyme from erthryocytes has received most attention. Human erythrocyte purine nucleoside phosphorylase has been highly purified (*96–99*), and a crystalline preparation has recently been described (*100*).

The major reactions catalyzed by purine nucleoside phosphorylase are of the type:

$$\text{Purine ribonucleoside (deoxyribonucleoside)} + P_i \rightleftharpoons \text{purine} + \text{ribose-1-P (deoxyribose-1-P)}$$

Although incomplete information is available on the substrate specificity of enzymes from different tissues, inosine, guanosine, and their deoxyribo-derivatives and xanthosine have been reported as substrates; the enzyme does not react with adenosine [however, see Korn and Buchanan (*101*); Tarr (*102*)]. An enzyme specific for guanosine and its deoxy derivative has been isolated from rabbit bone marrow (*95*), and it has been reported that purine nucleoside phosphorylase from beef liver forms 5-amino-4-imidazolecarboxamide ribonucleoside from ribose 1-phosphate and 5-amino-4-imidazolecarboxamide (*103*). It is likely that the reactions with ribo- and deoxyribonucleosides are catalyzed by the same enzyme, as indicated by constancy of activity ratios during purification (*100*). The equilibrium of the reaction favors nucleoside synthesis (*104*), and an equilibrium constant of 0.005 for the phosphorolytic reaction has been reported (*97*). The affinity of mammalian purine nucleoside phosphorylase for nucleosides is usually high, and K_m values between 10 and 58 μM have been reported for the enzyme from liver, erythrocytes, and fish muscle (*94, 99, 105–107*); a value of 210 μM was reported for phosphorolysis of guanosine by an enzyme from bone marrow (*95*). Animal tissue purine nucleoside phosphorylase is inhibited by *p*-chloromercuribenzoate and other reagents that react with thiol groups (*95–99*); and the reaction has been reported to have an ordered sequential reaction mechanism (*93, 108*).

Purine nucleoside phosphorylases from spleen (*92*), Ehrlich ascites-tumor cells (*86, 90*), and human erythrocytes (*99*) also catalyze a slow base-exchange reaction in the absence of P_i:

$$[^{14}C]\text{guanine} + \text{inosine} \rightleftharpoons [^{14}C]\text{guanosine} + \text{hypoxanthine}$$

Using a highly purified enzyme preparation from which traces of P_i had been removed on Sephadex columns, Kim *et al.* (*99*) showed that the transfer is direct and does not involve the intermediate formation of ribose 1-phosphate. Pinto and Touster (*90*) reported that the phosphorolytic and base transfer activities of the Ehrlich ascites-tumor cell enzyme can be separated by sucrose density-gradient centrifugation in the presence of mercaptoethanol; these workers also obtained evidence for the conversion of phosphorolytic into transfer activity when enzyme preparations are left at 4° in the presence of mercaptoethanol, and reported differential inhibition of the transfer activity by EDTA. However, Kim *et al.* (*99*) could not separate the phosphorolytic and base transfer activities of the erythrocyte enzyme by mercaptoethanol or EDTA treatment, and Abrams *et al.* (*92*) reported a constant ratio of the two activities during purification of phosphorylase from spleen and Ehrlich ascites-tumor cells. Although Krenitsky (*93*) could not show a base-exchange reaction with crystalline enzyme from calf spleen, an activity less than about 2% of that in the presence of P_i would not have been detected. Kim *et al.* (*99*) have shown that with an enzyme from human erythrocytes the transfer activity is only between 0.2 and 1% of the phosphorolytic activity.

Paterson and Simpson (*88*) used the uridine-supported synthesis of inosine from hypoxanthine in intact Ehrlich cells as a system to test potential nucleoside analogs and inhibitors. The most effective inhibitors of inosine synthesis were S^6-alkylated derivatives of 6-mercaptopurine and 6-thioguanine. However, these compounds did not inhibit nucleoside phosphorylase isolated from the tumor cells; presumably they interfere with the entry of nucleosides into the cell. The most likely mechanism of inosine synthesis here is formation of uracil and ribose 1-phosphate from uridine, catalyzed by pyrimidine nucleoside phosphorylase, followed by reaction of ribose 1-phosphate and hypoxanthine catalyzed by purine nucleoside phosphorylase; however, Paterson and Simpson (*87*) obtained indirect evidence that free ribose 1-phosphate is not an intermediate in the transfer process. Inhibition of human erythrocyte purine nucleoside phosphorylase has also been studied with intact cells and isolated enzyme by measuring the conversion of hypoxanthine into inosine (*106*). Xanthine, guanine, 6-mercaptopurine, and 6-thioguanine were strong competitive inhibitors, and adenine, 2,6-diaminopurine, 6-thioxanthine, allopurinol, and alloxanthine were only weak inhibitors, though the latter two compounds do react slowly with the phosphorylase (*109*); these workers also confirmed that 6-methylthioinosine inhibits nucleoside transport.

The inosine analog formycin B (3-(β-D-ribofuranosyl)pyrazolo[4,3-*d*]-6(*H*)-7-pyrimidone) is a strong competitive inhibitor of erythrocyte purine nucleoside phosphorylase (*107*). This inhibition is interesting because of the low toxicity of formycin B (*110*), and analogs of this type will be useful in studies of the physiological significance of purine nucleoside phosphorylases.

D. Purine Phosphoribosyltransferases

Purine phosphoribosyltransferases (Fig. 1, reactions 11–14) catalyse the direct reaction between a purine base and PRPP to form the corresponding 5′-phosphoribosyl derivative and PP_i. In general, our comments are restricted to enzymes from mammalian tissues. However, it should be mentioned that in bacteria there is evidence for separate phosphoribosyltransferases for adenine, hypoxanthine, guanine, and xanthine (*111*, *112*) while in animal tissues it is likely that there are only two enzymes, one reacting with adenine and 5-amino-4-imidazolecarboxamide and the other with hypoxanthine, guanine, and xanthine; evidence supporting these conclusions is given below.

Adenine phosphoribosyltransferase has been purified from beef liver (*113*, *114*) and Ehrlich ascites-tumor cells (*115*). The enzyme from beef liver reacts with adenine, 5-amino-4-imidazolecarboxamide, 8-aza-adenine, 2-aza-adenine, 4-aminopyrazolo[3,4-*d*]pyrimidine and 2-fluoroadenine, but not with hypoxanthine, guanine, or 6-mercaptopurine. An enzyme reacting with 2,6-diaminopurine has been isolated from L cells in culture (*116*). Kinetic studies by Hori and Henderson (*117*) suggested that the purified ascites tumor cell enzyme has a nonsequential mechanism; however, recent results indicate an ordered reaction mechanism (*118*). Adenine phosphoribosyltransferase from most animal tissues has a high affinity for adenine, and K_m values of 0.9–2.6 μM have been reported (*117*, *119–122*); Blair (*116*) found a value of 30 μM with enzyme from L cells.

Adenine phosphoribosyltransferase from Ehrlich ascites-tumor cells is inhibited by nucleoside monophosphates, particularly AMP and dAMP (*117*, *123–125*), but at low nucleotide concentrations stimulation of activity by AMP, GMP, and dAMP (*124*) and by nucleoside triphosphates has been reported (*126–128*). The significance of these effects in the regulation of enzyme activity remains to be established. The AMP analog, formycin-A 5′-phosphate strongly inhibits the enzyme from Ehrlich ascites-tumor cells (*129*).

The properties of adenine phosphoribosyltransferase in mouse liver change with time after birth (*127*). Enzyme extracted from the livers of mice between 17 and 51 days after birth (corresponding to the period of rapid increase in liver weight) had a low K_m for PRPP, was not stimulated by ATP, and was less inhibited by AMP than enzyme obtained before or after this period. These observations together with the report that both adenine and hypoxanthine-guanine phosphoribosyltransferase increase in specific activity after partial hepatectomy (*130*) implicate the phosphoribosyltransferases in the metabolism of rapidly growing cells. High activities of both enzymes have been found in mouse and rat liver and spleen, in ascites-tumor cells, and in rat bone marrow (*130*). The high level of hypoxanthine-guanine phosphoribosyltransferase in brain tissue (*130–132*) is of considerable interest in the metabolism of this organ and is discussed further in Section IV.

Although hypoxanthine-guanine phosphoribosyltransferase has been widely studied in animal tissues, there have been few studies with partially purified or purified preparations; kinetic results have been reported with enzyme from beef liver (*133*), Ehrlich ascites-tumor cells (*119*) and human erythrocytes (*134*); the enzyme has been purified from yeast (*135*). Evidence that a single enzyme is involved in the reaction with hypoxanthine and guanine includes similar rates of heat inactivation (*136*) and similar electrophoretic properties (*137*) with both substrates. Earlier observations (*119*) of different ratios of rates with hypoxanthine and guanine in enzyme fractions eluted from starch gel probably resulted from interference by gel components in the assay systems used. Moreover, both activities are lost together in humans with the Lesch-Nyhan syndrome (*138*) (cf. Section IV), and resistance to the growth-inhibitory effects of a number of hypoxanthine and guanine analogs results in loss of activity with both hypoxanthine and guanine (*139–141*). Henderson *et al.* (*137*) have also obtained kinetic evidence that a single enzyme from human erythrocytes reacts with both hypoxanthine and guanine. These authors concluded that hypoxanthine-guanine phosphoribosyltransferase, like adenine phosphoribosyltransferase, has an ordered reaction mechanism with the formation of ternary enzyme–substrate complexes. There is also evidence that the enzyme from human erythrocyte reacts slowly with xanthine (*136*). The enzyme from animal tissues also reacts with 6-mercaptopurine (*119, 120, 133, 142*), 8-azaguanine (*142*), and perhaps with other hypoxanthine and guanine analogs that act as growth inhibitors [cf. Brockman (*143*)].

K_m values for hypoxanthine and guanine of 11 and 2.5–3.3 μM, respectively, have been obtained with hypoxanthine-guanine phosphoribosyltransferase from ascites tumor cells (*119*), and similar values have been reported for the erythrocyte enzyme (*137*). However, Murray (*127*) found a K_m value for PRPP for the mouse-liver enzyme (reacting with hypoxanthine) of 28 μM compared with 240 μM reported by Henderson *et al.* (*137*). Hypoxanthine-guanine phosphoribosyltransferase from Ehrlich ascites-tumor cells is competitively inhibited by IMP and GMP, but not by AMP (*123*).

E. Purine Nucleoside Kinases

The major purine nucleoside kinase present in animal tissues catalyzes the phosphorylation of adenosine by ATP to form AMP and ADP (Fig. 1, reaction 15). The enzyme has been isolated from rabbit liver (*144*), Ehrlich ascites-tumor cells (*145, 146*), human tumor cells in culture (*147*), blood platelets (*148, 149*), rat liver and kidney (*150*), chick cells in culture (*151–154*), and human tissues (*155*).

It is generally found that adenosine kinase, in common with other kinases, has an absolute requirement for a bivalent cation. However, with adenosine as substrate, the enzyme from H. Ep. No. 2 cells in culture (*147*) and from Ehrlich ascites-tumor cells (*146*) did not require added Mg^{2+}. The ascites enzyme reacted with $MgATP^{2-}$ as substrate, but at a lower rate than with free ATP; uncomplexed Mg^{2+} was a noncompetitive inhibitor of the reaction. However, some Mg^{2+} tightly bound to the enzyme was required for enzyme activity; this Mg^{2+} could be released with *p*-chloromercuribenzoate and removed by dialysis to give an inactive enzyme. Full activity was restored by addition of 0.1 m*M* Mg^{2+} after removal of *p*-chloromercuribenzoate with dithiothreitol.

Adenosine is by far the best substrate for adenosine kinase and K_m values of 1.8 (*147*) and 2.8 μM (*146*) have been reported. However, many adenosine analogs also react with adenosine kinase, and generation of nucleotide analogs inside cells through the action of this enzyme is important in chemotherapy (Section IV). The enzyme from rabbit liver (*144*) reacts with analogs in which the 6-amino group is replaced by a methylamino group or by an H or Cl atom, but not with analogs where the amino group is replaced by a mercapto or hydroxyl group (thioinosine and inosine). Oxidation but not methylation at N-1 is permissible, and the 2-amino and 2-fluoro derivatives are good substrates. Replacement of H by CH_2OH at the C-1 of the ribose portion of adenosine (psicofuranine) prevents reaction, but 3′-deoxyadenosine is a substrate. Similar substrate specificity has been reported by Schnebli (*147*), who also showed the enzyme to react with the powerful antitumor agent 6-methylthioinosine to form the ribonucleotide (6-methylthioIMP) (also see references *145*, *155–159* and Section IV).

Although it is generally assumed that mammalian purine nucleoside kinase does not react with inosine, there have been some reports to the contrary (*145*, *160*, *161*). Pierre and LePage (*160*) described the conversion of inosine into IMP with an extract from Ehrlich ascites-tumor cells. The ascites line lacked hypoxanthine-guanine phosphoribosyltransferase, and so there was no possibility of ambiguity arising from free base formation. There is some doubt as to the physiological significance of this reaction, as even in the presence of high concentrations of Mg^{2+} (5 m*M*) inosine kinase was inhibited by high ATP concentrations (complete inhibition was obtained with 2 m*M* ATP).

A kinase specific for deoxyadenosine has been partially purified from calf thymus (*162*).

III. Supply of Purine Nucleosides and Bases

Two of the groups of enzymes discussed in Section II, the 5′-nucleotidases and the purine nucleoside phosphorylases, are involved in the production of purine fragments that can either be reutilized or catabolized to uric acid and excreted. Strong inhibition by ATP of AMP hydrolysis to adenosine and stimulation of AMP deaminase by ATP, together with the presence of adenosine deaminase and the absence of adenosine phosphorylase in animal tissues, would channel hydrolysis products to inosine. There is little evidence for the direct conversion of inosine into IMP, and this compound must first be cleaved to hypoxanthine before it can be utilized by hypoxanthine-guanine phosphoribosyltransferase. The following discussion outlines the available evidence suggesting that hypoxanthine (and possibly xanthine) is the major purine fragment available for utilization in mammalian tissue.

The importance of blood, and in particular the erythrocyte, as a vehicle for transport of purines has been evident since Lajtha and Vane (*163*) reported the dependence of bone-marrow cells on the liver for their purine supply. Intact rabbits incorporate [^{14}C]formate readily into DNA adenine and thymine of marrow cells, but hepatectomy or portal occlusion causes a 10- to 50-fold decrease in the ratio (incorporation into DNA adenine and thymine). In earlier work (*164*), addition of liver enzymes needed for *de novo* purine nucleotide biosynthesis increased 20- to 40-fold the incorporation of [^{14}C]formate into nucleic acid adenine and guanine of Ehrlich ascites cells *in vitro*, giving rates similar to those found *in vivo*. Marrow and a range of neoplastic cells showed similar effects, and it was suggested (*164*) that there might be two types of tissues, those that, like liver, can make sufficient purine nucleotide for their requirements, and those that, like bone marrow, depend on a supply of purine formed in other tissues. Lajtha and Vane (*163*) concluded that in the mammalian organism the liver is the main source of purines for the bone marrow cells and possibly for the cells of most peripheral tissue.

In these early experiments, the rapid incorporation of [^{14}C]adenine into nucleic acids [a consequence of the high activity and wide distribution of adenine phosphoribosyltransferase and the absence of adenine deaminase (cf. Section II, D)] drew attention to this purine as the transported base available for nucleotide formation. It now seems unlikely that free adenine (apart from that derived from dietary sources) is a significant extracellular precursor for mammalian purine utilization *in vivo*, although most studies have been with mice or rats, and species differences may exist. Henderson and LePage (*165*) found no free adenine or guanine in the plasma of rats several days after the injection of [^{14}C]adenine, although high levels of radioactivity were present in tissue and blood-cell adenine nucleotides, and rapid incorporation of radioactivity into the purines of nucleotides and nucleic acids of the tumor took place on subsequent implantation of Sarcoma 180. The main radioactive component of plasma was an unidentified "acidic purine" (λ_{max} 255 nm in acid) estimated to be present at a concentration of 0.1 m*M*. Little or no free adenine was detected in liver and intestine of mice 24 hours after injection of adenine; the main radioactive nonnucleotide constituents were hypoxanthine and an unidentified compound with similar chromatographic properties (*166*). The properties of both of these unknown compounds are consistent with those of xanthine. The specific activity of hypoxanthine was greater than that of *tissue* adenine nucleotides, but the specific activity of erythrocyte adenine nucleotides was not measured. Henderson and LePage (*167*) measured transfer of radioactivity from perfused rat liver, prelabeled with adenine, to red cells, and from washed prelabeled mouse red cells to mouse tissues. These workers suggested that red cell nucleotides were the transport form for purine utilization by other tissues and discussed the balance of purine uptake and release as a function of oxygen tension and energy supply. Henderson and LePage's results (*167*) emphasize the care necessary in comparing results obtained with different species. These authors prelabeled rat liver with [^{14}C]adenine and observed transfer of label to the adenine nucleotides of the red cells of perfused whole blood; this result could either have occurred by utilization of adenine or adenosine (provided from the liver) directly or by utilization of hypoxanthine or xanthine by the red cells followed by conversion of IMP into adenine nucleotides. Species differences occur in the ability of red cells to convert IMP into AMP; thus, in rabbit and mouse erythrocytes, [^{14}C]hypoxanthine is incorporated into both ATP and GTP (*168–170*), but in human erythrocytes it has been reported that there is virtually no conversion of hypoxanthine or guanine into ATP (*171*). With the results in rats (*167*), it would be necessary to show that the erythrocytes could convert IMP into AMP before direct utilization of adenine or adenosine could be discounted.

Henderson and LePage's proposals (*167*) have been confirmed and the detailed biochemical mechanisms have been clarified by recent studies of purine metabolism in rabbit erythrocytes *in vivo* (*169*) and *in vitro* (*172*) and in human erythrocytes *in vitro* (*173*). [^{14}C]Adenine could not be detected in plasma 5 minutes after intravenous injection, and at 30 minutes 15% of the injected adenine was in erythrocyte nucleotides, mainly as ATP. The erythrocytes took up adenine, guanine, hypoxanthine, and xanthine and converted them into nucleotides, but hypoxanthine and xanthine were the only purines released (*169*). The interconversion pathways involving AMP deaminase and GMP reductase permit conversion of adenine and guanine into IMP. Failure of the erythrocyte to release adenine and guanine is probably due in part to conversion of AMP into ATP and GMP into GTP. Phosphorylation of IMP and XMP to higher phosphates has not been shown, and the monophosphates are substrates of 5′-nucleotidase. The greater inhibitory effect of ATP on hydrolysis of AMP than of IMP by 5′-nucleotidase of Ehrlich cells (*30*) would also channel metabolism toward inosine and away from adenosine if it also operates with the red cell nucleotidase. Finally, the phosphorolysis of inosine and xanthosine but not adenosine by erythocyte nucleoside phosphorylase ensures that no matter what purine is taken up by the erythrocyte, the only products made available for utilization by peripheral tissues are hypoxanthine and xanthine. Although, *in vitro*, xanthine release sometimes exceeded hypoxanthine release (*172*), it was suggested that, *in vivo*, an adequate supply of glutamine, and the presence of XMP aminase, would lower the XMP pool and make hypoxanthine the main purine available for salvage. The recent report (*174*) that normal human plasma contains 0.04 m*M* hypoxanthine and 0.03 m*M* xanthine indicates the possibility that IMP → XMP → xanthosine → xanthine is a major metabolic pathway. These results also indicate the need for a reassessment of IMP and XMP as the sources of plasma urate. Jorgensen and Poulsen (*175*) found slightly lower levels of hypoxanthine and xanthine in plasma, but measurements made in samples deproteinized with $LiClO_4$ to avoid loss of purines (*174*, *176*) may be more reliable. Lower plasma levels (hypoxanthine + xanthine, 80–210 μg/100 ml, approximately 0.01 m*M*, in normal human subjects) have been reported recently (*177*). Conversion of human erythrocyte adenine nucleotides into hypoxanthine and, to a lesser extent, xanthine, had been observed earlier (*178*), as had the deamination of guanine to xanthine by human plasma and the net conversion of guanosine into xanthine in blood (*179*).

Liver, with its great capacity for *de novo* purine synthesis, is of primary interest as a source of purines, and a report that xanthine is probably the main 6-oxopurine present in the liver (*180*) is consistent with reports of this compound in plasma.

An efficient system to carry purines from organs with a purine surplus to organs with a purine requirement would need some regulation of erythrocyte metabolism to promote hypoxanthine and xanthine release at suitable points in the circulatory system. Although the mechanism is not known, Hershko *et al.* (*172*) confirmed Henderson and LePage's observation (*165*) that raised oxygen levels led to release of purines by erythrocytes.

It has been reported by Bertles and Beck (*181*) that loss of RNA during rabbit reticulocyte maturation is associated with the release of hypoxanthine, cytidine, and uracil in the extracellular fluid; no adenine or adenosine was detected. It has been suggested that these purine and pyrimidine fragments may be utilized for the nutrition of the neighboring bone marrow cells. There is also evidence from other tissues that a decrease in levels of adenine nucleotides is associated with release of hypoxanthine and inosine. Thus, Overgaard-Hansen (*182*) has shown that incubation of Ehrlich ascites-tumor cells with glucose or 2-deoxyglucose results in decreased ATP levels and a transient increase in AMP and ADP accompained by an increase in hypoxanthine and inosine; only traces of adenosine could be detected. 6-Oxopurine formation was apparently initiated by deamination of AMP to IMP, which was then hydrolyzed to inosine and split by phosphorylase. Similar results with ascites-tumor cells have been reported by Wu and Racker (*183*) and McComb and Yushok (*184*). In addition inosine and hypoxanthine are released from heart muscle when adenine nucleotide breakdown is induced with myocardial hypoxia (*31*, *185*).

Though much remains to be done to evaluate the significance of purine utilization, it seems reasonable to propose that mammalian cells with 6-oxopurine phosphoribosyltransferases (cf. Section II, D) have extracellular supplies of hypoxanthine and xanthine available for utilization, and that despite the high levels of adenosine kinase and adenine phosphoribosyltransferase present in many mammalian cells there is at present no evidence that these enzymes operate on extracellular sources of adenosine or adenine (apart from those from dietary sources). Nor is there direct evidence that these enzymes generate AMP using endogenous intracellular supplies of adenosine or adenine. This is mainly due to the difficulty of demonstrating intracellular levels of unphosphorylated adenine derivatives. 5-Amino-4-imidazolecarboxamide is a substrate of adenine phosphoribosyltransferase, and is incorporated into nucleic acid purines (*186*, *187*), but here too there is no reliable information on intracellular or plasma levels.

IV. Chemotherapy and Control of Growth

A. Conversion of Drugs into Active Metabolites

The existence of "salvage" pathways of purine nucleotide biosynthesis has a twofold significance in chemotherapy with antimetabolites of purines or purine nucleotides: the presence of the salvage enzymes permits intracellular conversion of these antimetabolites into their nucleotide derivatives, which in many cases seem to be the active forms of the drugs; also, the dependence of any cell type in an animal on supplies of preformed purines makes it particularly vulnerable to blocking of this supply. The enzymes clearly implicated in this activation of drugs are the purine phosphoribosyltransferases and adenosine kinase, though the possibility should not be overlooked that successive reactions of purine nucleoside phosphorylases and kinases could convert purine analogs into their nucleotide derivatives. An important distinction arises between those nucleotide analogs that are metabolically stable at the 5′-monophosphate level and function as analogs of AMP, IMP, or GMP and those that are converted in target cells into 5′-triphosphate derivatives and function as analogs of the natural nucleoside triphosphates, often serving as substrates and inhibitors of enzymes involved in nucleic acid biosynthesis. The literature of this subject is very extensive and has been dealt with in a number of reviews (*143*, *188–192*) and a recent book (*193*), so only a few selected examples will be discussed here to illustrate the main points.

6-Mercaptopurine, an excellent structural analog of hypoxanthine, is one of the most effective drugs at present available for treatment of a number of types of leukemia and related neoplastic conditions (*194*). The closely related drug Imuran (azathioprine; 6-(1-methyl-4-nitro-5-imidazolyl)mercaptopurine), generates 6-mercaptopurine in a nonenzymatic second-order reaction with glutathione and other thiols (*191*, *195*) and it, like 6-mercaptopurine, is extensively used as an immunosuppressive agent in control of homograft rejection and autoimmune disease [for references to earlier work see (*196*, *197*)]. 6-Mercaptopurine is rapidly converted, in sensitive cells, into its nucleotide, 6-thioIMP (*140*, *198*). Kinetic studies indicate that in Ehrlich ascites-tumor cells, where the level of 6-thioIMP may reach 0.5 μmoles per gram wet weight, the enzyme responsible for this conversion is hypoxanthine-guanine phosphoribosyltransferase. With enzyme from Ehrlich cells, V_{max} was 20% of that with guanine (*119*). Since 6-mercaptopuiine has a low K_m (11 μM in conditions where K_m for the natural substrates is 11 μM hypoxanthine and 3 μM guanine), the drug also serves as a powerful competitive inhibitor [K_i = 4–8 μM for hypoxanthine or guanine (*119*)] of these natural preformed purines for nucleotide biosynthesis. 6-ThioIMP is converted into other nucleotide derivatives that are of potential significance as metabolic inhibitors: it is oxidized to 6-thioXMP in Ehrlich cells (*199*) or methylated to 6-methylthioIMP. In H. Ep. No. 2 cells the concentration of methylated nucleotide was almost as high as that of 6 thioIMP (*200*). The observation that 6-mercaptopurine can be converted into 6-thioguanine residues of DNA in adenocarcinoma 755 (*201*) implies that 6-thioXMP can be metabolized through the reaction sequence of 2′-deoxyribonucleoside triphosphate formation to 6-thiodGTP and incorporated by DNA polymerase. 6-Thioguanine has antimetabolite properties and clinical uses similar to those of 6-mercaptopurine. Like 6-mercaptopurine, it is a substrate for hypoxanthine-guanine phosphoribosyltransferase. It has an even greater affinity than 6-mercaptopurine for this enzyme (K_i 2.6 μM with the Ehrlich cell enzyme) (*202*). The enzyme from human erythrocytes shows a similar relationship (*134*). In this case, the nucleotide generated in target cells is a substrate for kinase action and subsequent incorporation into nucleic acids (*203*).

Typical of the purine nucleoside analogs that are converted into nucleotides *in vivo* by kinase action is 6-methylthioinosine (6-methylmercaptopurine ribonucleoside). This compound is the subject of extensive investigation as an antineoplastic agent, particularly because of the marked synergism that it shows with 6-mercaptopurine (*204*, *205*) in mouse tumorlines.

The methylthio derivative is one of the most effective substrates of adenosine kinase, and very high intracellular levels of 6-methylthioIMP (ca. 1 m*M*) are generated in Ehrlich ascites-tumor cells (*158*). A major factor in development of these high levels of 6-methylthioIMP is the resistance of the nucleotide to kinase action. Whereas the action of adenylate kinase prevents high levels of AMP accumulating in cells (*80, 81*), this mechanism is unable to operate with 6-methylthioIMP, and consequently this powerful regulatory analog of AMP (cf. Section IV, B) can accumulate to levels never found for AMP *in vivo*.

An example of a purine nucleoside analog that is converted into a nucleotide analog by kinase action but readily further phosphorylated to the triphosphate level is arabinosyladenine (*206*; for a review, see *207*). The arabinosyl derivative of adenine, like that of cytosine, is a powerful inhibitor of DNA synthesis *in vivo*, and these compounds are of potential interest as antiviral agents and selective regulators of cell growth; it has recently been shown that arabinosyl analogs of ATP and CTP attach arabinosyl nucleotides to DNA in the presence of DNA polymerase purified from *Escherichia coli* (*208*), and it will be of interest to find whether this reaction or the implicated "pseudofeedback" inhibition of nucleotide biosynthesis (*207*) is a major factor in the inhibitory action of arabinosyl nucleosides.

Although very large numbers of purine and purine nucleoside analogs are known to inhibit cell function, and in most cases investigated, the conversion of these analogs into nucleotide derivatives has been demonstrated, there are relatively few cases where the enzymes involved have been characterized (cf. *192, 193* for reviews). However, it seems likely that in all cases it is the enzymes that permit reutilization of preformed purine that lead to lethal synthesis of nucleotide analogs from purine antimetabolites.

B. Control of Purine Nucleotide Biosynthesis by Drugs and Their Metabolites

In the case of an antimetabolite such as 6-mercaptopurine, there may be no single site of inhibition that accounts for the selective action of the drug against sensitive cells. In no case is there enough information to define exactly the mechanism of action of a purine antimetabolite, but the special cases of 6-mercaptopurine and 6-methylthioinosine are discussed to illustrate some of the principles involved and to show the relevance of the dual biosynthetic pathways, "salvage" and *de novo*, in design of effective compounds for artificial control of growth and cell function.

It is proposed that cells showing a high sensitivity to 6-mercaptopurine are those that depend heavily on hypoxanthine as a source of purine nucleotides. Exposure to 6-mercaptopurine inhibits conversion of hypoxanthine into IMP through competitive inhibition of hypoxanthine-guanine phosphoribosyltransferase. The lower the hypoxanthine level at the target cell surface, the more effectively will administered 6-mercaptopurine compete with it, and the higher will be the rate of 6-thioIMP synthesis relative to the rate of IMP synthesis. This will be the case whether the lower supply of hypoxanthine results from decreased supply from the main "donor" organs (e.g., the liver) or from interference with the transport system by which the erythrocyte can release hypoxanthine at one region in the circulatory system after taking it up at another region (cf. Section III). 6-ThioIMP, formed in the target cells, has multiple inhibitory effects. It is a powerful competitive inhibitor of IMP dehydrogenase; with the enzyme from Ehrlich cells, K_i is 4 μM for 6-thioIMP and K_m is 14 μM for IMP (*209*). 6-ThioIMP also inhibits adenylosuccinate synthetase and adenylosuccinate lyase from Ehrlich cells (*210*). Thus, 6-mercaptopurine can interfere not only with hypoxanthine salvage, but also, through its nucleotide derivative, with the pathways leading from hypoxanthine to AMP and GMP. The biological action of 6-mercaptopurine in animal cells has often been attributed to inhibition of phosphoribosyl pyrophosphate amidotransferase, the first enzyme of *de novo* purine synthesis. This assignment, based largely on studies of inhibition of the pigeon liver amidotransferase by 6-thioIMP (cf. *211*), seems unlikely in view of the very low sensitivity of mammalian enzymes to the analog nucleotide (50% inhibition at >1 mM 6-thioIMP (*212*, *213*)) and of the synergistic action of 6-mercaptopurine and azaserine, a known inhibitor of the *de novo* pathway (cf. *214*). However, this inhibition of *de novo* purine nucleotide biosynthesis may contribute to the total effect of the drug, particularly when reinforced by inhibition of the amidotransferase by 6-methylthioIMP, formed from 6-mercaptopurine as described in Section IV, A. 6-MethylthioIMP is one of the most effective inhibitors known of the amidotransferase from animal cells (*212*, *213*) mimicking the natural feedback nucleotides, and it has been suggested that there is a correlation between formation of this nucleotide analog and the apparent inhibition of *de novo* purine nucleotide biosynthesis *in vivo* by 6-mercaptopurine (*200*).

Further points of inhibition might arise at any of the numerous enzymatic reactions involved in conversion of 6-thioIMP into 6-thioguanine of DNA or as a consequence of incorporation or terminal attachment of this abnormal analog.

In the case of 6-methylthioinosine, although the metabolic fate of the drug may be simpler, its pattern of enzyme inhibition may be even more complex. The high levels of 6-methylthioIMP generated in sensitive cells are adequate, in view of known *in vitro* inhibitory effects, to account for the powerful inhibition of *de novo* synthesis observed *in vivo* (*215*). As an AMP analog resistant to kinase action, 6-methylthioIMP has many other potential sites of action as an artificial regulator of enzyme activity. AMP is one of the most significant natural regulatory compounds, activating reaction sequences that generate ATP and inhibiting sequences that consume ATP in biosynthesis, but these regulatory effects are subject to self-balancing controls since AMP, ADP, and ATP concentrations are kept at levels satisfying the equilibrium requirements of adenylate kinase (*8*, *81*). AMP analogs that are not substrates of adenylate kinase can override this "adenylate control system." 6-MethylthioIMP is an activator of phosphofructokinase and an inhibitor of fructose-1,6-diphosphosphatase (*202*), and to define the effects of 6-methylthioinosine in intact cells, it may be necessary to examine the numerous reactions regulated by AMP.

The synergism between 6-mercaptopurine and 6-methylthioinosine (*204*, *205*) is of special interest, since, while both are substrates for "salvage" enzymes, there is evidence that the former mainly inhibits "salvage" pathways, whereas the latter, as its nucleotide, inhibits mainly *de novo* pathways of purine nucleotide formation. It may be misleading to seek the source of the synergism at the level of multiple enzyme inhibitions in the target cell. If the suggestions made here are valid, the availability of hypoxanthine would be an important variable determining the effectiveness of 6-mercaptopurine, and inhibition of *de novo* purine nucleotide biosynthesis in liver by 6-methylthioIMP might decrease the release of hypoxanthine from liver to plasma, leading to lower levels of IMP and higher levels of 6-thio-IMP in target cells. Now that methods are available for measurement of hypoxanthine in plasma (*174*), it should be possible to determine whether these speculations are justified.

Combination chemotherapy is leading to greatly increased effectiveness in the use of antimetabolites to control growth and cell function and it seems likely that recognition of the dual dependence of cells on preformed purines and on simple precursors will be a useful guide in rational design of drug combinations.

C. Defects in Purine Utilization and Nucleotide Formation by *de Novo* Synthesis: Gout, the Lesch-Nyhan Syndrome, and Drug Resistance

Gout, which is expressed clinically in episodes of acute arthritis, or chronic arthritis, with the formation of uric acid calculi, has been associated with increased rates of *de novo* purine synthesis contributing to hyperuricemia in the majority of gouty patients (*216*). In other cases it may be due to increased nucleotide catabolism (*217*).

Even that section of the gouty population characterized by increased purine synthesis is composed of a heterogeneous group of genetic abnormalities. Deficiencies in glucose-6-phosphatase can cause increased ribose 5-phosphate to be made available for purine synthesis (*218*, *219*), and a study by Kelley *et al.* (*220*, *221*) has revealed a partial deficiency (90–99% loss) of hypoxanthine-guanine phosphoribosyltransferase in ten members of six families having these symptoms. The mutant enzyme of one of these families differed in its relative activity for each of the purine substrates, and those of two families studied differed in their heat stability. Physical and kinetic studies of adenine phosphoribosyltransferase in erythrocytes from these patients suggest that an altered or new enzyme is present. The apparent V_{max} of most of these preparations were also increased, but it could not be determined whether this was due to a change in the catalytic activity of the enzyme, or an increased amount of enzyme protein. The precise manner by which lack of hypoxanthine-guanine phosphoribosyltransferase gives rise to overproduction of uric acid is not yet known, although increased availability of PRPP has been implicated (*222–224*).

Azathioprine had no effect in two of those patients lacking the enzyme, whereas it was shown to suppress significantly the *de novo* purine synthesis in other gouty patients, two of whom were manifesting overproduction of uric acid, and three out of four of whom showed normal uric acid production (*225*). This suppression was taken as evidence that PRPP amidotransferase has normal sensitivity to feedback inhibition in patients who responded to the drug and therefore altered feedback sensitivity could not be the cause of increased purine synthesis.

The xanthine oxidase inhibitor, allopurinol, has also been used to control uric acid levels in gouty patients, and this, unlike azathioprine treatment, results in increased levels of hypoxanthine and xanthine (*177*). In most patients allopurinol also produces a decrease in total purine excretion (hypoxanthine + xanthine + uric acid). Suggested mechanisms for this effect (*226*) are feedback inhibition of *de novo* purine synthesis by allopurinol ribonucleotide, or increased feedback by increased IMP or XMP formed from hypoxanthine or xanthine resulting from inhibition of xanthine oxidase and consequent higher substrate levels. Support for this is given by the work of Pomales *et al.* (*227*, *228*), who showed that, in mice, allopurinol produces an increase in incorporation of labeled hypoxanthine into soluble nucleotides and into adenine and guanine of nucleic acids. In man, it has been shown that allopurinol enhances the re-use of exogenous hypoxanthine (*229*). A third suggested mechanism for inhibition of *de novo* synthesis by allopurinol is limitation of PRPP concentration by the formation of IMP or allopurinol ribonucleotide, although the failure of orotic acid to diminish purine synthesis *de novo* in man argues against this concept. In this connection, however, it is interesting to note the increased formation of PRPP in the red blood cells of some gouty patients, not suffering from any known specific enzymatic defect (*223*), and the increased purine biosynthesis in skin fibroblasts of similar gouty patients, associated with increased concentrations of PRPP (*217*). As might be expected, allopurinol had no effect on *de novo* purine synthesis in those patients lacking hypoxanthine-guanine phosphoribosyltransferase (*226*).

The lack of this enzyme is associated also with a sex-linked neurological disorder of children called the Lesch-Nyhan syndrome (*131, 138*). This syndrome is characterized by self-mutilation, mental retardation, marked overproduction of uric acid (6 times normal), and accelerated *de novo* biosynthesis of purines (200 times as much radioactive glycine incorporated into urinary uric acid as normal). The possible functions of the transferase and their relation to the Lesch-Nyhan syndrome are discussed by Henderson (*122, 230*).

Rosenbloom *et al.* (*231*) found two factors implicated as potential causes of increased purine biosynthesis in fibroblasts from such patients. One was the rise in PRPP concentration, with no change in its rate of synthesis. While there is little evidence that PRPP is limiting *in vivo*, the relative affinities of purine phosphoribosyltransferases and PRPP amidotransferase suggest that the former, if present, would have a competitive advantage for this substrate. The other factor was that hypoxanthine and xanthine caused stimulation of purine synthesis *de novo*. There was no evidence that this effect was caused by a decrease in the concentration of feedback-inhibitory guanine ribonucleotides because the concentration of purine ribonucleotides in mutant fibroblast cells was normal. These authors point out, however, that loss of transferase may lead to lowered concentrations of guanine ribonucleotides in basal ganglia, where the activity of the enzyme is normally 200 times greater than in fibroblasts.

Both azathioprine and allopurinol failed to suppress *de novo* purine biosynthesis in patients with the Lesch-Nyhan syndrome (*232*). Azathioprine did not reduce the plasma uric acid level or the urinary excretion of uric acid, whereas allopurinol dramatically reduced both, there being a simultaneous increase in uric acid precursors, so that the total excretion of urinary oxopurines remained constant. Rosenbloom *et al.* (*231*) pointed out that if high concentrations of oxopurine in cerebrospinal fluid cause abnormal functioning of the central nervous system, then treatment with allopurinol may prove detrimental. However, no change in the neurological symptoms was observed during allopurinol therapy (*233*). Also, gouty patients with partial deficiencies of hypoxanthine-guanine phosphoribosyltransferase can have as high oxopurine concentrations in their cerebrospinal fluid as individuals with the syndrome, but may still have no detectable neurological dysfunction (*230*). It appears more likely that a decrease in some necessary purine nucleotide such as GTP or increase in some unusual purine metabolite such as methylated xanthine is responsible for neurological damage (*233*, *234*).

Resistance to 6-mercaptopurine inhibition of purine biosynthesis in fibroblasts from patients with Lesch-Nyhan syndrome, because of a genetic lesion, can be considered an example of the frequently observed emergence of drug-resistance in previously susceptible cell populations, associated with a decreased conversion of purine bases or nucleosides to ribonucleotides (*143*, *189*, *235–237*). More recent examples of this phenomenon are the following: the inactivity of hypoxanthine and guanine and their thioanalogs as inhibitors of the formation of formylglycinamide ribonucleotide in thiopurine-resistant mammalian cells lacking hypoxanthine-guanine phosphoribosyltransferase (*238*); lack of adenine phosphoribosyltransferase, and possibly also some evidence for alterations in the specificity of this enzyme, associated with resistance to 2,6-diamopurine in L-strain mouse mouse cells (*239*), and with resistance to 2-fluoroadenine (*240*). Loss of adenosine kinase has been defined as the mechanism of resistance to many cytotoxic nucleosides, such as 6-methylthioinosine in H.Ep. No. 2 cells in culture (*159*), and as the reason for the inability of this ribonucleoside and of formycin A to inhibit the growth of a resistant Ehrlich ascites-tumor line (*158*).

Metabolic consequences of the resistance to azaguanine, due to loss of hypoxanthine-guanine phosphoribosyltransferase, have been suggested as a possible reason for the ability of A9 mouse fibroblast cells to suppress the malignant character of mouse tumor cells (*240a*).

Resistance due to a loss of hypoxanthine-guanine phosphoribosyltransferase, associated with a 2-3-fold increase in adenine phosphoribosyltransferase, has also been implicated in one case of leukemic leukocytes from a patient selected *in vivo* during a course of treatment with 6-mercaptopurine, although it was evident in 14 other resistant cases that there were other mechanisms of resistance as well (*241*). Indeed, detoxication, caused by cleavage of the sulfur group, decreased permeability to the analog, decreased sensitivity of an enzyme system to the inhibitor, or increased synthesis of the inhibited enzyme have all been proposed as methods by which a cell line may develop resistance [see Brockman (*235*)].

Despite the diversity of resistance mechanisms discussed here, the predominant cause of failure in chemotherapy with purine analogs seems to be a change in the nature of the target cell line's purine utilization systems resulting in failure to activate the drug and sometimes associated with increased use of *de novo* pathways. One of the main aims of current investigations of combination chemotherapy is to find drug combinations that will bring target cell populations to levels that can be controlled by the body's immunological defenses before drug-resistant cell populations arise. The sensitivity of immunologically active cells to the same drugs that are most effective in control of many pathological cells is one of the main barriers to an effective solution of this problem and is the basis for much of the present effort to find carrier groups that will deliver inhibitory nucleotides selectively in target cells without damaging normal cells that may closely resemble the target cells in biochemical properties such as dependence on supplies of preformed purines.

V. General Conclusions

Though this review is restricted to mammalian systems, the phenomenon of reutilization of preformed purines is much more general. The significance of the reutilization pathways is particularly clear in microorganisms, where mutation has frequently resulted in an absolute dependence on purines such as adenine. Although the enzymes involved have often not been characterized, the large number of studies in which radioactive purines are used as precursors for nucleic acid synthesis illustrates the general occurrence of the appropriate "salvage" enzymes.

The physiological significance of an enzyme is most conveniently studied in a mutant lacking that enzyme, and in the case of hypoxanthine utilization we have the opportunity, rare in mammalian systems, to study the consequences of lack of hypoxanthine-guanine phosphoribosyltransferase associated with the Lesch-Nyhan syndrome. This topic is, therefore, considered in some detail in the review. Though investigations are at an early stage and many aspects of the syndrome are still obscure, there has been increased recognition of the possible significance of purine reutilization pathways in detoxication mechanisms. In this respect the significance of adenine phosphoribosyltransferase in animal cells deserves detailed investigation. Activities in general are high, and the low K_m (cf. Section II) makes this an effective pathway to AMP, which, through AMP deaminase, can eventually form guanine nucleotides. However, the absence of known sources of endogenous adenine and the very low levels of adenine found in tissues raises the possibility that this very effective "adenine trap" serves to protect cells against free adenine rather than to maintain supplies of AMP.

There are striking analogies between the biosynthetic pathways leading to pyridine (*10*) and purine nucleotides, and many of the principles discussed by Chaykin (*10*) are equally relevant here. In both cases there are *de novo* and reutilization pathways, and the preformed heterocyclic compounds may be of dietary or endogenous origin. The molecules available for "salvage" have an amino functional group (nicotinamide or adenine) or an "oxy" functional group (nicotinate or hypoxanthine). Chaykin has discussed the special problems that arise in distinguishing between pathways directed toward provision of essential nucleotides, detoxication of preformed fragments, or conservation of valuable heterocyclic derivatives in storage forms.

Higher plants are particularly attractive systems for studying the significance of reutilization of preformed purines. As the meristematic cells vacuolate, taking on structural roles and often reaching a senescent state, there is an opportunity for purines no longer needed as components of nucleic acids or nucleotide coenzymes to be released in diffusible forms (nucleosides and bases) by the "salvage" enzymes discussed in Section II. The possibility arises that these preformed purines then establish gradients along roots and shoots, influencing patterins of cell division and morphogenesis. Enzymes leading to release and reutilization of purines are widely distributed among higher plants. Adenine phosphoribosyltransferase of soybean callus culture shows a marked rise on stimulation of cell division by kinetin (*242*).

The existence of purine reutilization pathways is relevant to studies of the mechanism of action of cytokinins. These N^6-alkyladenines show stimulation of plant cell division at levels as low as 10^{-8} to 10^{-9} M; among the most active compounds are those with N^6-isopentenyl substituents corresponding to those found in some species of tRNA (cf. Hall in this volume). From earlier work it had been proposed that natural or synthetic cytokinins were converted into nucleotides by purine utilization pathways and then incorporated into RNA [for a review, see Letham (*243*)]. A problem in this mechanism has been failure to demonstrate a kinase that will phosphorylate N^6-alkylAMP to N^6-alkylADP. It now seems more likely (*244*) that the converse sequence is involved; isopentenyl groups are attached to adenine at the tRNA level, and degradation of tRNA releases cytokinin nucleotides, nucleosides, and bases. There is indirect evidence that the bases may be the active form: N^6-benzyl-9-methyladenine, in which the nitrogen required for nucleotide formation is blocked, is a very active cytokinin, but shows no incorporation into nucleic acids (*245*). N^6-Benzyl-9-butyladenine, which is probably even more resistant to dealkylation than the 9-methyl derivative, is more effective than N^6-benzyladenine as a stimulator of cell division in soybean callus culture (*246*).

Nucleotide synthesis from preformed purines may be one of the most primitive biological reactions. Adenine and guanine are readily formed by nonenzymatic reactions in conditions thought likely to have existed prior to the origin of life (cf. *247*, *248*). It has been proposed that the availability of complex purines and amino acids facilitated the reactions leading to self-replicating cells, and if this is so, the primitive systems that made purines available for nucleic acid formation were the forerunners of the sophisticated systems that now permit enzymatic reutilization of preformed purines.

Studies of enzyme regulation emphasize the lack of information on the way in which observations on isolated enzymes may be related to metabolic processes. The problems of correlating the enzymology and physiology of purine utilization are particularly evident in multicellular organisms, but this is one of the most promising regions for studies of artificial control of growth and cell function and is likely to provide fundamental information on the natural mechanisms that regulate morphogenesis and cell division.

Acknowledgments

We wish to thank Dr. Arthur Kornberg for provision of facilities to complete this review while one of us (M. R. A.) was Visiting Professor in the Department of Biochemistry, Stanford University School of Medicine. The financial support of the Australian Research Grants Committee, The Reserve Bank of Australia, the University of Adelaide Anti-Cancer Foundation, and the U.S. Public Health Service is gratefully acknowledged.

References

1. A. Kornberg, *in* "The Chemical Basis of Heredity" (W. D. McElroy and B. Glass, eds.), p. 579. Johns Hopkins Press, Baltimore, Maryland, 1957.
2. A. Kornberg, *Advan. Enzymol.* **18,** 191 (1957).
3. A. Kornberg, *in* "Horizons in Biochemistry" (M. Kasha and B. Pullman, eds.), p. 251. Academic Press, New York, 1962.
4. D. Elson, *Ann. Rev. Biochem.* **34,** 449 (1965).
5. M. Laskowski, *Advan. Enzymol.* **29,** 165 (1967).
6. I. R. Lehman, *Ann. Rev. Biochem.* **36,** 645 (1967).
7. D. Shugar and H. Sierakowska, This series **7,** 369 (1967).
8. D. E. Atkinson, *Ann. Rev. Biochem.* **35,** 85 (1966).
9. E. R. Stadtman, *Advan. Enzymol.* **28,** 41 (1966).
10. S. Chaykin, *Ann. Rev. Biochem.* **36,** 149 (1967).
11. P. Reichard, *Advan. Enzymol.* **21,** 263 (1959).
12. A. Larsson and P. Reichard, This series **7,** 303 (1967).
13. J. M. Buchanan and S. C. Hartman, *Advan. Enzymol.* **21,** 199 (1959).
14. R. L. Blakley and E. Vitols, *Ann. Rev. Biochem.* **37,** 201 (1968).
15. L. A. Heppel, *in* "The Enzymes" (P. D. Boyer, H. Lardy, and K. Myrbäck, eds.), 2nd. ed., Vol. 5, p. 49. Academic Press, New York, 1961.
16. R. Itoh, A. Mitsui, and K. Tsushima, *J. Biochem., Tokyo* **63,** 165 (1968).
17. R. Itoh, A. Mitsui, and K. Tsushima, *Biochim. Biophys. Acta* **146,** 151 (1967).
18. G. de Lamirande, C. Allard, and A. Cantero, *J. Biophys. Biochem. Cytol.* **4,** 373 (1958).
19. G. de Lamirande, C. Allard, and A. Cantero, *Cancer Res.* **18,** 952 (1958).
20. P. Fritzson, *Europ. J. Biochem.* **1,** 12 (1967).
21. C. C. Widnell and J. C. Unkeless, *Proc. Natl. Acad. Sci. U.S.* **61,** 1050 (1968).
22. E. Sulkowski, W. Björk, and M. Laskowski, *J. Biol. Chem.* **238,** 2477 (1963).
23. P. L. Ipata, *Biochem. Biophys. Res. Commun.* **27,** 337 (1967).
24. P. L. Ipata, *Biochemistry* **7,** 507 (1968).

25. M. S. Center and F. J. Behal, *Arch. Biochem. Biophys.* **114,** 414 (1966).
26. Z. Ahmed and J. L. Reis, *Biochem. J.* **69,** 386 (1958).
27. J. Lisowski, *Biochim. Biophys. Acta* **113,** 321 (1966).
28. H. L. Segal and B. M. Brenner, *J. Biol. Chem.* **235,** 471 (1960).
29. A. R. P. Paterson and A. Hori, *Can. J. Biochem. Physiol.* **41,** 1339 (1963).
30. A. W. Murray and B. Friedrichs, *Biochem. J.* **111,** 83 (1969).
31. R. Burger and J. M. Lowenstein, *J. Biol. Chem.* **242,** 5281 (1967).
32. H. P. Baer, G. I. Drummond, and E. L. Duncan, *Mol. Pharmacol.* **2,** 67 (1966).
33. L. A. Heppel, J. Hurwitz, and B. L. Horecker, *J. Am. Chem. Soc.* **79,** 630 (1957).
34. W. D. McElroy, *in* "Methods in Enzymology" (S. P. Colowick and N. O. Kaplan, eds.), Vol. VI, p. 203. Academic Press, New York, 1963.
35. R. C. Hartenstein and I. Fridovich, *J. Biol. Chem.* **242,** 740 (1967).
36. L. G. Howell and I. Fridovich, *J. Biol. Chem.* **242,** 4930 (1967).
37. H. M. Kalckar, *J. Biol. Chem.* **167,** 461 (1947).
38. A. Roush and E. R. Norris, *Arch. Biochem.* **29,** 124 (1950).
39. R. Currie, F. Bergel, and R. C. Bray, *Biochem. J.* **104,** 634 (1967).
40. B. R. Baker, *J. Med. Chem.* **10,** 59 (1967).
41. S. Kumar, K. K. Tewari, and P. S. Krishnan, *Biochem. J.* **95,** 797 (1965).
42. S. Kumar, V. Josan, K. C. S. Sanger, K. K. Tewari, and P. S. Krishnan, *Biochem. J.* **102,** 691 (1967).
43. V. Josan and P. S. Krishnan, *Biochem. Biophys. Res. Commun.* **31,** 299 (1968).
44. R. Wolfenden, *J. Am. Chem. Soc.* **88,** 3157 (1966).
45. T. K. Sharpless and R. Wolfenden, *in* "Methods in Enzymology" (L. Grossman and K. Moldave, eds.), Vol. XII, p. 126. Academic Press, New York, 1967.
46. R. Wolfenden, T. K. Sharpless, and R. Allan, *J. Biol. Chem.* **242,** 977 (1967).
47. T. G. Brady and W. O'Connell, *Biochem. Biophys. Acta* **62,** 216 (1962).
48. M. Rockwell and M. H. Maguire, *Mol. Pharmacol.* **2,** 574 (1966).
49. N. Pfrogner, *Arch. Biochem. Biophys.* **119,** 141 (1967).
50. G. Ronca, C. Bauer, and C. A. Rossi, *Europ. J. Biochem.* **1,** 434 (1967).
51. P. Andrews, *Biochem. J.* **91,** 222 (1964).
52. N. Pfrogner, *Arch. Biochem. Biophys.* **119,** 147 (1967).
53. J. G. Cory and R. J. Suhadolnik, *Biochemistry* **4,** 1729 (1965).
54. H. P. Baer, G. I. Drummond, and J. Gillis, *Arch. Biochem. Biophys.* **123,** 172 (1968).
55. B. M. Chassy and R. J. Suhadolnik, *J. Biol. Chem.* **242,** 3655 (1967).
56. S. Fredericksen, *Arch. Biochem. Biophys.* **113,** 383 (1966).
57. A. Askari and J. E. Franklin, *Biochem. Biophys. Acta* **110,** 162 (1965).
58. A. Razin and J. Mager, *Israel J. Med. Sci.* **2,** 614 (1966).
59. A. Askari and S. N. Rao, *Biochim. Biophys. Acta* **151,** 198 (1968).
60. J. A. Muntz, *J. Biol. Chem.* **201,** 221 (1953).
61. H. Weil-Malherbe and R. H. Green, *Biochem. J.* **61,** 218 (1955).
62. J. Mendicino and J. A. Muntz, *J. Biol. Chem.* **233,** 178 (1958).
63. B. Cunningham and J. M. Lowenstein, *Biochem. Biophys. Acta* **96,** 535 (1965).
64. B. Setlow, R. Burger, and J. M. Lowenstein, *J. Biol. Chem.* **241,** 1244 (1966).
65. B. Setlow and J. M. Lowenstein, *J. Biol. Chem.* **242,** 607 (1967).
66. B. Setlow and J. M. Lowenstein, *J. Biol. Chem.* **243,** 3409 (1968).
67. Y. P. Lee, *J. Biol. Chem.* **227,** 987 (1957).
68. K. L. Smiley, A. J. Berry, and C. H. Suelter, *J. Biol. Chem.* **242,** 2502 (1967).
69. Y. P. Lee, *J. Biol. Chem.* **227,** 993 (1957).
70. Y. P. Lee, *J. Biol. Chem.* **227,** 999 (1957).
71. K. L. Smiley and C. H. Suelter, *J. Biol. Chem.* **242,** 1980 (1967).

72. R. Wolfenden, Y. Tomozawa, and B. Bamman, *Biochemistry* **7,** 3965 (1968).
73. M. R. Atkinson and A. W. Murray, *Biochem. J.* **104,** 10C (1967).
74. L. D. Smith and D. E. Kizer, *Federation Proc.* **27,** 792 (1968).
75. D. E. Kizer, B. A. Howell, J. A. Clouse, and B. C. Shirley, *Cancer Res.* **26,** 1376 (1966).
76. Y. P. Lee and M. H. Wang, *J. Biol. Chem.* **243,** 2260 (1968).
77. A. W. Murray and M. R. Atkinson, unpublished results.
78. B. S. Vanderheiden, *Biochem. Biophys. Res. Commun.* **21,** 265 (1965).
79. H. Klenow, *in* "Biochemists Handbook" (C. Long, ed.), p. 411. Spon, London, 1961.
80. P. Mandel, This series **3,** 299 (1964).
81. H. A. Krebs, *Proc. Roy. Soc.* **B159,** 545 (1964).
82. B. Setlow and J. M. Lowenstein, *J. Biol. Chem.* **243,** 6216 (1968).
83. M. Freidkin and H. M. Kalckar, *in* "The Enzymes" (P. D. Boyer, H. Lardy, and K. Myrbäck, eds.), Vol. 5, p. 237. Academic Press, New York, 1961.
84. E. Robins, D. E. Smith, and R. E. McCaman, *J. Biol. Chem.* **204,** 927 (1953).
85. E. D. Korn and J. M. Buchanan, *J. Biol. Chem.* **217,** 183 (1955).
86. A. M. Gotto, A. W. Meikle, and O. Touster, *Biochem. Biophys. Acta* **80,** 552 (1964).
87. A. R. P. Paterson and A. I. Simpson, *Can. J. Biochem.* **43,** 1693 (1965).
88. A. R. P. Paterson and A. I. Simpson, *Can. J. Biochem.* **43,** 1701 (1965).
89. A. R. P. Paterson, *Can. J. Biochem.* **43,** 257 (1965).
90. B. Pinto and O. Touster, *J. Biol. Chem.* **241,** 772 (1966).
91. V. E. Price, M. C. Otey, and P. Plesner, *in* "Methods in Enzymology" (S. P. Colowick and N. O. Kaplan, eds.), Vol. II, p. 448. Academic Press, New York, 1955.
92. R. Abrams, M. Edmonds, and L. Libenson, *Biochem. Biophys. Res. Commun.* **20,** 310 (1965).
93. T. A. Krenitsky, *Mol. Pharmacol.* **3,** 526 (1967).
94. H. L. A. Tarr and J. E. Roy, *Can. J. Biochem.* **45,** 409 (1967).
95. E. W. Yamada, *J. Biol. Chem.* **236,** 3043 (1961).
96. F. M. Huennekens, E. Nurk, and B. W. Gabrio, *J. Biol. Chem.* **221,** 971 (1956).
97. K. K. Tsuboi and P. B. Hudson, *J. Biol. Chem.* **224,** 879 (1957).
98. C. H. De Verdier and B. J. Gould, *Biochim. Biophys. Acta* **68,** 333 (1963).
99. B. K. Kim, S. Cha, and R. E. Parks, *J. Biol. Chem.* **243,** 1763 (1968).
100. R. P. Agarwal and R. E. Parks, *Federation Proc.* **27,** 585 (1968).
101. E. D. Korn and J. M. Buchanan, *J. Biol. Chem.* **217,** 183 (1955).
102. H. L. A. Tarr, *Can. J. Biochem. Physiol.* **36,** 517 (1958).
103. E. D. Korn, F. C. Charalampous, and J. M. Buchanan, *J. Am. Chem. Soc.* **75,** 3610 (1953).
104. M. R. Atkinson and R. K. Morton, *in* "Comparative Biochemistry" (M. Florkin and H. S. Mason, eds.), Vol. II, p. 1. Academic Press, New York, 1960.
105. M. Friedkin and H. M. Kalckar, *J. Biol. Chem.* **184,** 437 (1950).
106. T. A. Krenitsky, G. B. Elion, A. M. Henderson, and G. H. Hitchings, *J. Biol. Chem.* **243,** 2876 (1968).
107. M. R. Sheen, B. K. Kim, and R. E. Parks, *Mol. Pharmacol.* **4,** 293 (1968).
108. B. K. Kim, S. Cha, and R. E. Parks, *J. Biol. Chem.* **243,** 1771 (1968).
109. T. A. Krenitsky, G. B. Elion, R. A. Strelitz, and G. H. Hitchings, *J. Biol. Chem.* **242,** 2675 (1967).
110. H. Umezawa, T. Sawa, Y. Fukagawa, G. Koyama, M. Murase, M. Hamanda, and T. Takeuchi, *J. Antibiotics (Tokyo) Ser. A* **18,** 178 (1965).
111. R. W. Brockman, C. S. Debavadi, P. Stutts, and D. J. Hutchison, *J. Biol. Chem.* **236,** 1471 (1961).

112. G. P. Kalle and J. S. Gots, *Biochim. Biophys. Acta* **53,** 166 (1961).
113. J. G. Flaks, M. J. Erwin, and J. M. Buchanan, *J. Biol. Chem.* **228,** 201 (1957).
114. J. K. Roy, C. A. Haavik, and R. E. Parks, *Proc. Am. Assoc. Cancer Res.* **3,** 146 (1960).
115. M. Hori and J. F. Henderson, *J. Biol. Chem.* **241,** 1406 (1966).
116. D. G. R. Blair, *Can. J. Biochem.* **45,** 435 (1967).
117. M. Hori and J. F. Henderson, *J. Biol. Chem.* **241,** 3404 (1966).
118. R. E. A. Gadd and J. F. Henderson, *Federation Proc.* **27,** 639 (1968).
119. M. R. Atkinson and A. W. Murray, *Biochem. J.* **94,** 64 (1965).
120. M. R. Atkinson and A. W. Murray, *Biochem. J.* **94,** 71 (1965).
121. J. F. Henderson, H. R. Miller, W. N. Kelley, F. M. Rosenbloom, and J. E. Seegmiller, *Can. J. Biochem.* **46,** 703 (1968).
122. J. F. Henderson, *Federation Proc.* **27,** 1053 (1968).
123. A. W. Murray, *Biochem. J.* **100,** 671 (1966).
124. M. Hori, R. E. A. Gadd, and J. F. Henderson, *Biochem. Biophys. Res. Commun.* **28,** 616 (1967).
125. A. Razin and J. Mager, *Israel J. Chem.* (*Proc.*) **5,** 121 (1967).
126. A. W. Murray, *Biochem. J.* **103,** 271 (1967).
127. A. W. Murray, *Biochem. J.* **104,** 675 (1967).
128. A. W. Murray and P. C. L. Wong, *Biochem. J.* **104,** 669 (1967).
129. J. F. Henderson, A. R. P. Paterson, I. C. Caldwell, and M. Hori, *Cancer Res.* **27,** 715 (1967).
130. A. W. Murray, *Biochem. J.* **100,** 664 (1966).
131. F. M. Rosenbloom, W. N. Kelley, J. Miller, J. F. Henderson, and J. E. Seegmiller, *J. Am. Med. Assoc.* **202,** 175 (1967).
132. M. E. Balis, *Federation Proc.* **27,** 1067 (1968).
133. L. N. Lukens and K. A. Herrington, *Biochim. Biophys. Acta* **24,** 432 (1957).
134. T. A. Krenitsky, R. Papaioannou, and G. B. Elion, *J. Biol. Chem.* **244,** 1263 (1969).
135. R. L. Miller and A. L. Bieber, *Biochemistry* **7,** 1420 (1968).
136. W. N. Kelley, F. M. Rosenbloom, J. F. Henderson, and J. E. Seegmiller, *Biochem. Biophys. Res. Commun.* **28,** 340 (1967).
137. J. F. Henderson, L. W. Brox, W. N. Kelley, F. M. Rosenbloom, and J. E. Seegmiller, *J. Biol. Chem.* **243,** 2514 (1968).
138. J. E. Seegmiller, F. M. Rosenbloom, and W. N. Kelley, *Science* **155,** 1682 (1967).
139. R. W. Brockman, L. L. Bennett, Jr., M. S. Simpson, A. R. Wilson, J. R. Thomson, and H. E. Skipper, *Cancer Res.* **19,** 856 (1959).
140. R. W. Brockman, *Cancer Res.* **20,** 643 (1960).
141. R. W. Brockman, G. G. Kelley, P. Stutts, and V. Copeland, *Nature* **191,** 469 (1961).
142. J. L. Way and R. E. Parks, *J. Biol. Chem.* **231,** 467 (1958).
143. R. W. Brockman, *Advan. Cancer Res.* **7,** 129 (1963).
144. B. Lindberg, H. Klenow, and K. Hansen, *J. Biol. Chem.* **242,** 350 (1967).
145. K. J. Pierre, A. P. Kimball, and G. A. LePage, *Can. J. Biochem.* **45,** 1619 (1967).
146. A. W. Murray, *Biochem. J.* **106,** 549 (1968).
147. H. P. Schnebli, D. L. Hill, and L. L. Bennett, *J. Biol. Chem.* **242,** 1997 (1967).
148. H. Holmsen and M. C. Rozenberg, *Biochim. Biophys. Acta* **155,** 326 (1968).
149. M. C. Rozenberg and H. Holmsen, *Biochim. Biophys. Acta* **155,** 342 (1968).
150. H. E. Franz, M. Franz, and K. Decker, *Z. Physiol. Chem.* **327,** 243 (1962).
151. C. Scholtissek, *Biochim Biophys. Acta* **145,** 228 (1967).
152. C. Scholtissek, *Biochim. Biophys. Acta* **145,** 238 (1967).
153. C. Scholtissek, *Biochim. Biophys. Acta* **155,** 14 (1968).

154. C. Scholtissek, *Biochim. Biophys. Acta* **158,** 435 (1968).
155. D. H. W. Ho, J. K. Luce, and E. Frei, *Biochem. Pharmacol.* **17,** 1025 (1968).
156. I. C. Caldwell, J. F. Henderson, and A. R. P. Paterson, *Can. J. Biochem.* **44,** 229 (1966).
157. H. T. Shigeura and S. D. Sampson, *Nature* **215,** 419 (1967).
158. I. C. Caldwell, J. F. Henderson, and A. R. P. Paterson, *Can. J. Biochem.* **45,** 735 (1967).
159. L. L. Bennett, H. P. Schnebli, M. H. Vail, P. W. Allan, and J. A. Montgomery, *Mol. Pharmacol.* **2,** 432 (1966).
160. K. J. Pierre and G. A. LePage, *Proc. Soc. Exptl. Biol. Med.* **127,** 432 (1968).
161. H. L. A. Tarr *Can. J. Biochem.* **42,** 1535 (1964).
162. V. Krygier and R. L. Momparler, *Biochim. Biophys. Acta* **161,** 578 (1968).
163. L. G. Lajtha and J. R. Vane, *Nature* **182,** 191 (1958).
164. R. M. S. Smellie, R. Y. Thomson, R. Goutier, and J. N. Davidson, *Biochim. Biophys. Acta* **22,** 585 (1956).
165. J. F. Henderson and G. A. LePage, *Cancer Res.* **19,** 67 (1959).
166. E. L. Bennett and B. J. Krueckel, *Biochim. Biophys. Acta* **17,** 515 (1955).
167. J. F. Henderson and G. A. LePage, *J. Biol. Chem.* **234,** 3219 (1959).
168. B. A. Lowy, M. K. Williams, and I. M. London, *J. Biol. Chem.* **236,** 1439 (1961).
169. J. Mager, A. Hershko, R. Zeitlin-Beck, T. Shoshani, and A. Razin, *Biochim. Biophys. Acta* **149,** 50 (1967).
170. L. J. Fontenelle and J. F. Henderson, *Biochim. Biophys. Acta* **177,** 175 (1969).
171. B. A. Lowy, M. K. Williams, and I. M. London, *J. Biol. Chem.* **237,** 1622 (1962).
172. A. Hershko, A. Razin, T. Shoshani, and J. Mager, *Biochim. Biophys. Acta* **149,** 59 (1967).
173. J. Mager, A. Dvilansky, A. Razin, E. Wind, and G. Izak, *Israel J. Med. Sci.* **2,** 297 (1966).
174. P. J. Orsulak, W. Haab, and M. D. Appleton, *Anal. Biochem.* **23,** 156 (1968).
175. S. Jorgensen and H. E. Poulsen, *Acta Pharmacol. Toxicol.* **11,** 223 (1955).
176. R. Currie, *Nature* **205,** 1212 (1965).
177. S. Goldfinger, J. R. Klinenberg, and J. E. Seegmiller, *J. Clin. Invest.* **44,** 623 (1965).
178. S. Jorgensen and H. E. Poulsen, *Acta Pharmacol. Toxicol.* **11,** 287 (1955).
179. S. Jorgensen, *Acta Pharmacol. Toxicol.* **12,** 303 (1956).
180. E. Reid and B. M. Stevens, *Biochem. J.* **68,** 367 (1958).
181. J. F. Bertles and W. S. Beck, *J. Biol. Chem.* **237,** 3770 (1962).
182. K. Overgaard-Hansen, *Biochim. Biophys. Acta* **104,** 330 (1965).
183. R. Wu and E. Racker, *J. Biol. Chem.* **234,** 1036 (1959).
184. R. B. McComb and W. D. Yushok, *Cancer Res.* **24,** 198 (1964).
185. R. M. Berne, *Physiol. Rev.* **44,** 1 (1964).
186. C. S. Miller, S. Gurin, and D. W. Wilson, *Science* **112,** 654 (1950).
187. L. G. Lajtha, W. D. Noyes, and R. Oliver, *Exptl. Cell Res.* **16,** 471 (1959).
188. J. F. Henderson and H. G. Mandel, *Advan. Pharmacol.* **2,** 297 (1963).
189. R. W. Brockman and E. P. Anderson, *Ann. Rev. Biochem.* **32,** 463 (1963).
190. G. B. Elion and G. H. Hitchings, *Advan. Chemotherapy* **2,** 91 (1965).
191. G. B. Elion, *Federation Proc.* **26,** 898 (1967).
192. J. J. Fox, K. A. Watanabe, and A. Bloch, This series **5,** 251 (1966).
193. M. E. Balis, "Antagonists and Nucleic Acids." North-Holland Publ., Amsterdam, 1968.
194. D. A. Karnofsky, *in* "Drugs of Choice, 1968–1969" (W. Modell, ed.), p. 540. Mosby, St. Louis, Missouri, 1967.

195. A. H. Chalmers, P. R. Knight, and M. R. Atkinson, *Australian J. Exptl. Biol. Med. Sci.* **45,** 681 (1967).
196. R. Schwartz, "Symposium on Immunosuppressive Drugs," *Federation Proc.* **26,** 877 (1967).
197. M. C. Berenbaum, *J. Clin. Pathol.* (*Suppl.*) **20,** 471 (1967).
198. A. R. P. Paterson, *Can. J. Biochem. Physiol.* **37,** 1011 (1959).
199. M. R. Atkinson, G. Eckermann, and J. Stephenson, *Biochim. Biophys. Acta* **108,** 320 (1965).
200. P. W. Allan and L. L. Bennett, *Proc. Am. Assoc. Cancer Res.* **9,** 2 (1968).
201. J. P. Scannell and G. H. Hitchings, *Proc. Soc. Exptl. Biol. Med.* **122,** 627 (1966).
202. A. W. Murray and G. R. Donaldson, unpublished results.
203. G. A. LePage and I. G. Junga, *Mol. Pharmacol.* **3,** 37 (1967).
204. M. C. Wang, A. I. Simpson, and A. R. P. Paterson, *Cancer Chemotherapy Rept.* **51,** 101 (1967).
205. F. M. Schabel, W. R. Laster, and H. E. Skipper, *Cancer Chemotherapy Rept.* **51,** 111 (1967).
206. J. J. Brink and G. A. LePage, *Can. J. Biochem.* **43,** 1 (1965).
207. S. S. Cohen, This series **5,** 1 (1966).
208. M. R. Atkinson and A. Kornberg, unpublished results.
209. M. R. Atkinson, R. K. Morton, and A. W. Murray, *Biochem. J.* **89,** 167 (1963).
210. M. R. Atkinson, R. K. Morton, and A. W. Murray, *Biochem. J.* **92,** 398 (1964).
211. R. J. McCollister, W. R. Gilbert, Jr., D. M. Ashton, and J. B. Wyngaarden, *J. Biol. Chem.* **239,** 1560 (1964).
212. D. L. Hill and L. L. Bennett, *Biochemistry* **8,** 122 (1969).
213. B. S. Tay, R. M. Lilley, A. W. Murray, and M. R. Atkinson, *Biochem. Pharmacol.* **18,** 936 (1969).
214. J. F. Henderson, *Biochem. Pharmacol.* **12,** 551 (1963).
215. L. L. Bennett, R. W. Brockman, H. P. Schnebli, S. Chumley, G. J. Dixon, F. M. Schabel, E. A. Dulmadge, H. E. Skipper, J. A. Montgomery, and H. J. Thomas, *Nature* **205,** 1276 (1965).
216. J. B. Wyngaarden, *in* "The Metabolic Basis of Inherited Disease" (J. B. Stanbury, J. B. Wyngaarden, and D. S. Frederickson, eds.), 2nd ed., p. 667. McGraw-Hill, New York, 1966.
217. J. F. Henderson, F. M. Rosenbloom, W. N. Kelley, and J. E. Seegmiller, *J. Clin. Invest.* **47,** 1511 (1968).
218. F. P. Alepa, R. R. Howell, J. R. Klinenberg, and J. E. Seegmiller, *Am. J. Med.* **42,** 58 (1967).
219. S. Jakovcic and L. B. Sorensen, *Arthritis Rheumat.* **10,** 129 (1967).
220. W. N. Kelley, F. M. Rosenbloom, J. F. Henderson, and J. E. Seegmiller, *Proc. Natl. Acad. Sci. U.S.* **57,** 1735 (1967).
221. W. N. Kelley, *Federation Proc.* **27,** 1047 (1968).
222. O. W. Jones, D. M. Ashton, and J. B. Wyngaarden, *J. Clin. Invest.* **41,** 1805 (1962).
223. A. Hershko, C. Hershko, and J. Mager, *Israel J. Med. Sci.* **4,** 939 (1968).
224. F. M. Rosenbloom, J. F. Henderson, I. C. Caldwell, W. N. Kelley, and J. E. Seegmiller, *J. Biol. Chem.* **243,** 1166 (1968).
225. W. N. Kelley, F. M. Rosenbloom, and J. E. Seegmiller, *J. Clin. Invest.* **46,** 1518 (1967).
226. W. N. Kelley, F. M. Rosenbloom, J. Miller, and J. E. Seegmiller, *New Engl. J. Med.* **278,** 287 (1968).

227. R. Pomales, S. Bieber, R. Friedman, and G. H. Hitchings, *Biochim. Biophys. Acta* **72,** 119 (1963).
228. R. Pomales, G. B. Elion, and G. H. Hitchings, *Biochim. Biophys. Acta* **95,** 505 (1965).
229. R. W. Rundles, E. N. Metz, and H. R. Silberman, *Ann. Internal Med.* **64,** 229 (1966).
230. J. F. Henderson, *Federation Proc.* **27,** 1075 (1968).
231. F. M. Rosenbloom, W. N. Kelley, J. M. Miller, J. F. Henderson, and J. E. Seegmiller, *J. Clin. Invest.* **46,** 1110 (1967).
232. L. B. Sorensen and P. J. Benke, *Nature* **213,** 1122 (1967).
233. L. Sweetman, *Federation Proc.* **27,** 1055 (1968).
234. L. Sweetman, *Federation Proc.* **27,** 1109 (1968).
235. R. W. Brockman, *Cancer Res.* **25,** 1596 (1965).
236. D. J. Hutchison, *Advan. Cancer Res.* **7,** 235 (1963).
237. A. R. P. Paterson, *Can. Cancer Conf.* **5,** 417 (1963).
238. R. W. Brockman and S. Chumley, *Biochim. Biophys. Acta* **95,** 365 (1965).
239. D. G. R. Blair, *Can. J. Biochem.* **46,** 217 (1968).
240. L. L. Bennett, M. H. Vail, S. Chumley, and J. A. Montgomery, *Biochem. Pharmacol.* **15,** 1719 (1966).
240a. H. Harris, O. J. Miller, G. Klein, P. Worst, and T. Tachibana, *Nature* **223,** 363 (1969).
241. J. D. Davidson and T. S. Winter, *Cancer Res.* **24,** 261 (1964).
242. P. B. Nicholls and A. W. Murray, *Plant Physiol.* **43,** 645 (1968).
243. D. S. Letham, *Ann. Rev. Plant Physiol.* **18,** 349 (1967).
244. A. W. Galston and P. J. Davies, *Science* **163,** 1288 (1969).
245. H. Kende and J. E. Taveres, *Plant Physiol.* **43,** 1244 (1968).
246. M. R. Atkinson and D. C. Elliott, unpublished results.
247. A. I. Oparin, *Advan. Enzymol.* **27,** 347 (1965).
248. H. H. Pattee, *Advan. Enzymol.* **27,** 381 (1965).

Ribosome Specificity of Protein Synthesis *in Vitro*

ORIO CIFERRI[1] AND
BRUNO PARISI

*Department of Genetics,
University of Pavia,
Pavia, Italy*

I. Introduction 121
II. Role of Ribosomes in the Synthesis of Protein 122
III. Protein Synthesis *in Vitro* by Heterologous Systems 125
IV. Attempts to Characterize the Sites Responsible for Ribosome Specificity 129
V. Ribosome Specificity, Cellular Organelles, and the Regulation of Protein Synthesis 135
VI. Ribosome-Specific Inhibitors of Protein Synthesis 137
VII. Conclusions 139
References 141
Note Added in Proof 144

I. Introduction

It is now well established that there exist two broad categories of ribosomes that may be exemplified by the ribosomes from bacteria and those from animal tissues. Among the many differences between them, the most evident is perhaps the sedimentation coefficient, that of the bacterial ribosomes being approximately 70 S ("70 S type") while that of animal ribosomes is approximately 80 S ("80 S-type"). Besides structural differences, these 70 S and 80 S ribosomes appear also to differ functionally. Ribosomes of a given size seem to be unable to catalyze one or more of the reactions required for protein synthesis *in vitro* when incubated with supernatant (polymerizing)[2] enzymes prepared from organisms endowed with ribosomes of a different size.[3]

[1] Present address: Botanical Institute, University of Cagliari, Cagliari, Italy.

[2] To clarify the somewhat confused terminology, in this paper, we use three different terms, randomly employed, to define three different types of preparation, and three levels of purification: (a) *Supernatant enzymes* are all enzymes and factors required for protein synthesis present in the "soluble" fraction (e.g., a 105,000 × *g* supernatant), including tRNA and aminoacyl-tRNA synthetases. (b) *Polymerizing enzymes* are partially purified preparations of (a) that still contain all factors required for amino acid polymerization onto ribosomes. (c) *Transfer factors* are the separated transfer enzymes [e.g., T_u, T_s and G from bacteria (*1–3*), TF-I and TF-II from animal tissues (*4–6*), etc.].

[3] We recently proposed (*7*) the use of the term "ribosome specificity" in place of

The aim of this article is to describe and discuss some recent results in this area, the goal being to understand how the protein-synthesizing machineries of different organisms function. Such results supplement the data already available, but so far insufficient, on the differences that exist between the two types of ribosomes. It has been said that the "basic divergence in cellular structure, which separates the bacteria and blue-green algae from all other cellular organisms, probably represents the greatest single evolutionary discontinuity to be found in the present-day living world" (*9*). However there is no convincing explanation why evolution has separated eukaryotic from prokaryotic organisms to such an extent that macromolecules and macromolecular complexes (e.g., the ribosomes) from one such organism are no longer recognized by the counterparts from another organism. Is the higher complexity of the 80 S ribosomes justified by the relations, structural and functional, that exist in eukaryotic organisms between ribosomes and the endoplasmic reticulum? Or is such a higher complexity related to the existence of different "states" of such ribosomes in relation to different forms, or "states" of mRNA (long-lived, masked forms, etc.)?

The study of ribosome specificity in protein synthesis may answer, at least in part, some of such questions. The observations discussed in this article deal mainly with the function rather than the structure of ribosomes. It has been suggested that if one cannot study structure, then he should study function and, failing the second objective, then study correlation. We feel, however, that the study of structure and function should go in parallel. After all, if one does not know what a watch is made for, very little may he learn on how to measure time by demolishing the machine. On the other hand, the man who knows how and why a watch ticks may guess how it is made inside.

II. Role of Ribosomes in the Synthesis of Protein

Ribosomes are the site of protein synthesis. On the ribosomes, mRNA, aminoacyl-tRNA's and polymerizing enzymes interact to assemble sequentially the amino acids into the polypeptide chain specified by the template mRNA.

As already mentioned, ribosomes may be divided into two broad categories according to their size, those with a sedimentation coefficient of approximately 80 S (80 S ribosomes) and those with a sedimentation

"species specificity" previously used since the size of the ribosome is one, perhaps the only, factor clearly correlated with the activity of mixed preparations. We now realize that the term "ribosome specificity" was employed by Nathans and Lipmann in the paper in which they first described the phenomenon (*8*).

coefficient of approximately 70 S (70 S ribosomes) (*10–12*). Cytoplasmic ribosomes of eukaryotic organisms are characteristic of the former group, while the latter group encompasses ribosomes of prokaryotic organisms and ribosomes from chloroplasts and mitochondria (*13–15*). It should be emphasized that a sharp distinction between these two groups is perhaps an oversimplification since, within the same group, ribosomes appear to have different sedimentation coefficients and differences are evident in the sedimentation coefficients of the ribosomal subunits and of ribosomal RNA's (*16–19*). In addition, besides a general paucity of data for all types of ribosomes from a spectrum of representative organisms, and especially in the case of organelle ribosomes, no definite conclusions on the types (particularly the subtypes) of ribosomes may be drawn until the procedures for isolation, purification, and analysis of the different types of ribosomes are standardized. Notwithstanding all these limitations, a broad distinction between prokaryotic and organelle (70 S type) ribosomes and eukaryotic cytoplasmic (80 S type) ribosomes seems to be justified and acceptable as a working hypothesis.

With the development of cell-free systems capable of incorporating radioactive amino acids into compounds insoluble in hot trichloroacetic acid, it became possible to study the mechanisms and the reactions leading to the synthesis *in vitro* of the polypeptide chain. In addition, the polymerization of a single species of amino acid directed by a synthetic homopolyribonucleotide (*20*) has further simplified the study of the reactions involved in the synthesis of proteins. Table I illustrates a paradigm of the components required for the translation of natural mRNA (or of synthetic heteropolyribonucleotides) and of a synthetic homopolyribonucleotide. It is evident that, at least in the case of *Escherichia coli* ribosomes, the requirements for the polymerization of phenylalanine directed by poly U are simpler than those for the translation of certain synthetic heteropolyribonucleotides or of natural mRNA's (such as viral RNA). Obviously, as compared with the synthesis of a protein *in vivo*, the synthesis of polyphenylalanine is most probably an oversimplification. Nevertheless, the poly U-phenylalanine system has provided extremely useful insights into the mechanism of the synthesis of the peptide bond and on the functioning of the ribosomal machinery. Furthermore, by employing poly U, it is possible to overcome the difficulties that may arise in connection with other factors, including initiation factors and *N*-formylmethionyl-tRNA, required for the translation of added natural mRNA, but not of poly U (*21–24*). Likewise, if phenylalanyl-tRNA is used in place of phenylalanine and unesterified tRNA, any interference due to species specificity in the reaction between tRNA and aminoacyl-tRNA synthetase(s) is eliminated. It is well known [for a recent review, see (*25*)] that lack of esterification of an

TABLE I

REQUIREMENTS FOR PROTEIN SYNTHESIS *in Vitro* DIRECTED BY DIFFERENT TYPES OF mRNA's

Translation of natural mRNA or synthetic heteropolyribonucleotides		Translation of poly U	
Ribosomes		Ribosomes	
Transfer enzymes		Transfer enzymes	
mRNA		Poly U	
tRNA's	or aminoacyl-tRNA's	tRNA	or phenylalanyl-tRNA
Amino acids		Phenylalanine	
Aminoacyl-tRNA synthetases		Phenylalanyl-tRNA synthetase	
ATP		ATP	
Formyl groups donors	or *N*-formylmethionyl-tRNA[a]	—	
Formylating enzymes			
Initiation and termination factors[a]		—	
GTP		GTP	
Mg^{2+}, K^+ (or NH_4^+), SH^- groups		Mg^{2+}, K^+ (or NH_4^+), SH^- groups	

[a] Requirements demonstrated up to now only in the case of 70 S ribosomes.

TABLE II
EFFECT OF HOMOLOGOUS AND HETEROLOGOUS tRNA OR PHENYLALANYL-tRNA ON POLYPHENYLALANINE SYNTHESIS BY THE *Escherichia coli* AND CASTOR BEAN SEEDLINGS SYSTEMS

Ribosomes and polymerizing enzymes from	tRNA from	Phenylalanyl-tRNA from	Phenylalanine incorporated, % of the homologous system[a]
E. coli	*E. coli*	—	100
	Castor bean	—	5.4
	—	*E. coli*	100
	—	Castor bean	79.1
Castor bean	Castor bean	—	100
	E. coli	—	95.6
	—	Castor bean	100
	—	*E. coli*	119.2

[a] Incorporation of phenylalanine from either phenylalanyl-tRNA or phenylalanine plus tRNA directed by poly U. Data recalculated from Parisi *et al.* (*30*).

amino acid with its cognate tRNA may occur if tRNA and aminoacyl-tRNA synthetase are prepared from different organisms. As shown in Table II, it is now evident (*8, 26–31*) that heterologous[4] aminoacyl-tRNA's may be utilized efficiently for protein synthesis by a ribosomal system even in cases in which the ribosomal system is unable to utilize the unesterified tRNA (preparations of supernatant or polymerizing enzymes still contain the enzymes required for the esterification of amino acids with tRNA).

Most of the data discussed in the present article were obtained with the poly U-phenylalanine system. It is not known whether the conclusions reached can be extrapolated to the biosynthesis of more complex polypeptides and, eventually, of whole proteins.

III. Protein Synthesis *in Vitro* by Heterologous Systems[4]

Attempts to exchange ribosomes and polymerizing, or supernantant, enzymes from different organisms are reported in Table III. The results in this table illustrate the incorporation of amino acids from aminoacyl-tRNA's as directed by the endogenous mRNA present on the ribosomes. Liver ribosomes incorporate amino acids when the polymerizing enzymes

[4] The adjective "heterologous" is employed to designate any component prepared from a different organism (e.g., heterologous tRNA). Likewise, we use the term "heterologous system" for any system for protein synthesis *in vitro* in which ribosomes and polymerizing enzymes are prepared from different organisms (e.g., *E. coli* ribosomes and yeast polymerizing enzymes).

TABLE III
RIBOSOME SPECIFICITY OF PROTEIN SYNTHESIS *in Vitro* DIRECTED BY ENDOGENOUS mRMA EMPLOYING AMINOACYL-tRNA[a]

Ribosomes from	Polymerizing or supernatant enzymes from: Mammalian liver	Rat kidney	Yeast	*Escherichia coli*	*Lactobacillus arabinosus*	*Azotobacter vinelandii*	*Proteus shermanii*
Rat	+ (*8, 26, 31*)	+ (*26*)	+ (*26*)	− (*8, 26, 31*)	− (*26*)	− (*26*)	− (*26*)
E. coli	− (*8, 26, 31*)	− (*26*)	− (*26*)	+ (*8, 26, 31*)	+ (*26*)	+ (*26*)	+ (*26*)

[a] +, active system; −, lack of activity. The symbol "−" has been arbitrarily set to indicate that the heterologous system has an activity always less than 20% of that of the homologous system. However, in general, the activity was in the order of a few percent of the homologous system. Likewise "+" indicates that the heterologous system has an activity of at least 70% of the homologous system. Minor discrepancies may be due to the fact that, in certain experiments, polymerizing enzymes were present in less than saturating amounts. In addition, incomplete removal of all transfer factors from certain preparations of ribosomes may give rise to active systems even with heterologous noncompatible polymerizing enzymes. For instance, if *E. coli* ribosomes are not extensively washed with NH_4Cl (*2*), activity may be evident when polymerizing enzymes from yeast are employed (*32*). Furthermore, it must be remembered that certain preparations from eukaryotic organisms may contain ribosomes of the 70 S and/or transfer factors present in the cytoplasmic organelles (see later discussion).

are prepared from yeast or from kidney or liver of different animals, but not when the polymerizing enzymes are prepared from bacteria. The phenomenon is reciprocal, and *E. coli* ribosomes promote the polymerization of amino acids *in vitro* in the presence of bacterial polymerizing enzymes, but not with enzymes from mammalian tissues or yeast. More results are available on the activity of heterologous systems in which poly U is used as mRNA (Table IV).

Even in the case of the poly U–phenylalanine system, ribosomes and polymerizing enzymes are interchangeable only when both are prepared either from eukaryotic organisms (whether animals or plants) or from prokaryotic organisms.

As previously mentioned, the origin of the aminoacyl-tRNA has no effect on the activity, or lack of activity, of heterologous systems since homologous systems utilize efficiently aminoacyl-tRNA of different origins even when the species specificity in the reaction between tRNA and aminoacyl-tRNA synthetase prevents the utilization of the unesterified tRNA. Nevertheless, there are many reports on the existence of ribosome specificity in experiments in which unesterified tRNA was employed (*38–44*); such results should, however, be confirmed by the demonstration that lack of activity of heterologous mixtures does not arise from an incompatibility between supernatant enzymes and heterologous tRNA.

A more detailed study of the specificity in protein synthesis utilized preparations from castor bean seedlings and *E. coli* (*30*). The lack of activity of heterologous systems (*E. coli* ribosomes and polymerizing enzymes from castor bean seedlings, and *vice versa*) came not from either a selective inactivation of ribosomes by heterologous polymerizing enzymes or from the presence of specific inhibitory activities in heterologous polymerizing enzymes or ribosomes. Likewise, the binding of phenylalanyl-tRNA to ribosomes directed by poly U (*45*) was not inhibited by polymerizing enzymes heterologous to the ribosomes.

The evidence therefore indicates that lack of protein synthesis *in vitro* in some heterologous systems arises specifically from a lack of interaction between ribosomes of a given size and one or more of the transfer enzymes prepared from organisms endowed with ribosomes of a different size. Barring the presence of specific inhibitors, active heterologous systems may be obtained only when both ribosomes and polymerizing enzymes are prepared from either eukaryotic or prokaryotic organisms. Since prokaryotic organisms contain ribosomes of the 70 S type and eukaryotic organisms ribosomes of the 80 S type in the cytoplasm, the sedimentation coefficient of ribosomes appears to be the factor more clearly correlated with the activity of heterologous systems. The 70 S type of ribosome is active in *in vitro* protein synthesis with polymerizing enzymes from organisms

TABLE IV

RIBOSOME SPECIFICITY OF POLYPHENYLALANINE SYNTHESIS DIRECTED BY POLY U EMPLOYING PHENYLALANYL-tRNA[a]

Ribosomes from	Polymerizing or supernatant enzymes from									
	Mammals		Insects			Higher plants		Microorganisms		
	Liver	Human cells[b]	*Tenebrio molitor*	*Bombyx mori*	*Drosophila melano-gaster*	Wheat embryos	Castor bean seedlings	*Saccha-romyces* spp.	*Escherichia coli*	*Bacillus* spp.
Mammals										
Liver	+	+ (*34*)	+ (*36*)	+ (*35*)		+ (*33*)	+ (*30*)	+ (*30*)	− (*30, 33, 36*)	
Human cells[b]	+ (*34*)	+						+ (*34*)	− (*34*)	
Insects										
Tenebrio molitor	+ (*36*)		+						− (*36*)	
Drosophila melanogaster					+			+ (*37*)	− (*37*)	
Higher plants										
Wheat embryos	+ (*33*)					+			− (*33*)	
Castor bean seedlings	+ (*30*)						+	+ (*30*)	− (*30*)	− (*30*)
Microorganisms										
Saccharomyces spp.	+ (*30*)	+ (*34*)					+ (*30*)	+	− (*30*)	− (*28, 30*)
E. coli	− (*30, 33, 35*)	− (*34*)	− (*36*)	− (*35*)		− (*33*)	− (*30*)	− (*30*)	+	+ (*30*)
Bacillus spp.							− (*30*)	− (*28*)	+ (*28, 30*)	+

[a] +, active system; −, lack of activity. For explanation of symbols, see Table III.   [b] Grown *in vitro* (EUE).

endowed with the 70 S type of ribosomes, but not with polymerizing enzymes from organisms containing the 80 S type, and *vice versa*. Thus, relying on the evidence so far available, the specificity of the reaction(s) between the polymerizing enzymes and the ribosomes of the two types appears the most likely explanation for the phenomenon.

IV. Attempts to Characterize the Sites Responsible for Ribosome Specificity

The detailed mechanism of ribosome specificity may be approached by studying either the ribosomes or the polymerizing enzymes. Indeed, the whole ribosome or only one of the two subunits, or even only a few of the ribosomal proteins, may be responsible for ribosome specificity. One approach to the problem is to prepare "hybrid" ribosomes, i.e., ribosomes formed by interaction of subunits from 70 S (50 S + 30 S) ribosomes from a prokaryotic organism (e.g., *E. coli*) and the complementary subunits from 80 S (60 S + 40 S) ribosomes from a eukaryotic organism (e.g., yeast). Such hybrid ribosomes (50 S + 40 S and 60 S + 30 S) could be challenged with polymerizing enzymes from organisms endowed with 70 S or 80 S ribosomes and, provided polymerizing activity was evident, would permit the identification of the specific ribosome subunit. However, up to now no interspecific recombination of this type has been achieved [e.g., (*46*)]. There are several reports that 70 S ribosomes may easily be split into monomeric subunits and then recombined to produce biologically active ribosomes (e.g., *47*). Furthermore, it is also possible to prepare active hybrid ribosomes between 30 S and 50 S ribosomal subunits from different prokaryotic organisms (*48–50*). In contrast, it has been very difficult to reconstitute biologically active 80 S ribosomes even from monomeric subunits prepared from the same ribosomes. However, two recent reports indicate that it is now possible to reconstitute active particles from the separated subunits of mammalian and yeast ribosomes (*51*, *52*) and attempts to make hybrid ribosomes of this type are now therefore conceivable.

An alternative approach relies on the fractionation of the transfer factors from an eukaryotic organism (e.g., yeast) and a prokaryotic (e.g., *E. coli*) and the assay of protein synthesis *in vitro* in the presence of mixtures of complementary transfer factors in part homologous and in part heterologous to the ribosomes.

The activity of *E. coli* ribosomes with partially fractionated *E. coli* transfer factors (T and G) (*32*, *53*) and unfractionated yeast enzymes are reported in Table V. Such data confirm that yeast enzymes cannot substitute for all *E. coli* transfer factors; however, in the presence of *E. coli*

TABLE V
EFFECT OF TRANSFER FACTORS FROM *Escherichia coli* AND YEAST ON POLYPHENYLALANINE SYNTHESIS BY *E. coli* RIBOSOMES

Additions	Phenylalanine incorporated[a] (pmoles/assay)	Incorporation (%)
—	0.090	1.7
E. coli T factor	0.260	5.0
E. coli G factor	0.086	1.6
E. coli T factor + *E. coli* G factor	5.192	100
Yeast enzymes (75 μg)	0.120	2.3
Yeast enzymes (150 μg)	0.181	3.5
E. coli T factor + yeast enzymes (75 μg)	1.075	20.7
E. coli T factor + yeast enzymes (150 μg)	2.033	39.1
E. coli G factor + yeast enzymes (75 μg)	0.373	7.2
E. coli G factor + yeast enzymes (150 μg)	0.101	1.9
E. coli T factor + *E. coli* G factor + yeast enzymes (75 μg)	5.109	98.4
E. coli T factor + *E. coli* G factor + yeast enzymes (150 μg)	5.660	109.0

[a] Incorporation of phenylalanine from phenylalanyl-tRNA directed by poly U. Data from Ciferri *et al.* (*7*).

T factor, yeast polymerizing enzymes are capable of promoting protein synthesis. Thus it appears that yeast enzymes contain a fraction that may substitute for *E. coli* G factor.

An attempt has been made to separate the yeast fraction that complements with *E. coli* T factor on *E. coli* ribosomes. By chromatography on TEAE-cellulose (*54*), two protein fractions were separated from yeast (*32*). Although such fractions appeared to have lost polymerizing activity on yeast ribosomes even when combined, there is a clear complementation between one peak of yeast enzymes and *E. coli* T factor but not with *E. coli* G factor (*7*). In the presence of *E. coli* T factor, increasing amounts of the complementary yeast fraction stimulate a proportional increase of phenylalanine polymerization, which reaches approximately the same level as the homologous system from *E. coli* (Fig. 1).

Since all the results so far reported clearly point out an interchangeability among *E. coli* and yeast enzymes for G activity but not for T activity, an attempt was made to ascertain if only T_u or T_s [the two transfer factors that make up T (*1*, *2*)] is ribosome-specific. When only *E. coli* T_s factor is employed, the yeast fraction appears to be endowed also with T_u activity for *E. coli* ribosomes (Table VI).

These results indicate that crude or partially purified preparations of yeast transfer enzymes may substitute for transfer factors G and T_u of

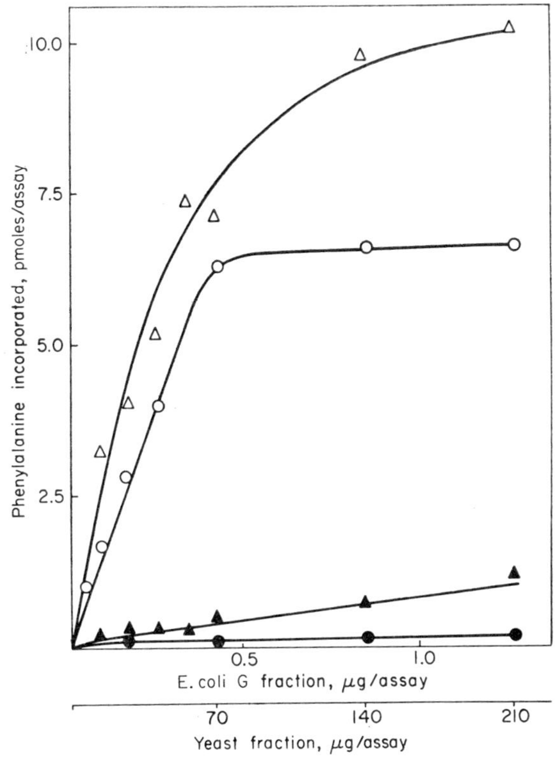

FIG. 1. Effect of increasing concentrations of *E. coli* G factor or of a partially purified yeast fraction on polyphenylalanine synthesis by *E. coli* ribosomes in the presence of *E. coli* T factor. Incorporation of phenylalanine from phenylalanyl-tRNA directed by poly U. Redrawn from (7). ○——○, *E. coli* T factor + *E. coli* G factor; ●——●, *E. coli* G factor; △——△, *E. coli* T factor + yeast fraction; ▲——▲, yeast fraction.

TABLE VI

PRESENCE OF T_U- AND G-LIKE ACTIVITIES FOR *Escherichia coli* RIBOSOMES OF A PARTIALLY PURIFIED FRACTION FROM YEAST POLYMERIZING ENZYMES

Additions		Phenylalanine incorporated, % of the homologous system[a]
Factors from *E. coli*	Yeast fraction	
$T_u + T_s + G$	—	100
$T_s + G$	—	1.2
$T_s + G$	+	94
T_s	+	101.5
—	+	11

[a] The complete system was composed of *E. coli* ribosomes and *E. coli* T ($T_u + T_s$) and G transfer factors. Incorporation of phenylalanine from phenylalanyl-tRNA directed by poly U. Data recalculated from Ciferri *et al.* (7).

E. coli. On the contrary, *E. coli* T_s factor cannot be substituted for by any fraction, or even crude preparations, from yeast.

These observations can be interpreted in at least two ways. Perhaps the most obvious possibility is that, since only the T_s fraction is ribosome specific, the T_u and G fractions are functionally equal in *E. coli* and in yeast and do not discriminate between the two types of ribosomes. From such an interpretation, one could speculate that, if evolution from prokaryotes to eukaryotes is accepted, in yeast only transfer factor T_s and that part of the ribosome that reacts with T_s has evolved to such an extent that it is no longer interchangeable with *E. coli* T_s factor. One then wonders if in more evolved, and hence more differentiated, organisms ribosome specificity is also more differentiated than between microorganisms such as *E. coli* and yeast. Transfer factors from rat liver appear to behave in the same way as the transfer factors from yeast; when *E. coli* ribosomes are used, rat liver transfer factor T_2 may substitute for *E. coli* G factor while transfer factor T_1 cannot be used in place of *E. coli* T factor (*55*).

Since it can be shown that the complementary transfer factors are interchangeable among eukaryotes (*35, 54, 56*) and the same is true for the transfer factors from prokaryotic organisms (*2, 3*), it can be surmised that eukaryotic organisms contribute to prokaryotic ribosomes at least T_u- and G-like activities.

A second possibility also exists, namely, that in yeast and rat liver there are two G- and T_u-like activities, one specific for 70 S type ribosomes and one for 80 S type ribosomes.[5] Indeed, the following experiments indicate that there is some evidence to support this hypothesis. Transfer factors have been fractionated from *Prototheca zopfii* (*37*), a nonphotosynthetic unicellular alga, considered to be a natural achloric mutant of *Chlorella* (*57, 58*). From the previous results, one would expect the polymerizing enzymes prepared from this organism to be active only on 80 S ribosomes, like those from yeast or rat liver. On the contrary, as shown in Table VII, *P. zopfii* polymerizing enzymes are active on both 80 S and 70 S ribosomes (i.e., *E. coli* and yeast ribosomes) while *P. zopfii* ribosomes are active only with yeast polymerizing enzymes and no activity is obtained with *E. coli* polymerizing enzymes. This last observation would indicate that *P. zopfii* extracts contain only 80 S type ribosomes, a contention that is strengthened

[5] We employ the term T-like and G-like activities for the separated complementary factors from polymerizing enzymes of eukaryotic organisms assuming, in the light of existing evidence (*56*), that the transfer factors from such organisms may be equated, at least functionally, to the bacterial transfer factors T and G. Thus, for instance, transfer factor T_1 (TF_1) from rabbit reticulocytes appears to have a function similar to that of *E. coli* T factor, the so-called "aminoacyl-tRNA, GTP binding factor." Similarly, transfer factor T_2 (TF_2) may be considered to be equivalent to *E. coli* G factor, the "GTP-dependent peptidyl-tRNA translocase."

TABLE VII
SPECIFICITY OF *P. zopfii* RIBOSOMES AND POLYMERIZING ENZYMES

Ribosomes from	Polymerizing enzymes from	Phenylalanine incorporated, % of the homologous system[a]
Saccharomyces cerevisiae	*S. cerevisiae*	100
	E. coli	3.2
	P. zopfii	113.8
Escherichia coli	*S. cerevisiae*	1.6
	E. coli	100
	P. zopfii	147.9
Prototheca zopfii	*S. cerevisiae*	93.7
	E. coli	10.7
	P. zopfii	100

[a] Incorporation of phenylalanine from phenylalanyl-tRNA directed by poly U (*37*).

by ultracentrifugation analysis of the ribosomes, which shows only one sharp peak with an $s^{\circ}_{20,w}$ of approximately 77 (*59*). Clearly, then, the significance of the polymerizing activity from *P. zopfii* on *E. coli* ribosomes comes into question. Is there one set of polymerizing enzymes that are not ribosome-specific, or are there two complete sets of polymerizing enzymes that are ribosome-specific? Or do there exist two T_s transfer factors that are ribosome-specific, and only one T_u and one G factor that are not ribosome-specific?

Transfer factors of *P. zopfii* can be separated by chromatography on a DEAE-Sephadex column (*37*). As shown in Fig. 2, two G-like activities are recovered in distinct fractions, one specific for yeast ribosomes, the other for *E. coli* ribosomes. T-like activities for *E. coli* and yeast ribosomes overlap at least partially in the first tubes, and it has not been possible to establish whether they are two distinct enzymes. However, the fact that *P. zopfii* T-like activity on *E. coli* ribosomes is inactivated faster than on yeast ribosomes also supports the hypothesis that two T factors are present.

As Fig. 2 shows, yeast ribosomes and *P. zopfii* T-like factor are active only when *P. zopfii* G_1 factor or the yeast G-like fraction (*7*) are added. With *P. zopfii* G_2 factor (active on *E. coli* ribosomes) or with the G factor from *E. coli*, the system shows no activity. On the other hand, the *P. zopfii* G_1 factor, active with yeast ribosomes, does not support activity in a system containing *E. coli* ribosomes and *E. coli* T factor.

Therefore, *P. zopfii* appears to have two complete sets of specific transfer factors, one for *E. coli* and one for yeast ribosomes. Since *E. coli* transfer factors do not substitute for any of the yeast transfer factors on yeast ribosomes, it must be surmised that bacteria have only one set of transfer

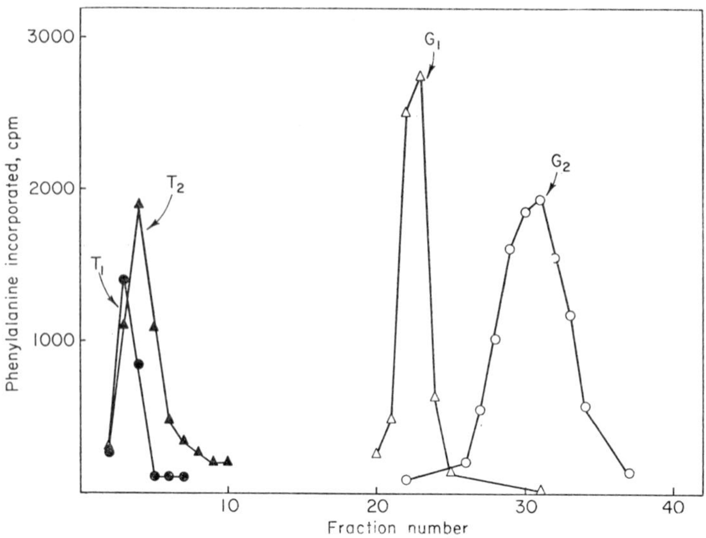

FIG. 2. Fractionation of ribosome-specific T- and G-like activities from *Prototheca zopfii* polymerizing enzymes. Phenylalanine incorporation from phenylalanyl-tRNA directed by poly U. Separation on a DEAE-Sephadex A-50 column by elution with a gradient between 0.15 and 0.35 *M* KCl (*37*). For each fraction the following assays were performed: ●——●, *E. coli* ribosomes + *E. coli* G factor; ○——○, *E. coli* ribosomes + *E. coli* T factor; ▲——▲, yeast ribosomes + fraction number 22 or yeast fraction (*7*); △——△, yeast ribosomes + fraction number 5. For the sake of clarity, the terms T and G have been employed also for the activities on yeast ribosomes (see also footnote 5). In addition to the reported complementation data, T-like activity on yeast ribosomes was confirmed (*37*) by assaying the enzyme-dependent binding of phenylalanyl-tRNA to ribosomes (*56*).

factors, all of which are strictly ribosome-specific. Such results could support an interpretation of ribosome specificity which is different from that previously reported (*7*), namely that prokaryotic organisms have only one set of transfer factors specific for 70 S ribosomes, while eukaryotic organisms contain transfer factors specific for 80 S ribosomes, as well as one or more of the transfer factors specific for 70 S ribosomes.

As mentioned previously, recent evidence indicates that cellular organelles, such as mitochondria and chloroplasts, contain 70 S type ribosomes (*13–15*) and that such ribosomes appear to catalyze protein synthesis *in vitro* in the presence of *E. coli* polymerizing enzymes (*60, 61*). It may therefore be surmised that the presence in *P. zopfii* of polymerizing enzymes active on *E. coli* ribosomes may arise from the presence of polymerizing enzymes specific for organelle ribosomes that are already present in the cytoplasm or that have been released in the cell sap during preparation of the cell-free extracts. The presence of polymerizing enzymes specific for organelle ribosomes may justify the activity on prokaryotic ribosomes of supernatant preparations from eukaryotic organisms even in those cells, such as *P. zopfii*, that have lost by mutation one type of organelle, since it is well established that the genetic determinants of some organelle proteins are of nuclear origin (*62*).

V. Ribosome Specificity, Cellular Organelles, and the Regulation of Protein Synthesis

Since cell organelles contain ribosomes of the 70 S type (*13–15*), eukaryotic cells should possess transfer factors specific for both 80 S and 70 S types. The most suitable material for studying the polymerizing enzymes responsible for organelle protein synthesis is therefore the green plant; being endowed with both mitochondria and chloroplasts, photosynthetic plants should contain a higher quantity, or a higher proportion, of transfer factors specific for 70 S type ribosomes. Prokaryotic and eukaryotic unicellular algae are an *ad hoc* material for such work since, besides being an extremely homogeneous material, they may be grown heterotrophically or autotrophically. In addition, the photosynthetic apparatus may be induced on dark-grown cells just by exposure to light while permanently heterotrophic lines are available or easily produced. With these types of algae it is possible to study the relationship that may exist between chloroplast formation and the induction of transfer factors specific for chloroplast ribosomes. Analogous studies may perhaps be performed in the case of mitochondria by employing aerobically and anaerobically grown cells of yeast and *petite* mutants of yeast or *poky* mutants of *Neurospora crassa*.[6]

Table VIII shows preliminary results on the polymerizing activity of *E. coli* and yeast ribosomes in the presence of polymerizing enzymes prepared from a prokaryotic alga, *Nostoc muscorum*, and light-grown eukaryotes such as *Euglena gracilis*, *Chlorella vulgaris*, and *Prototheca zopfii*. Polymerizing enzymes prepared from green cells of *E. gracilis* and *C. vulgaris* catalyze protein synthesis *in vitro* on both types of ribosomes while the extract prepared from autotrophically grown cells of the prokaryote *N. muscorum* is active only on *E. coli* ribosomes. These results are in good agreement with the hypothesis of the presence of polymerizing activity on both types of ribosomes in eukaryotic green cells and only on *E. coli* ribosomes in the prokaryotic algae containing only 70 S ribosomes (*11*). The finding that polymerizing enzymes from eukaryotic green algae are active on both types of ribosomes does not indicate the presence of two

[6] Although no experimental evidence is available, it may be predicted on the basis of the similarity of the size of the ribosomes that transfer factors from mitochondria and chloroplasts are interchangeable.

TABLE VIII
RIBOSOMES SPECIFICITY OF POLYMERIZING ENZYMES FROM ALGAE

Ribosomes from	Polymerizing enzymes from	Phenylalanine incorporated, % of the homologous system[a]
Escherichia coli	*E. coli*	100
	Nostoc muscorum	40.1
	Prototheca zopfii	91.0
	Chlorella vulgaris	50.0
	Euglena gracilis	49.6
Saccharomyces cerevisiae	*S. cerevisiae*	100
	N. muscorum	6.4
	P. zopfii	100.5
	C. vulgaris	81.5
	E. gracilis	131.0

[a] Incorporation of phenylalanine from phenylalanyl-tRNA directed by poly U (*37, 63*).

sets of ribosome-specific transfer factors or of only one set of enzymes active on both types of ribosomes. However, the fact that in *P. zopfii* two sets of transfer enzymes may be separated (Fig. 2) favors the hypothesis of two groups of transfer factors. This hypothesis is supported by the data reported in Fig. 3 on the polymerizing activity on the two types of ribosomes in cell-free extracts prepared from dark- and light-grown *E. gracilis* and from a permanently bleached mutant of the same organism (*64*). Indeed, the extracts prepared from the three types of cells are active on yeast ribosomes while activity on *E. coli* ribosomes is displayed only by the extracts prepared from light-grown cells. It may then be tentatively concluded that induction of the photosynthetic apparatus brings about a corresponding induction (or activation) of one or more of the transfer factors specific for 70 S ribosomes. Indeed, such factor(s) cannot be detected in the extracts prepared from cells grown under conditions that prevent the formation of a complete photosynthetic apparatus (dark and glucose) nor in those of a permanently bleached mutant, even when grown in the light. The activity of such extracts on *E. coli* ribosomes may, however, be restored by the addition of *E. coli* T factor thus indicating that such extracts are already endowed with G-like activity for 70 S ribosomes, as reported for yeast (see Section IV) (*59*). The results obtained with *P. zopfii* (Table VIII) seem at variance with the latter conclusions since the alga is a naturally occurring achloric eukaryote and should not display activity on *E. coli* ribosomes. However, these results may be explained by the release of specific transfer factors from chloroplast precursors or mitochondria (as may be the case in certain strains of yeast). Alternatively, the presence of

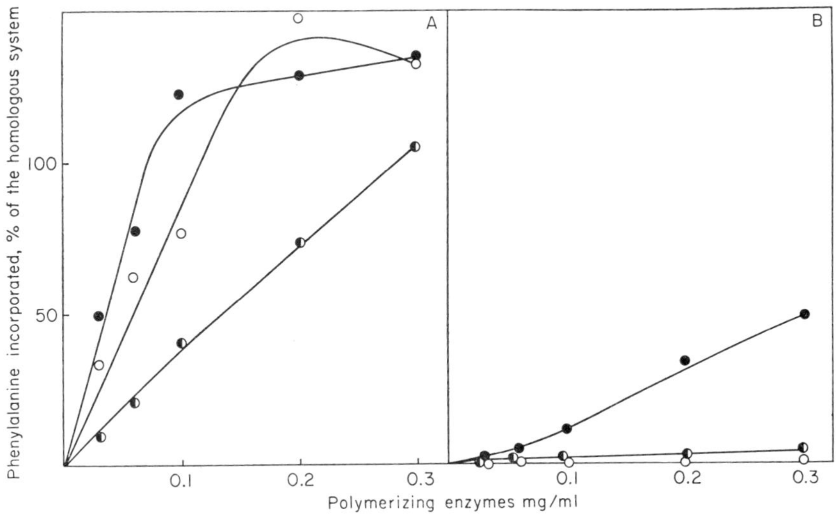

FIG. 3. Ribosome specificity of polymerizing enzymes from light- or dark-grown cells and from an achloric mutant of *Euglena gracilis*. Incorporation of phenylalanine from phenylalanyl-tRNA directed by poly U on yeast ribosomes (A) and *Escherichia coli* ribosomes (B) (*63*). ○——○, dark-grown cells; ◐——◐, achloric mutant; ●——●, light-grown cells.

such enzymes could be related to the loss of a mechanism for the control of the synthesis of these enzymes in *P. zopfii* in contrast to the stringent control in dark-grown cells that are still light-inducible, or in recently produced white mutants of green cells. Even more simply, organisms as different as *Euglena* and *Chlorella* may differ in the degree of autonomy of their chloroplasts (*65*) as far as the synthesis of the organelle itself and/or the synthesis of accessory enzymes, such as the transfer enzymes, are concerned.

VI. Ribosome-Specific Inhibitors of Protein Synthesis

A number of drugs, especially antibiotics, that inhibit protein synthesis *in vitro* act selectively on systems endowed with either 70 S or 80 S ribosomes (*66*). Little is known on the precise site of action of many of these compounds, but it is evident that to be ribosome-specific such drugs must interact selectively with sites that are different on the two types of ribosomes, or with the transfer factors specific for each type of ribosome.

Some of these antibiotics have been successfully employed in the selective inhibition *in vivo* of protein synthesis by cytoplasmic ribosomes (80 S) or by mitochondrial or chloroplast ribosomes (70 S). Among the better known ones are streptomycin and other aminoglycosidic antibiotics that inhibit protein synthesis presumably by inducing an ambiguity in the translation mechanisms (*miscoding*) which is easily reproduced *in vitro* (*67, 68*). Streptomycin and related antibiotics seem to induce a miscoding effect on bacterial ribosomes but not on cytoplasmic ribosomes from eukaryotic organisms (*69*). It is perhaps interesting to speculate that an *ante litteram* exploitation of the ribosome specificity of streptomycin may be considered to be a bleaching, probably by interference with chloroplast development and/or function, of *Euglena* and other algae (*70*). Chloramphenicol seems to interfere with the utilization of aminoacyl-tRNA by inhibiting the enzyme responsible for peptide synthesis (peptidyl transferase) on 70 S ribosomes only (*66, 71–73*). This antibiotic inhibits selectively the synthesis of mitochondrial enzymes in intact yeast cells (*74, 75*) and protein synthesis by isolated yeast mitochondria (*76*) without affecting cytoplasmic protein synthesis. Furthermore, chloramphenicol inhibits protein synthesis *in vitro* by mitochondrial ribosomes from *N. crassa* but not by cytoplasmic ribosomes from the same organism (*61*). Likewise, it inhibits the biosynthesis of some chloroplast enzymes as well as chloroplast protein synthesis (*77–79*) without impairing cytoplasmic function or protein synthesis by cytoplasmic ribosomes (*80*). Macrolide antibiotics, such as erythromycin, inhibit only bacterial (*66*) and yeast mitochondrial (*76*) protein synthesis, presumably by preventing the binding of aminoacyl-tRNA and/or mRNA to ribosomes (*66*).

On the other hand, cycloheximide, other glutarimide antibiotics and some related ipecac alkaloids, such as emetine, inhibit protein synthesis *in vitro* and *in vivo* only in systems containing 80 S ribosomes (*81, 82*). We have recently demonstrated that pederin, the toxic substance isolated from the insect *Paederus fuscipes*, appears to inhibit protein synthesis *in vitro* only in the case of 80 S ribosomes (*34, 83*). At a concentration of only 0.2 μg/ml, pederin inhibits the incorporation of phenylalanine in the cell-free extracts prepared from human cells grown *in vitro*, rat liver, *S. cerevisiae*, *P. zopfii*, and *C. vulgaris*, while it is ineffective at concentrations of 40 μg/ml on systems prepared from the bacteria *B. subtilis* and *E. coli* and the blue-green alga *N. muscorum* (*37*). The remarkable potency and selectivity of the drug both *in vitro* and *in vivo* (*84*) makes pederin and attractive tool for studies of protein synthesis *in vitro* and *in vivo* in eukaryotic organisms.

The use of some of these compounds in the selective inhibition of protein synthesis *in vivo* by eukaryotic-type ribosomes (80 S) or by prokaryotic-type ribosomes (70 S), suggests that the exploitation of the different specificity of these and other compounds may be of considerable interest. Such compounds may indeed help to elucidate the mechanisms for the regulation of protein synthesis in eukaryotic organisms and the relations that may exist in the cells *in toto* between protein synthesis by cytoplasm and by organelles.

VII. Conclusions

All the evidence at present indicates that ribosomes of the 70 S-type (prokaryotic organisms) are unable to catalyze one or more of the reactions required for protein synthesis *in vitro* when incubated with polymerizing enzymes obtained from the cytoplasm of eukaryotic organisms endowed with cytoplasmic ribosomes of the 80 S-type. (The possible exception represented by *P. zopfii* polymerizing enzymes is discussed in Section IV). The same appears to be true for ribosomes of the 80 S type and polymerizing enzymes from organisms containing the 70 S type of ribosomes. In contrast to the cytoplasm, organelles of eukaryotic cells contain ribosomes of the 70 S type that appear to respond to *E. coli* polymerizing enzymes (*60, 61*). At least in the case of *N. crassa* mitochondria, the organelle ribosomes respond to organelle or *E. coli* polymerizing enzymes, but not to cytoplasmic polymerizing enzymes. Likewise cytoplasmic ribosomes respond to cytoplasmic or rat liver polymerizing enzymes, but not to those prepared from mitochondria or *E. coli* (*61*). If the incompatibility *in vitro* reflects a similar situation *in vivo*, the phenomenon could represent a regulatory mechanism for protein synthesis in eukaryotic cells. Protein synthesis by genetically and metabolically semiautonomous cellular organelles could thus be uncoupled from protein synthesis in cell cytoplasm by the incompatibility between the cytoplasmic polymerizing enzymes and organelle ribosomes and *vice versa*. Such a contention appears to be strengthened by the finding that cell-free extracts prepared from dark-grown cells of *Euglena* or light-grown cells of a permanently bleached strain of the same organism, known to be totally devoid of or with an incomplete photosynthetic apparatus, display no activity on 70 S ribosomes. It is interesting to note that permanently bleached cells of *Euglena* do not show a band of satellite DNA (presumably chloroplast DNA) (*85*). This could indicate that one or more of the chloroplast polymerizing enzymes are coded for by the DNA of the organelle. If, on the other hand, the polymerizing enzymes for organelle protein synthesis were of nuclear origin, then the nucleus could control protein synthesis also in cellular organelles. Of course, this does not rule out other possibilities for a nuclear control over protein synthesis in the organelles (*62*).

The specificity of *E. coli* ribosomes appears to be only partial, since two out of the three transfer factors (T_u and G) may be provided by crude or partially purified preparations of polymerizing enzymes obtained from eukaryotic organisms such as yeast (*7*) or rat liver (*55*). At present the possibility cannot be excluded that the preparations from the eukaryotic organisms are contaminated by the polymerizing enzymes, which act on the mitochondrial ribosomes. However, since the polymerizing enzymes from such organisms do not catalyze protein synthesis on ribosomes from *E. coli*, it might be surmised that one of the mitochondrial enzymes (most probably T_s) is absent or inactivated in the extracts.

Before speculating on the significance of the partial ribosome specificity, it is important to determine unambiguously whether eukaryotic cells contain different sets of enzymes for each type of ribosome (cytoplasmic, mitochondrial, chloroplastic) as is the case for tRNA and aminoacyl-tRNA synthetase at least in *N. crassa* mitochondria (*86*), or whether the cells contain only one type of the nonspecific transfer factor (such as T_u and G) and two, or three, different types of the specific factor (T_s). One could surmise that if the function of the nonspecific factors is the same for all types of ribosomes, there should be only one type of T_u and G per cell, as evolution should have disposed of a meaningless redundancy. The evidence obtained from *P. zopfii* seem to contradict such a teleological hypothesis; for the modulation of protein synthesis in the cytoplasm and in the organelles, two G factors (and two T_u factors), each specific for one type of ribosome, may provide a selective advantage as compared to that of a more streamlined scheme composed of only one G and one T_u factor.

Recently there has been a revival of the hypothesis that organelles may have evolved from a symbiotic relationship between a "primitive" eukaryotic cell and a prokaryotic cell (*87*) (perhaps an ancestral bacterium in the case of mitochondria and an ancestral blue-green alga in the case of chloroplasts). If such is the case, ribosome specificity in protein synthesis could be explained on the basis of a different biological origin of the different types of ribosomes.

If ribosome specificity is of any importance in the regulation of protein synthesis, then the evidence from the partial activity of tumor ribosomes in response to *E. coli* polymerizing enzymes (*31*, *88*) may be significant. Further work on this subject is, however, required before drawing any conclusion,[7] as the transition from normal to malignant cells is accompanied, or caused, by profound alterations in the structure and function of the cell.

Whatever the reasons or the causes for ribosome specificity, the phenomenon appears to be widespread and reproducible. Ribosome specificity

[7] In our laboratory, ribosomes from ascites tumor cells do not respond to *E. coli* polymerizing enzymes nor to purified *E. coli* transfer factors for the incorporation of phenylalanine directed by poly U (*63*). On the contrary, Canning and Griffin (*31*) studied the incorporation of various amino acids under the direction of the endogenous mRNA.

TABLE IX
A TENTATIVE CLASSIFICATION OF RIBOSOME SPECIFICITY OF PROTEIN SYNTHESIS *in Vitro*[a]

Ribosomes from	Polymerizing enzymes from			
		Eukaryotes		
	Prokaryotes	Cytoplasm	Mitochondria	Chloroplasts
Prokaryotes	+	−	+	(+)
Eukaryotes				
Cytoplasm	−	+	−	(−)
Mitochondria	+	−	+	(+)
Chloroplasts	+	(−)	(+)	(+)

[a] +, active system; (+), active system (tentative); −, lack of activity; (−), lack of activity (tentative).

sharply dichotomizes between ribosomes of the prokaryotic and eukaryotic types rather than between ribosomes from unicellular and multicellular organisms or from plants and animals (Table IX). Its study may contribute to the elucidation of the structural and functional differences between the two types of ribosomes. In addition, it may be a tool of some value in comparative studies of the mechanisms for initiation of protein synthesis, peptide bond formation, peptide translocation, etc., in ribosomes of the eukaryotic and prokaryotic types. A clearer picture of the phenomenon and of its significance will emerge only when it becomes possible to purify and separate the transfer factors from cytoplasmic and organelle polymerizing enzymes and possibly to prepare active "hybrid" ribosomes from the complementary subunits of the cytoplasmic and organelle ribosomes. The goal is perhaps not too distant; we hope that it will not sound preposterous to propose a time scale of just a few years. [*See* Note Added in Proof, p. 144.]

ACKNOWLEDGMENTS

The experimental work performed in the Authors' laboratory was supported by grants from the Consiglio Nazionale delle Ricerche. The Authors are grateful to Drs. Antonia Perani, Orsola Tiboni, and Maria Grandi for their skillful and enthusiastic participation in the experimental work and to Dr. J. Van Etten for constructive criticism of the manuscript. In addition, they wish to acknowledge the technical assistance of Messers. G. Di Pasquale and E. Garbagnoli and the untiring cooperation of Miss Olga Cignoli in the preparation of the manuscript.

REFERENCES

1. Y. Nishizuka and F. Lipmann, *Proc. Natl. Acad. Sci. U.S.* **55,** 212 (1966).
2. J. Lucas-Lenard and F. Lipmann, *Proc. Natl. Acad. Sci. U.S.* **55,** 1562 (1966).

3. A. S. Skoultchi, Y. Ono, H. M. Moon, and P. Lengyel, *Proc. Natl. Acad. Sci. U.S.* **60,** 675 (1968).
4. R. Arlinghaus, J. Shaeffer, and R. Schweet, *Proc. Natl. Acad. Sci. U.S.* **51,** 1291 (1964).
5. E. Gasior and K. Moldave, *J. Biol. Chem.* **240,** 3346 (1965).
6. L. Skogerson and K. Moldave, *Arch. Biochem. Biophys.* **125,** 497 (1968).
7. O. Ciferri, B. Parisi, A. Perani, and M. Grandi, *J. Mol. Biol.* **37,** 529 (1968).
8. D. Nathans and F. Lipmann, *Proc. Natl. Acad. Sci. U.S.* **47,** 497 (1961).
9. R. Stanier, M. Doudoroff and E. Adelberg, "The Microbial World," 2nd ed., p. 85. Prentice-Hall, Englewood Cliffs, New Jersey, 1963.
10. M. Petermann, "The Physical and Chemical Properties of Ribosomes," pp. 12 and 15. Elsevier, Amsterdam, 1964.
11. M. M. Taylor and R. Storck, *Proc. Natl. Acad. Sci. U.S.* **52,** 958 (1964).
12. P. O. P. Ts'O, *Ann. Rev. Plant Physiol.* **13,** 45 (1962).
13. E. N. Svetailo, I. I. Philippovich, and N. H. Sissakian, *J. Mol. Biol.* **24,** 405 (1967).
14. N. K. Boardman, R. I. B. Francki, and S. G. Wildman, *J. Mol. Biol.* **17,** 470 (1966).
15. H. Küntzel and H. Noll, *Nature* **215,** 1340 (1967).
16. E. Stutz and H. Noll, *Proc. Natl. Acad. Sci. U.S.* **57,** 774 (1967).
17. A. H. Reisner, J. Rowe, and H. M. Macindoe, *J. Mol. Biol.* **32,** 587 (1968).
18. U. E. Loening, *J. Mol. Biol.* **38,** 355 (1968).
19. H. Noll, *Nature* **215,** 360 (1967).
20. M. W. Nirenberg and J. H. Matthaei, *Proc. Natl. Acad. Sci. U.S.* **47,** 1588 (1961).
21. G. Brawerman and J. M. Eisenstadt, *Biochemistry* **5,** 2784 (1966).
22. M. Revel and F. Gros, *Biochem. Biophys. Res. Commun.* **25,** 124 (1966).
23. W. Stanley, Jr., M. Salas, A. Wahba, and S. Ochoa, *Proc. Natl. Acad. Sci. U.S.* **56,** 290 (1966).
24. J. Lucas-Lenard and F. Lipmann, *Proc. Natl. Acad. Sci. U.S.* **57,** 1050 (1967).
25. G. D. Novelli, *Ann. Rev. Biochem.* **36,** 449 (1968).
26. R. Rendi and S. Ochoa, *J. Biol. Chem.* **237,** 3707 (1962).
27. G. Von Ehrenstein and F. Lipmann, *Proc. Natl. Acad. Sci. U.S.* **47,** 941 (1961).
28. C. F. Heredia and H. O. Halvorson, *Biochemistry* **5,** 946 (1966).
29. T. Okamoto, *J. Biochem.* **55,** 19 (1964).
30. B. Parisi, G. Milanesi, J. L. Van Etten, A. Perani, and O. Ciferri, *J. Mol. Biol.* **28,** 295 (1967).
31. L. Canning and A. C. Griffin, *Biochim. Biophys. Acta* **103,** 522 (1965).
32. B. Parisi, A. Perani, M. Grandi, and O. Ciferri, *Giorn. Bot. It.* **102,** 317 (1968).
33. J. E. Allende and M. Bravo, *J. Biol. Chem.* **241,** 5813 (1966).
34. A. Perani, B. Parisi, L. De Carli, and O. Ciferri, *Biochim. Biophys. Acta* **161,** 223 (1968).
35. K. Shimura, S. Ejiiri, and T. Kobayashi, *Proc. 7th Intern. Congr. Biochem., Tokyo Symp.* **II-4,1,** 121 (1967).
36. J. Ilan and F. Lipmann, *Acta Biochem. Pol.* **13,** 353 (1966).
37. O. Tiboni, unpublished results.
38. D. Grünberger, *Collection Czechv. Chem. Commun.* **29,** 2400 (1964).
39. W. Zillig, D. Schachtschabel, and W. Krone, *Z. Physiol. Chem.* **318,** 100 (1960).
40. A. G. So and E. W. Davie, *Biochemistry* **2,** 132 (1963).
41. K. M. Downey, A. G. So, and E. W. Davie, *Biochemistry* **4,** 1703 (1965).
42. E. Krajewska and W. Szer, *European J. Biochem.* **2,** 250 (1967).
43. B. Parisi and O. Ciferri, *Biochemistry* **5,** 1638 (1966).

44. J. Van Etten, B. Parisi, and O. Ciferri, *Nature* **212,** 932 (1966).
45. M. W. Nirenberg and P. Leder, *Science* **145,** 1399 (1964).
46. S. Lederberg and J. M. Mitchinson, *Biochim. Biophys. Acta* **55,** 104 (1962).
47. W. Gilbert, *J. Mol. Biol.* **6,** 374 (1963).
48. S. Lederberg and V. Lederberg, *Exptl. Cell Res.* **25,** 198 (1961).
49. M. Takeda and F. Lipmann, *Proc. Natl. Acad. Sci. U.S.* **56,** 1875 (1966).
50. F. N. Chang, C. J. Sih, and B. Weisblum, *Proc. Natl. Acad. Sci. U.S.* **55,** 431 (1966).
51. T. E. Martin and I. G. Wool, *Proc. Natl. Acad. Sci. U.S.* **60,** 569 (1968).
52. S. S. Rao and A. P. Grollman, *Biochem. Biophys. Res. Commun.* **29,** 696 (1967).
53. J. E. Allende, N. W. Seeds, T. W. Conway and H. Weissbach, *Proc. Natl. Acad. Sci. U.S.* **58,** 1566 (1967).
54. D. Richter and F. Klink, *Biochemistry* **6,** 3569 (1967).
55. L. Felicetti, personal communication.
56. L. Felicetti and F. Lipmann, *Arch. Biochem. Biophys.* **125,** 548 (1968).
57. M. W. Beijerinck, *Rec. Trav. Bot. Néerl.* **1,** 14 (1904).
58. E. E. Butler, *Science* **120,** 274 (1954).
59. B. Parisi, unpublished results.
60. J. M. Eisenstadt and G. Brawerman, *Biochemistry* **5,** 2777 (1966).
61. H. Küntzel, *FEBS Letters* **4,** 140 (1969).
62. A. Gibor and S. Granick, *Science* **145,** 890 (1964).
63. A. Perani, unpublished results.
64. L. Provasoli, S. Hutner, and J. J. Pintner, *Cold Spring Harbor Symp. Quant. Biol.* **16,** 113 (1951).
65. J. A. Schiff and M. H. Zeldin, *J. Cellular Physiol.* **72,** Suppl. 1, 103 (1968).
66. D. Vazquez and R. E. Monro, *Biochim. Biophys. Acta* **142,** 155 (1967).
67. J. Davies, W. Gilbert, and L. Gorini, *Proc. Natl. Acad. Sci. U.S.* **51,** 883 (1964).
68. J. Davies, L. Gorini, and B. D. Daves, *Mol. Pharmacol.* **1,** 93 (1965).
69. I. B. Weinstein, M. Ochoa, Jr., and S. M. Friedman, *Biochemistry* **5,** 3332 (1966).
70. L. Provasoli, S. Hutner, and A. Schatz, *Proc. Soc. Exptl. Biol. Med.* **69,** 279 (1948).
71. O. Jardetzky and G. R. Julian, *Nature* **201,** 397 (1964).
72. R. Rendi and S. Ochoa, *J. Biol. Chem.* **237,** 3711 (1964).
73. E. Cundliffe and K. McQuillen, *J. Mol. Biol.* **30,** 237 (1967).
74. M. Huang, D. R. Biggs, G. D. Clark-Walker, and A. W. Linnane, *Biochim. Biophys. Acta* **114,** 334 (1966).
75. G. D. Clark-Walker and A. W. Linnane, *J. Cell Biol.* **34,** 1 (1967).
76. G. D. Clark-Walker and A. W. Linnane, *Biochem. Biophys. Res. Commun.* **25,** 8 (1966).
77. A. W. Linnane and P. R. Stewart, *Biochem. Biophys. Res. Commun.* **27,** 511 (1967).
78. D. Spencer, *Arch. Biochem. Biophys.* **111,** 381 (1965).
79. R. M. Smillie, D. Graham, M. R. Dwyer, A. Grieve, and N. F. Tobin, *Biochem. Biophys. Res. Commun.* **28,** 604 (1967).
80. J. M. Eisenstadt and G. Brawerman, *J. Mol. Biol.* **10,** 392 (1964).
81. H. L. Ennis and M. Lubin, *Science* **146,** 1474 (1964).
82. A. P. Grollman, *Proc. Natl. Acad. Sci. U.S.* **56,** 1867 (1966).
83. O. Tiboni, B. Parisi,and O. Ciferri, *Giorn. Bot. It.* **102,** 337 (1968).
84. A. Brega, A. Falaschi, L. De Carli, and M. Pavan, *J. Cell Biol.* **36,** 485 (1968).
85. J. Leff, M. Mandel, H. T. Epstein, and J. A. Schiff, *Biochem. Biophys. Res. Commun.* **13,** 126 (1963).
86. W. E. Barnett and D. H. Brown, *Proc. Natl. Acad. Sci. U.S.* **57,** 452 (1967).

87. L. Sagan, *J. Theoret. Biol.* **14,** 225 (1967).
88. A. C. Griffin, *Advan. Cancer Res.* **10,** 83 (1967).

Note Added in Proof

Recently it has been demonstrated that it is possible to separate from *S. cerevisiae* two G-like activities, each specific for only one type of ribosomes (*1*). Moreover, within the group of the 70 S ribosomes the factors required for peptide chain initiation (*2*) appear to be interchangeable. Thus there is a functional interchangeability of initiation factors between *E. coli* and *B. subtilis* ribosomes. In addition, the factors from *E. coli* catalyze the reactions for *in vitro* peptide chain initiation on mitochondrial ribosomes from *N. crassa* and chloroplast ribosomes from *E. gracilis* (*3*).

1. O. Tiboni, B. Parisi, A. Perani, and O. Ciferri, *J. Mol. Biol.* in press.
2. M. Salas, M. J. Miller, A. J. Wahba, and S. Ochoa, *Proc. Natl. Acad. Sci. U.S.* **57,** 1865 (1967).
3. F. Sala, unpublished results.

Synthetic Nucleotide-peptides

ZOE A. SHABAROVA

Department of Chemistry,
Laboratory of Bioorganic Chemistry,
Moscow State University, USSR

I. Introduction 145
II. Nomenclature of Nucleotide-peptides 147
III. Synthesis of Nucleotide-(P-N)-peptide Phosphoramidates and Nucleotide-(P-O)-seryl(or threonyl)-peptide Phosphoesters . . 148
IV. Properties of Nucleotide-peptide Bonds in Synthetic Models . . 151
A. Nucleotide-(P-N)-peptide Phosphoramidates 151
B. Nucleotide-(5′ → *O*)-seryl(or threonyl)peptide Phosphoesters . 167
C. Specific Cleavage of Nucleotide-peptide Bonds 169
D. Concluding Remarks 169
V. Nucleotide-(P-N)-peptide Phosphoramidates as Models of "Active Nucleotides" 170
A. Possible Models of Nucleotide Activation by Nucleotidyl Transferases 170
B. Ribonucleoside-5′-phosphoamidase 172
VI. Conclusion 178
References 179

I. Introduction

Since the discovery of the relationship between the functions and biosynthesis of proteins and nucleic acids, a new class of compounds, the nucleotide-peptides, has become the object of extensive investigation. Nucleotide-peptides are compounds with a covalent bond between a nucleotide (or oligonucleotide) and an amino acid (or peptide). In some cases, these compounds are intermediate structures formed in the course of transfer of amino acids (and, possibly, of nucleotides); hence, studies of them can furnish information about the mechanisms of these transformations. They may also be responsible for the structural arrangement of biopolymers and also may give very interesting evidence about the processes in which they take part.

Nucleotide-peptides have been isolated from various sources [see the review of A. Bogdanov (*1*)]. They differ not only in the structure and length of the nucleotide and peptide chains of which they are composed, but, chiefly, in the nature of the linkage determining the type of nucleotide-peptide, its reactivity and, consequently, its functions.

As far as the structure and functions are concerned, the best-known nucleotide-peptides are those with anhydride-type linkages (aminoacyladenylates), with ester nucleotide-peptide bonds (*O*-aminoacyl-tRNA's) as well as the peptide derivatives of UDP-*N*-acetylmuramic acid.

In addition, there are many nucleotide-peptides with unidentified linkages between the two main portions. Such compounds, isolated from acid-soluble fractions of various cells, include the peptide derivatives of nucleoside 5′-phosphates (*2–5*), amino acid and peptide derivatives of nucleoside polyphosphates (*6–9*), as well as peptide derivatives of oligonucleotides (*10*).

Of special interest are the nucleotide-peptide fragments found in high polymer nucleic acids. Highly purified DNA (*11–14*) and ribosomal RNA preparations (*15, 16*), isolated from various sources, contain small amounts of peptides (0.2–0.5%) believed to be covalently bound to the nucleic acid. This conclusion is usually based on the fact that the peptides do not separate from DNA or RNA in the usual deproteinization procedures, even the most drastic ones. In the case of ribosomal RNA, more direct confirmation is available, for nucleotide-peptide fragments have been isolated from RNA alkaline hydrolyzates (*17–19*). Some properties of these compounds indicate that the peptides have phosphoamide bonds with nucleotides (*20*). For DNA-peptides, two types of nucleotide-peptide linkage have been suggested: phosphoester and phosphoamide bonds (*21*) to the hydroxy or amino groups of hydroxyamino (amino) acids, respectively.

It has been suggested (*11, 12*) that peptides play the role in DNA of "stitching together" single polynucleotide chains. It was also assumed that one end of the peptide is linked to the polynucleotide by a phosphoester bond to N-terminal serine, the other end by an ester bond to the carboxyl of the C-terminal amino acid. This suggestion is supported by a number of factors: the molecular weight of DNA decreases on treatment with reagents cleaving the ester bonds, and *O*-phosphoserine was found among the products of DNA acid hydrolysis.

However, no direct confirmation of the above assumption (isolation of the respective nucleotide-peptide fragments, analysis of terminal peptide groups released on DNA depolymerization, etc.) is available. Hence there is continuing doubt concerning the hypothesis of Bendich *et al.* (*11, 12*) (see Discussion in References *12, 22*).

Peptides may become attached to ribosomal RNA in the course of ribosome biosynthesis (*16*), but the significance of this finding remains obscure.

One more aspect of nucleotide-peptides that calls for some attention is the nature of the chemical interaction between nucleotides and enzymes catalyzing various transformations of mononucleotides (RNA- and DNA-polymerase, ligases and some others). Very little is known about these interactions although some suggestions have been made about the formation of covalently bonded complexes between mononucleotides and enzymes (*23*) resembling the reactions involved in the transfer of inorganic phosphate (*24–26*). Also of interest is the recent evidence of a ligase-adenylate intermediate (*27*, *28*). It is well established that there exist various specific noncovalent interactions between nucleotides and enzymes catalyzing their transfer. The study of covalent and noncovalent interactions will certainly contribute to the elucidation of such important biochemical processes as the synthesis of nucleic acids, the structural organization and functional characteristics of ribosomes, etc.

Difficulties arising in the course of elucidating the nature of nucleotide-peptide bonds and the biological role of nucleotide-peptides are due to the very scanty knowledge of such compounds. Investigators engaged in their study are usually inclined to resort to remote chemical analogies without taking into account specific features in the behavior of the nucleotide-peptides due to their component nucleotides and amino acids, which have different functional groups and heterocyclic bases. Hence a vast field of investigation is open to nucleotide-peptide chemistry, which is certain to gain importance in the nearest future. Reviews dealing with various aspects of the chemistry and biochemistry of nucleotide-peptides with ester and anhydride bonds by H. G. Zachau and H. Feldmann (*29*) in this series and by M. R. Hoagland (*30*) have appeared.

The present article deals with synthetic nucleotide-peptides, with phosphoamide and phosphoester bonds.

II. Nomenclature of Nucleotide-peptides

A phosphoric acid in which one hydroxyl group is replaced by an amino group to yield $(HO)_2P(O)\text{-}NH_2$ is phosphoramidic acid.[1] The salts of the esters of this compound are phosphoramidates, and the P-N link is an amide (or phosphoamide) bond. Whether the N is that of a single amino acid or a peptide, we use the term nucleotide-(P-N)-peptide to describe the resulting phosphoramidate. Nucleotide-(P-N)-peptides are thus peptides in which the N of an amino group is phosphorylated by a phosphoric ester (nucleotide).

Compounds involving esterification of nucleotide phosphoric acid with the hydroxyl of serine or threonine are phosphoesters; hence we refer to them as nucleotide-(P-O)-seryl(or threonyl)-peptide esters and the bond to the amino acid hydroxyl as an ester (or phosphoester) bond.

[1] "The Naming and Indexing of Chemical Compounds" (Introduction to Subject Index of Vol. 56 of *Chemical Abstracts*), para. 407; Handbook for Chemical Society Authors, 1960 (Chapter 6, Rules 3 and 4, examples).

The individual names contain, in order, the name of the nucleotide (nucleotidyl) radical, the number identifying the location of the P atom on the pentose (2′, 3′, or 5′), the locant (*N* or *O*) to which the *P* is connected, and the name of the amino acid (or peptide). Thus, uridylyl-(5′ → *N*)-phenylalanine; uridylyl-(3′ → N^{ϵ})-lysine; uridylyl-(5′ → *O*)-serylglycine; 5′-phospho-polyuridylyl-(5′ → *N*)-phenylalanylglycine.

The compound formed by linking the diesterified phosphorus of a dinucleoside monophosphate (or of an oligonucleotide) and the N or O of an amino acid contains a triply substituted phosphate. We refer to the link as an internucleotide-(P-N)-peptide (phospho)amide bond or an internucleotide-(P-O)-peptide (phospho)ester bond. Specification of all three radicals attached to the P atom is accomplished by replacing one in parentheses[2]; alternatively, it could be placed above or below, as in branched-peptide[2] or -saccharide[3] symbolism. Thus the phosphoamide (P-N) derivative of uridylyl-(3′ → 5′)-uridine, or UpU, is uridylyl-3′(→5′-uridine) → *N*-phenylalanine or uridylyl-3′(→*N*-phenylalanine)→5′-uridine.

Symbolic representation of these substances utilizes the conventional three-letter symbols.[4] Thus the above become: Urd5′*P* → *N*Phe; Urd3′*P* → N^{ϵ}Lys; Urd5′*P* → *O*Ser-Gly; (poly U)5′*P* → *N*Phe-Gly; Urd3′*P*(→ 5′Urd) → *N*Phe or Urd3′*P*(→*N*Phe) → 5′Urd. Others are: Cyd-(2′Ac,5′Thp)3′*P* → *N*PheOMe; dThd3′*P*(→5′Ado) → *N*Phe.

III. Synthesis of Nucleotide-(P-N)-peptide Phosphoramidates and Nucleotide-(P-O)-seryl(or threonyl)-peptide Phosphoesters

The synthesis of four types of compounds that may serve as possible models of nucleotide-peptides with phosphoamide and phosphodiester linkages has been described, and the compounds are shown in Fig. 1.

The methods of synthesis of nucleotide-(P-N)-peptides are quite similar to those of the peptides, the only difference being that instead of the carboxyl component there is a nucleotide (Fig. 2). Consequently, the most important factor is the choice of effective ways of activating the phosphoric acid residue in the nucleotide.

Methods of activating the phosphate groups in the mononucleotides as well as terminal phosphate groups in oligo- and polynucleotides have

[2] IUPAC–IUB Tentative Rules, "Abbreviated Designation of Amino Acid Derivatives and Peptides" [*J. Biol. Chem.* **241,** 2491 (1966), and elsewhere], para. 4.2.

[3] IUPAC–IUB Tentative Rules, "Abbreviations and Symbols for Chemical Names of Special Interest in Biological Chemistry" [*J. Biol. Chem.* **241,** 527 (1966), and elsewhere], para. 3.4.

[4] See footnote 3 reference, Section 5.3; also Abbreviations and Symbols, p. ix in this volume.

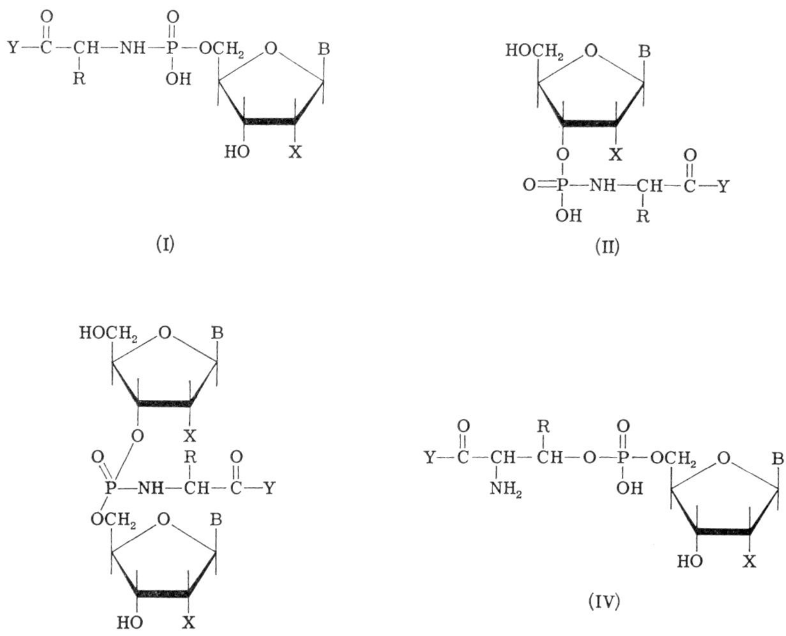

(I) (II) (III) (IV)

X = H or OH
B = Pyrimidine or purine bases
Y = OH, O-Alk or NH—CH(R)—COOH (peptide)

FIG. 1

been studied fairly extensively. The synthesis of the respective nucleotide-(P-N)-peptides (compounds of type I and II, Fig. 1) is most effectively accomplished by the carbodiimide method (*31–34*) and that of mixed anhydrides (*35–45*) (anhydrides of nucleotides and diphenylphosphoric acid are usually used). The chlorophosphate method is a modification of choice for the latter procedure; here the chloroanhydride of the nucleotide

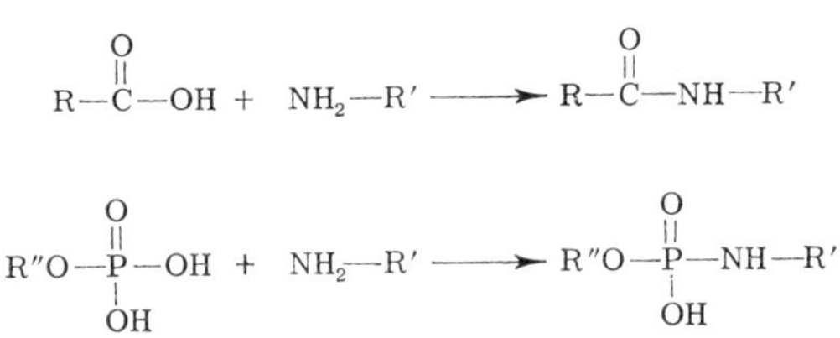

R and R′ = Amino acid (peptide) residues
R″ = Nucleoside (polynucleotide)

FIG. 2

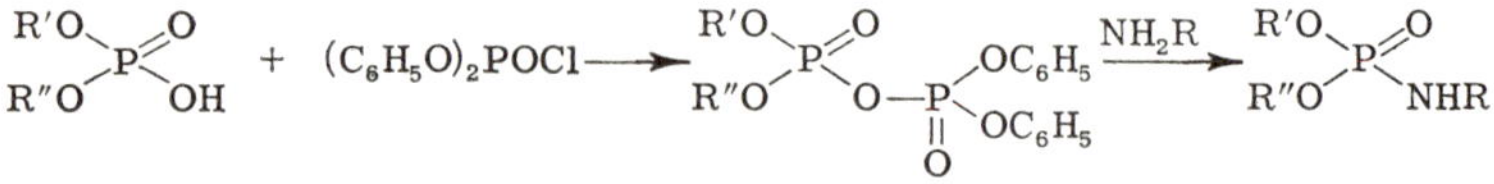

R = Amino acid (or peptide) residue
R′ and R″ = Nucleoside

FIG. 3

is used (*46–49*). For the preparation of nucleotide-peptides of type III (Fig. 1), formed by the internucleotide phosphorus, activation of the internucleotide phosphate group by preparation of the respective anhydride with diphenylphosphoric acid (*50–52*) is employed (Fig. 3).

The synthesis of nucleotide-(P-O)-peptides (compounds of type IV, Fig. 1) is essentially similar to that of the internucleotide bond, the only difference being the use of a hydroxyamino acid instead of the nucleoside component (Fig. 4).

In this case, the activation of the nucleotide was achieved either by way of the mixed anhydride with diphenylphosphoric acid (*41*, *53*), or the chloroanhydride (*54*), or by means of dicyclohexylcarbodiimide (*54*).

But the similarity between the formation of the nucleotide-(P-N)- and -(P-O)-peptide linkages and peptide and internucleotide bonds is not complete—the synthesis of nucleotide-peptides has peculiarities of its own due to the participation in the reaction of monomers (or polymers) belonging to different classes of natural compounds.

Compounds of type I have been synthesized that are derivatives of ribonucleoside 5′-phosphate (*31*, *32*, *35–38*, *41–44*, *53*, *55*, *56*), of uridine 5′-phosphates modified in the uracil residue (*45*), of ribonucleoside 3′,5′-diphosphates (*57*), of deoxythymidine 5′-phosphate (*39*, *43*), and of 5′-phosphopolyuridylic acid (*34*); they contain the amine as an amide fragment (*56*), amino acids (*31*, *32*, *34–37*, *39*, *55*, *57*), hydroxyamino acids (*41*, *44*, *53*), β-alanine (*56*), and peptides (*38*, *40*, *43*), sometimes containing serine

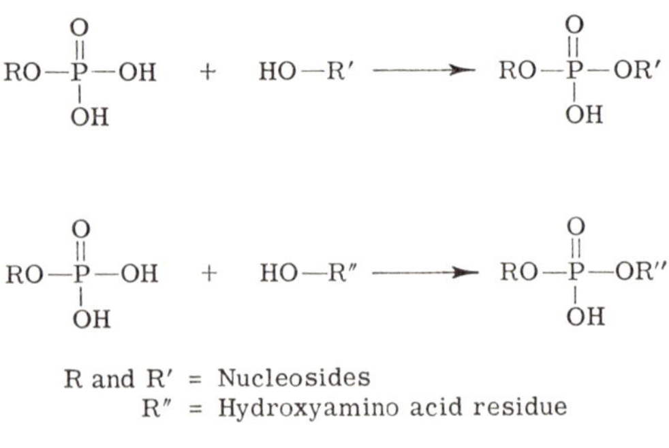

R and R′ = Nucleosides
R″ = Hydroxyamino acid residue

FIG. 4

(*42*, *43*) and threonine (*42*, *43*). Of the compounds belonging to type II, there have been synthesized phenylalanine derivatives of ribonucleoside 3′-phosphates (*36*, *58*) and deoxythymidine 3′-phosphate (*39*); of the compounds of type III: amide (*52*), amino acid (*50–52*), and peptide (*52*) derivatives of uridylyl-(3′ → 5′)-uridine (*50*), deoxythymidylyl-(3′ → 5′)-deoxythymidine (*50*), deoxythymidylyl-(3′ → 5′)-adenosine (*52*), deoxythymidylyl-(3′ → 5′)-uridine (*51*); of the compounds of type IV: derivatives of deoxythymidine 5′-phosphate (*54*) and uridine 5′-phosphate (*41*, *53*, *54*) containing serine (*41*, *53*, *54*) and peptides of serine (*53*) and threonine (*53*).

IV. Properties of Nucleotide-peptide Bonds in Synthetic Models

A. Nucleotide-(P-N)-peptide Phosphoramidates

The polyfunctional nature of the nucleotide-peptides and the abundance of single bonds allow for great variation in the relative localization of the parts of the molecule and intramolecular interactions of the groups not directly connected. These interactions may have a great effect on the behavior of the P-N bonds because of the specific properties of the phosphorus atom (i.e., the $p_\pi - d_\pi$ bonding with atoms not directly connected to it) as well as the properties of the nitrogen atom (i.e., its ability to protonize not only at the expense of the external protons, but also from the proton-donating groups of the same molecule).

The following is a discussion of the experimental evidence confirming existence of such interactions. Table I contains the relevant data.

1. Mononucleotide Phosphoramidates

Investigation of nucleotidyl-(5′ → *N*)-peptides with an amino acid residue containing no additional functional groups shows that the 5′-phosphoamide bond is readily hydrolyzed in acid but is stable in neutral and alkaline solutions (*37*, *59–61*), as is the phosphoamide bond in the simple amides and esters of phosphoric acid (*62–64*). In a mildly acid medium (pH 2.5–5.0), the hydrolytic stability of the phosphoamide bond in the nucleotidyl-(5′ → *N*)-peptides depends on the nature of the nucleotide heterocyclic base (*37*, *55*, *59*). Investigation of the structure of nucleotidyl-(5′ → *N*)-amino acids suggests that the base-phosphoaminoacyl residue interaction is effected at the expense of H-bonds. The schemes shown in Fig. 5 are suggested for the possible H-bonds in the methyl esters of uridylyl-(5′ → *N*)-phenylalanine (a) (*37*), guanylyl-(5′ → *N*)-phenylalanine (b) (*55*), inosinyl-(5′ → *N*)-phenylalanine (c) (*55*) and adenylyl (5′ → *N*)-phenylalanine (d).

Nucleotides are known to have base *anti*-orientation (*65–68*); in some

TABLE I
HYDROLYSIS OF SYNTHETIC NUCLEOTIDE-PEPTIDES

No.	Name and type of compound	$t_{1/2}$[a]	Percent hydrolysis of nucleotide-peptides at 37°, 60 min: In acid: [H$^+$]	In acid: P–N	In acid: P–05′	At pH 3–10: P–N	At pH 3–10: P–05′	At pH > 10: pH[l]	At pH > 10: P–N	At pH > 10: P–05′	References
1	Ado5′*P* → *N*PheOMe (I)		0.1	100		0	0	13	0	0[o]	(*37, 71*)
2	Ado5′*P* → *N*Phe (I)		0.1	50	50	0	0		0		(*60, 71*)
3	Ado5′*P* → *N*Phe-Val-GlyOMe (I)	86		+		0	0	13	0	0[o]	(*60*)
4	dThd5′*P* → *N*PheOMe (I)	3.5	0.1	100		0	0		Same		(*39*)
5	dThd3′*P* → *N*PheOMe (II)	3	0.1	100		dThd3′:5′*P* formed[i]			Same		(*39*)
6	dThd5′*P* → *N*Ser-GlyOEt (I)		0.1	50	10	0		13		10[m]	(*43*)
			1	50	50						
7	dThd3′*P*(→5′Ado) → *N*(*n*Bu) (III)		0.2	5[b]	0[z]	0 (pH 3–5, 37°, 2 hr)		14	0[b,x]		(*72*)
8	dThd3′*P*(→5′Ado) → *N*(Bzl) (III)		0.2	5[b]	0[z]	Same			Same		(*72*)
9	dThd3′P(→5′Ado) → *N*Phe (III)		0.2	10[b]	50(40)[c]	Same			Same		(*72*)
10	dThd3′*P*(→5′Ado) → *N*Ala-PheOMe (III)		0.1	20[d]		Same		14	0	80[n]	(*72*)
			1	40[b]							
11	Urd(2′,5′-Ac$_2$)3′*P* → *N*PheOMe (II)		0.1	100		—		10.8	70 (24 hr)		(*37*)
12	Guo5′*P* → *N*AlaOMe (I)	120		+		50 (pH 4, 24 hr)			0		(*55*)
13	Guo5′*P* → *N*PheOMe (I)	24		+		Same (21h)			0		(*55, 59*)
14	Guo5′*P* → *N*Phe-GlyOMe (I)	57		+		Same (24h)			0		(*59*)
15	Guo5′*P* → *N*Phe-ValOMe (I)			+		Same (25h)			0		(*59*)
16	Guo5′*P* → *N*Phe-Val-AlaOMe (I)	127		+		Same (27h)			0		(*59*)
17	Ino5′*P* → *N*AlaOMe (I)	151		+		40 (pH 4, 80 hr)			0		(*55*)
18	Ino5′*P* → *N*PheOMe (I)	50		+		Same			0		(*55*)
19	Urd5′*P* → *N*(Ethanolamine) (I)		0.1	100[e]	0	0			0		(*56*)
20	Urd5′*P* → *N*Phe (I)		0.1	50	50[f]	0			0		(*56, 71*)
21	Urd5′*P* → *N*PheOMe (I)	20	0.05	100[h]		50 (pH 4, 45 hr)		13	0[o]		(*37, 45, 57, 59*)
22	Urd5′*P* → *N*(βAla) (I)		0.1	100	0	0			0		(*56*)
23	Urd5′*P* → *N*Ser(*O*Thp)OMe (I)		1	20	70	+ (pH 4)	0	13	0[o]		(*44*)
			0.1	70	30						
24	Urd5′*P* → *N*SerOMe (I)			Same as above		Same		13.7	90[o]		(*44*)
25	Urd5′*P* → *O*SerOMe (IV)		1	75% UMP				10.8	40[p]		(*53*)
26	Urd5′*P* → *O*Ser(*N*Cbz)OMe (IV)		1	100[g]				10.8	40% UMP		(*53*)

27	Urd5′*P* → *N*Gly-AlaOMe (I)		1(0.1)	100(7)		0		13	0[q]		(*43*)
28	Urd5′*P* → *N*Gly-SerOMe (I)		1	100		0		13		10[m]	(*43*)
29	Urd5′*P* → *N*Ser-GlyOEt (I)		1	50	50	0		13		10[m]	(*43*)
			0.1	70	20						
30	Urd5′*P* → *N*Thr-GlyOEt (I)		1	20	80	0		13		10[m]	(*43*)
			0.1	40	50						
31	Urd5′*P* → *O*Ser-GlyOEt (IV)		1	0[y]		0[y]		10.8	40[r]		(*53*)
32	Urd5′*P* → *O*Ser(*N*Cbz)-GlyOEt (IV)		1	0[y]		0[y]		10.8	0[s,t]		(*53*)
33	Urd5′*P* → *O*Thr(*N*Cbz)-GlyOEt (IV)		1	0[y]		0[y]		10.8		0[t]	(*53*)
34	Urd5′*P* → *N*Phe-Val-GlyOMe (I)	155		+					0		(*59*)
35	Urd5′*P* → *N*Thr-Gly-PheOMe (I)		1	20	80	0		13	20	10[o]	(*43*)
			0.1	30	0						
36	Urd5′*P* → *N*Thr-Gly-Phe-GlyOMe (I)		1	20	80	0		13	30	15[o]	(*43*)
			0.1	0	0						
37	Urd(5-*n*BuNH)-5′*P* → *N*PheOMe (I)	48	1	100		0			0		(*45*)
38	Urd(5-Br)-5′*P* → *N*PheOMe (I)	40	1	100		0		~11[u]	+[v]		(*45*)
			0.05	75[h]							
39	Urd(5Me)-5′*P* → *N*PheOMe (I)	24	1	100		0			0		(*45*)
40	Urd(5Morph)-5′*P* → *N*PheOMe (I)	42	1	100		0			0		(*45*)
41	Urd(5-OH)-5′*P* → *N*PheOMe (I)	17	1	100		Rearr. pH 10, 20°,2 hr[j]		14	0[o]		(*45*)
42	Urd(3′-*P*)-5′*P* → *N*Phe (I)		0.1	30	70						(*71*)
43	Urd(3′-*P*)-5′*P* → *N*PheOMe (I)	60	0.1	100	0	0			0		(*57, 71*)
44	Urd(2′:3′-*P*)-5′*P* → *N*PheOMe (I)	105	1	100		0			0		(*57, 71*)
45	Urd-3′:5′*P* → *N*PheOMe		0.1		25	0	0[k]	13		+[w]	(*75*)
46	Urd(2′:3′-*P* → *N*PheOMe)5′*P* → *N*PheOMe		0.01	100		3′(2′)*P*Urd5′*P* → *N*PheOMe formed		3′(2′)*P*Urd5′*P* → *N*Phe formed			(*57, 75*)

[a] Half-time, in minutes, for hydrolysis given under "acid" in 0.05 *N* HCl at 37°C.
[b] 2 hours.
[c] P-03′ bond.
[d] 6 hours.
[e] Products are Urd5′*P*(70%) and Urd5′*P*-*P*5′Urd(17%).
[f] 60% in *N* HCl.
[g] Urd(10%) and Urd5′*P*(90%).
[h] 1.5 hours.
[i] In Me_2 formamide, 70°, 5 hours.
[j] Forms 5-hydroxyuridine 5′:O⁵-cyclic phosphate.
[k] Also P-03′ bond.
[l] pH 13 is 0.1 *N* NaOH; pH 14 is 1 *N* NaOH.
[m] Ester bond, 90%.
[n] Internucleotide bond 80% (P-O5′ 40%, P-O3′ 60%).
[o] Ester bond, 100%.
[p] 40% UMP, 20% Urd5′*P* → *N*Ser.
[q] Ester bond, 26%.
[r] Forms Urd5′*P* → *N*Ser-Gly, 40%.
[s] Forms UMP 7%.
[t] Ester bond 55%.
[u] 0.5 *N* NH_4OH, 12 hours, 20°.
[v] Forms BrUrd(5′:*O*⁴-cyclic)*P*; ester bond cleaved only in 1 *N* NaOH.
[w] Also P-03′ bond.
[x] Internucleotide bond cleaved.
[y] Phosphoester bond.
[z] Internucleotide bond stable.

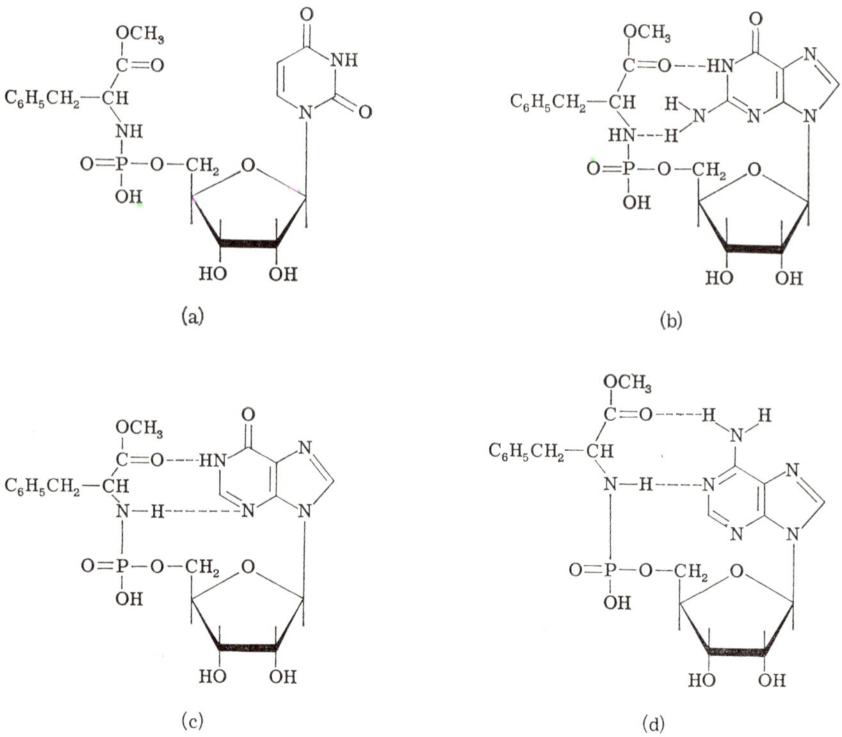

FIG. 5

nucleotide-peptides, however (structures b, c, and d), the barrier between the *anti*- and *syn*-conformation is compensated by formation of H-bonds.

Infrared and ultraviolet spectroscopic analyses provide some evidence confirming the structures suggested (*37*), and investigation of hydrolytic stability of nucleotide-(P-N)-peptides (*37*, *55*, *59*) gives a more detailed picture of their intramolecular interactions.

a. Acid Hydrolysis. In the amino acid derivative of guanosine 5′-phosphate (the "b" structure), the phosphoamide nitrogen is protonized under almost neutral conditions; this apparently accounts for its higher hydrolysis rate as compared to that of the uridine 5′-phosphate derivatives. On the other hand, inosinyl-(5′ → *N*)-phenylalanine and uridylyl-(5′ → *N*)-phenylalanine have approximately equal stability, as might be expected (see structures "a" and "c"). With a rise in temperature (*55*, *59*) or gradual increase in acidity (*37*, *55*, *59*), the P-N bonds in the phenylalanine derivatives of various nucleotides are hydrolyzed at similar rates, probably because of the decrease of the contribution of specific H-bonds.

The presence of other functional groups in the nucleotide-peptide molecule may cause changes in the properties of the P-N bond. For example, an additional phosphate group introduced into the 3′ (or 2′) position of the nucleotide fragment leads to some stabilization of the P-N bond, if the pH is lower than 2.5 (*57*). Under these conditions, the 5′-phosphoramidate and 3′(or 2′)-phosphate groups are supposed to be close to each other which could be due to the *gg*-conformation of the C5′-O5′ bond (*69*) and C2′-*endo* puckering of the furanose ring (*66*, *70*). At higher pH values this interaction does not take place because of the electrostatic repulsion of the dissociated phosphate groups.

If the nucleotide fragment is more complex, as in the case of 5′-phospho-polyuridylyl-(5′ → *N*)-phenylalanine, the stability of the P-N bond decreases, especially at pH values higher than 2.5 (*34*). The structure of the polymer derivative evidently not only precludes the P-N bond stabilizing interaction between the phosphoramidate fragment and the internucleotide phosphate group, but is certain to result in intramolecular protonization of the amide nitrogen by some proton-donating groups of the polynucleotide chain.

On the other hand, with the transition from a single amino acid to a peptide, the stability of the P-N bond increases. A study of the hydrolysis rates of the guanosine 5′-phosphate derivatives at different pH values and temperatures leads to the conclusion that steric hindrance of the protonization of phosphoramidate nitrogen is the decisive factor for the increased stability observed (*59*).

The stability of the P-N bond in the amino acid derivatives of 5′-UMP also increases on introducing in the uracil 5-position such substituents as bromo, morpholino, and *n*-butylamino groups (*45*). Since the electron influence of the substituent in position 5 cannot be transmitted to the 5′-phosphoramidate center along the chain, this result may also be considered as proof of such nucleotide conformation where the heterocyclic base is brought closer to the 5′-phosphoramidate unit so that the change in the base structure affects the reactivity of the P-N bond.

b. Intramolecular Catalysis of Acid Hydrolysis. Consideration of the hydrolytic stability of nucleotide-(5′ → *N*)-peptides (the structures of type I) indicates that the differences in stability of the P-N linkages are only quantitative. However, if the peptide fragment has some additional functional groups, an entirely new mode of nucleotide bond cleavage becomes possible. For example, (Fig. 6) nucleotidyl-(5′ → *N*)-amino acids with free carboxyls (*71*) are hydrolyzed by acid not only at the P-N linkage to form a nucleotide and an amino acid, but also at the phosphoester bond, yielding a nucleoside, inorganic phosphate, and the amino acid whereas acid hydrolysis of the same compounds but with blocked carboxyls, or of deoxythymidylyl-(3′ → *N*)-phenylalanine (*72*) and uridylyl-(5′ → *N*)-β-alanine (*56*), results only in cleavage of the P-N linkage (Fig. 7).

The above evidence leads to the conclusion that hydrolysis of the

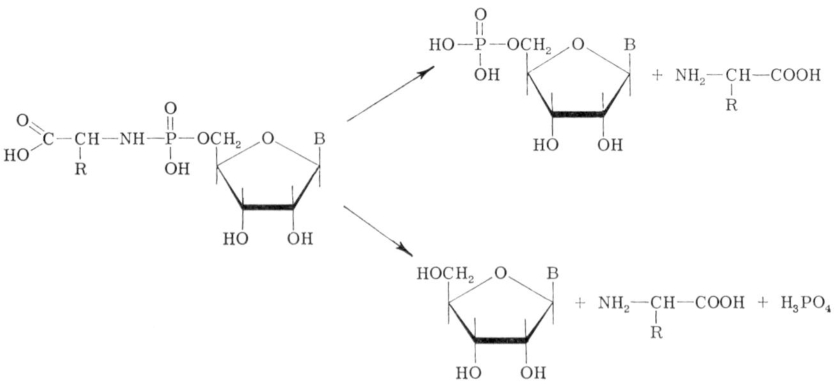

FIG. 6

phosphoester bond of the type I (Fig. 1) phosphoramidates bearing free carboxyls is due to the larger inductive effect of the COOH group as compared to the $COOCH_3$ group, resulting in reduced protonation of the amide nitrogen (*64*) and more active attack at the phosphate-bonded carbon atom. The absence of this reaction in the case of deoxythymidylyl-(3′ → *N*)-phenylalanine is probably accounted for by the nature of the C3′ atom in the sugar ring.

The direction of the cleavage of the nucleotide-(P-N)-peptide is determined to a great extent by the neighboring hydroxyl group. Nucleotidyl-(5′ → *N*)-serine esters (*44*) and those of nucleotidyl-(5′ → *N*)-seryl(or threonyl)-peptides (*43*) with the free hydroxyl group, on treatment with acid, hydrolyze to break not only the P-N bond, forming a nucleotide and a hydroxyamino acid (peptide), but also the phosphoester P-O5′ linkage to form a nucleoside and *O*-phosphoserine or *O*-phosphoseryl(or threonyl) peptides. In this case, one of the possible routes of hydrolysis of the P-O5′ linkage is by way of the cyclic phosphoramidate intermediate (Fig. 8).

Formation of *O*-phosphoserine or similar compounds on nucleophilic

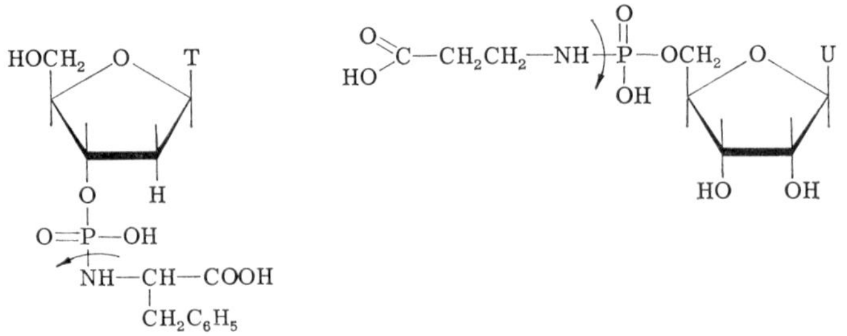

FIG. 7

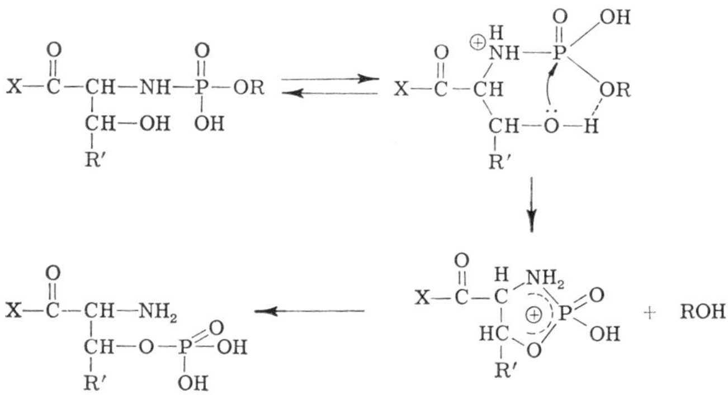

R = Nucleoside residue
R′ = H or CH_3
X = Alkoxy group or amino acid (or peptide) residue

FIG. 8

substitution at the phosphorus atom with participation of the serine (threonine) hydroxy group, testifies to the fact that it is the nucleoside that must leave. If the amino group is the leaving one, the intermediate will be nucleotidyl-(5′ → *O*)-serine ester (or that of the respective peptide). However, such compounds (see Section IV, B) are hydrolyzed in acid to form a nucleotide and a serine (peptide) ester—i.e., the cleavage of the serine-phosphorus linkage takes place; consequently, they cannot be the intermediates from which *O*-phosphoserine or similar compounds are formed.

This fact, however, does not exclude the amino group from also being the leaving one during the nucleophilic attack of the phosphorus atom in nucleotidyl-(5′ → *N*)-seryl(or threonyl)-peptides by the hydroxyamino acid hydroxyl. In this case, the nucleotide appears in the hydrolyzate not only as a result of the hydrolysis of the phosphoamide linkage (attack of the phosphorus atom by a molecule of water), but also as the result of the intramolecular nucleophilic substitution at the phosphorus atom.

However, the acid hydrolysis of uridylyl-(5′ → *N*)-ethanolamine occurs only at the P-N bond (no nucleoside or *O*-phosphoethanol amine were detected in the reaction mixture) (*56*). Apparently the nucleophilic substitution at the phosphorus atom with the participation of the neighboring hydroxyl leading to the cleavage of the P-O5′ linkage occurs only if facilitated by the inductive influence of other groups connected with nitrogen. It is quite possible that the conformation of the molecule is of importance for this process. For example, it is well known that, in aqueous solutions, serine favors a conformation with all the three functional groups close to each other (*73*).

c. Intramolecular Catalysis in Alkaline Hydrolysis. Intramolecular interaction of the carboxyl and hydroxyl groups with the phosphoramidate moiety is also observed in alkaline solution: uridylyl-(5′ → *N*)-serine under mild conditions is readily cleaved to uridine 5′-phosphate and serine. Such unusual lability of the phosphoamide linkage in alkali may be accounted for by intramolecular catalysis in which two groups—the dissociated carboxyl and nondissociated hydroxyl—take part simultaneously (Fig. 9).

In compounds with blocked hydroxyl (*44*), Urd5′*P* → *N*Ser(*O*Thp), or carboxyl (*43*), Urd5′*P* → *N*Ser(or Thr)-Gly, or with a distant hydroxyl group (*43*), Urd5′*P* → *N*Gly-Ser, the P-N linkage is stable in alkali. Furthermore, a neighboring carboxyl itself does not cause the alkaline destruction of the nucleotide-(P-N)-peptides (*44, 71*), whereas a neighboring hydroxyl leads to the cleavage of the nucleotide-peptide bond in the alkaline medium by way of the "a" route (*43*) (Fig. 10).

Thus, Urd5′*P* → *N*Ser(or Thr)-Gly, dThd5′*P* → *N*Ser-Gly, Urd5′*P* → *N*Thr-Gly-Phe and Urd5′*P* → *N*Thr-Gly-Phe-Gly are hydrolyzed to nucleosides (*43*). It is of interest that in case of the tri- and, particularly, tetrapeptide derivatives, the hydrolyzate contains also a nucleotide, which is an indication of the cleavage occurring not only at the P-O5′ linkage (route "a") but also the P-N linkage (route "b").

It is quite likely that the "a" and "b" mechanisms are also the result of intramolecular catalysis, i.e., the "a" route is the cleavage of the phosphoester bond (P-O5′) during the nucleophilic substitution at the phosphorus atom with the participation of the adjacent hydroxyl (intramolecular transesterification accompanied by the departure of the nucleoside); the "b" route is the cleavage of the P-N bond due to mutual catalytic action of the neighboring hydroxyl and the terminal carboxyl (the latter may come close to the phosphoramidate center in the tri- and tetrapeptide derivatives).

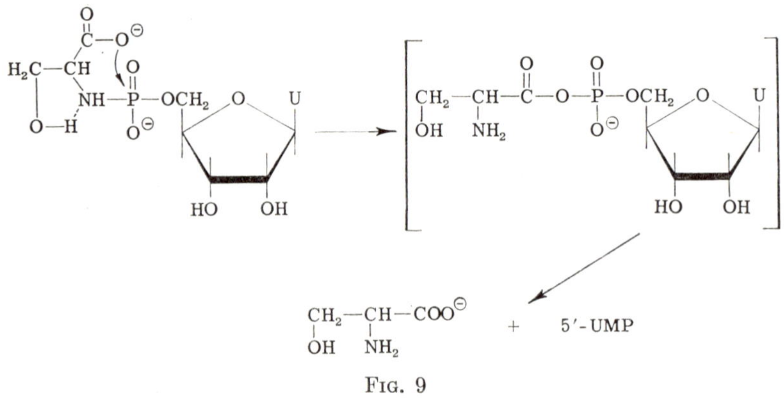

FIG. 9

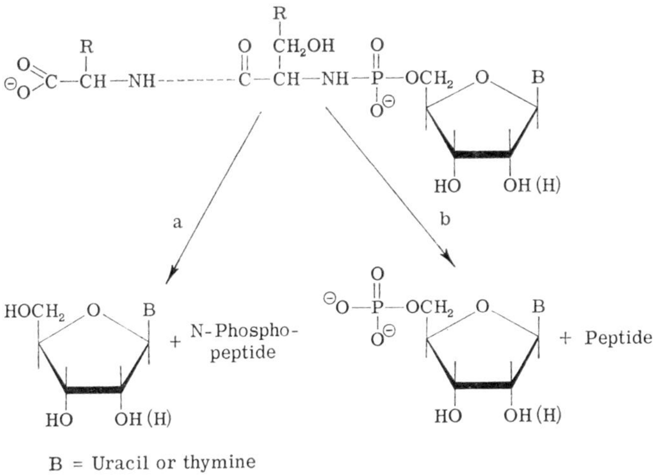

FIG. 10

This mode of cleavage differs from that of the above-mentioned uridylyl-(5′ → *N*)-serine (*44*); in the former, the phosphorus atom is attacked by the distant, not by the neighboring, carboxyl group.

Analogous intramolecular action of the favorably disposed hydroxyl group is clearly demonstrated in ribonucleoside 3′-phosphoamino acids where the position of the hydroxyl near the phosphoramidate group is fixed by the sugar ring. The amino acid derivatives of deoxythymidine 3′- and 5′-phosphates display practically the same behavior in aqueous solutions regardless of the position of the phosphoramidate residue: they are stable at pH > 3 (*39*). In contrast, the amino acid derivatives of ribonucleoside 3′-phosphates may exist under these conditions only when the 2′-hydroxyl group is blocked (*37*). The role of the 2′-hydroxyl is confirmed by the formation of a nucleoside 2′:3′-cyclic phosphate during the removal of the acetyl group under conditions in which the isolated P-N bond is stable (pH 10.0) (*37*) (Fig. 11).

The cyclic phosphate and the amino acid (and its ester) are accumulated in the reaction mixture in the equimolecular ratios. In this case, the simultaneous participation of the COOH group, formed during this alkaline treatment, cannot be excluded. However, the similar P-N bond cleavage in 5′-acetyl-2′-tetrahydropyranyl-cytidylyl-(3′ → *N*)-phenylalanine-OMe (*58*) after removal of the 2′-Thp group at pH 4.7 (where the $COOCH_3$ esteric bond is stable) is not in complete agreement with this mutual participation hypothesis.

These results imply that with the fixed position of the hydroxyl in relation to the phosphoramidate group, the nucleophilic attack of the

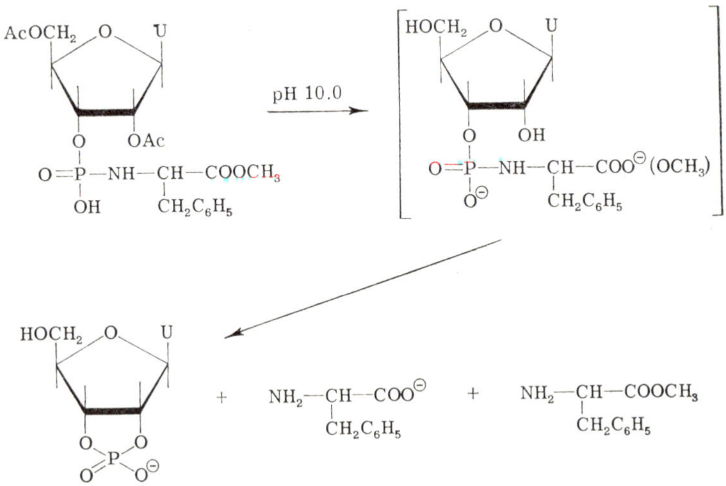

FIG. 11

former group may result in the departure of not only the nucleoside (in this case $3' \rightarrow 2'$ migration of the phosphoamino acid residue takes place), but also the amino acid. This testifies to the phosphoramidate being active enough to be converted into the diester of the phosphoric acid.

It is quite possible that the unsuccessful attempts to use nucleotide amidates for the synthesis of the internucleotide linkage may be due not to insufficient activity of the phosphorus, but to the fact that, in the course of the intermolecular interaction, the acidic catalysis (the amide nitrogen should be protonated) and the nucleophilic attack by the hydroxyl do not occur simultaneously. Intramolecularly, when the respective hydroxyl group is in a favorable position, it may play the role of both the general acid catalyst and of the nucleophilic agent.

The evidence obtained in the studies of synthetic nucleotide-peptides of types I and II (Fig. 1) allows us to draw the following conclusions about the behavior of natural nucleotide-peptides with phosphoamide bonds formed by the 5′- or 3′-phosphate of ribo- and deoxyribonucleotide (RNA- and DNA-peptides of types I and II):

1. The nucleotide-(P-N)-peptide phosphoamidate bond must be labile in acid and stable in neutral and alkaline solutions.

2. If the carboxyl group (the α-carboxyl of the amino acid forming the nucleotide-P-N-peptide bond) is close to the P-N bond, the latter should be cleaved in acid solution to form the N-terminal amino group and either a nucleotide or a nucleoside and inorganic phosphate. In deoxyribonucleotide-($3' \rightarrow N$)-peptides, if their phosphoamide centers should have

α-COOH groups, only P-N linkages are hydrolyzed in acid. All these compounds must be stable in alkaline solution.

3. Nucleotide-(P-N)-peptides in which the P-N bond is formed by an N-terminal seryl or threonyl will be cleaved by acid in two directions: to yield the terminal amino and phosphate groups of peptide and nucleotide (or polynucleotide), respectively, and to form *O*-phosphoseryl(or threonyl)-peptides and the corresponding nucleoside (or polynucleotide having no phosphate residue at the end previously attached to the peptide). Consequently the presence of *O*-phosphohydroxyamino acids should not be interpreted as unequivocal proof (*11, 12*) of the fact that, in the initial nucleotide-peptide, the fragments were linked through the hydroxyamino acid hydroxyl. On the contrary, as shown below in Section IV, B, the hydrolysis of such a linkage in acid is usually accompanied by the release of the serine (or threonine) hydroxyl group, the phosphorus remaining attached to the nucleoside. It is noteworthy that nucleotide-(P-N)-seryl(or threonyl)-peptides must be unstable in alkaline solution regardless of whether their nucleotide-peptide bond is of the phosphoamide type. The cleavage will occur at the phosphoester bond to form the respective *N*-phosphoseryl(or threonyl)-peptide and the corresponding nucleoside (or polynucleotide without the terminal phosphate).

4. When the nucleotide-(P-N)-peptide is formed with an amino acid bearing two free functional groups—the hydroxyl and the carboxyl—its P-N linkage is very easily cleaved in alkali. Lability of the P-N bond in such compounds is expected not only because of the presence of the free carboxyl in the hydroxyamino acid forming the P-N bond. The carboxy group may belong to any other amino acid (for example, to aspartic or glutamic acid) situated "inside" the peptide chain, if it is close to the P-N linkage as a result of the secondary structure.

5. Ribonucleotide-($3' \rightarrow N$)-peptides (RNA-peptides of type II) are also labile in alkaline and neutral media. The prolonged existence of such nucleotidepeptides *in vivo* is rather doubtful, unless the 2′-hydroxyl group of the nucleotide forming the P-N linkage is in some way blocked (e.g., methylated). The respective DNA-peptides should be stable in alkali, unless their N-terminal amino acids are serine or threonine.

2. Dinucleoside Monophosphate Phosphoramidates

One of the possible types of covalent linkage between the "residual" protein and the polynucleotide fragment in DNA-peptides is the phosphoamide bond formed by an internucleotide phosphorus atom and the amino group of the N-terminal amino acid of the peptide. This is confirmed by the ratio of the peptide and polynucleotide subunits (*74*), and by the stability of the natural DNA-peptides in alkaline solutions (*21*), etc.

The simplest models of such a "knot" of linkages are amidates of the mixed dinucleoside phosphates in which the internucleotide phosphorus is bound to an amino acid or a peptide, e.g., dThd3′*P*(→ 5′Ado) → *N*R, Fig. 12.

The study of the stability of the phosphoamide unit in such amidates reveals (*72*) that there are no considerable differences in the properties of phosphoamide linkages formed by an internucleotide phosphorus and those formed by the terminal 5′-phosphate residue of mononucleotides and polynucleotides (see Section IV, A, 1). The P-N bond in the dinucleoside phosphoramidates is hydrolyzed under acid conditions and is stable under neutral and alkaline conditions (*72*). But the properties of this bond are in some way affected by the phosphodiester structure of the dinucleoside phosphate.

This influence is revealed first of all in the fact that the dThd3′*P* → 5′Ado phosphoramidates are much more stable in acid than the mononucleotide analogs (see Table I). The greater stability of the internucleotide phosphoamide bond can be accounted for both by steric hindrance (the P-N center is localized between two sugar planes) and by greater $p_\pi - d_\pi$ overlap of nitrogen and phosphorus compared with derivatives formed by monoesterified phosphate. A similar increased stability of the P-N linkages has been also observed with the phenylalanine amidates of uridine 3′:5′-cyclic phosphate (*75*) and 5-methyluridine 5′-methylphosphate (*76*) in which the phosphoamide linkage is formed by a disubstituted phosphate.

The mode of cleavage of the deoxythymidylyl-(3′-5′)-adenosine phosphoramidates in acid (Fig. 13) depends upon the structure of the amidate residue (*72*). Derivatives of the simple amines (butylamine, benzylamine, and others) as well as the peptide derivatives (for example, dThd3′*P*-(→ 5′Ado) → *N*Ala-Phe) cleave under these conditions at the phosphoamide linkage to form the dinucleoside phosphate and amine or peptide, respec-

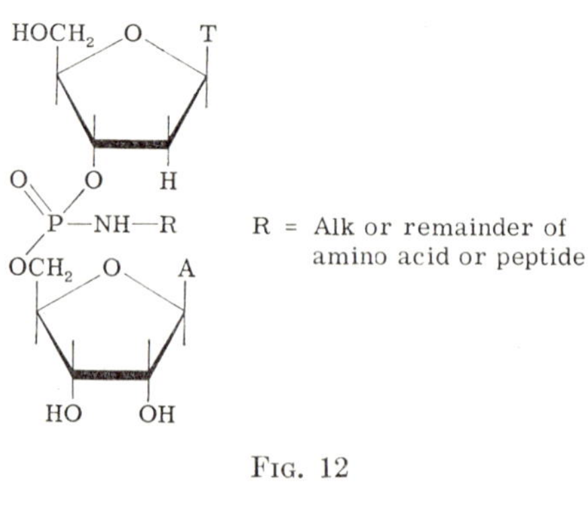

Fig. 12

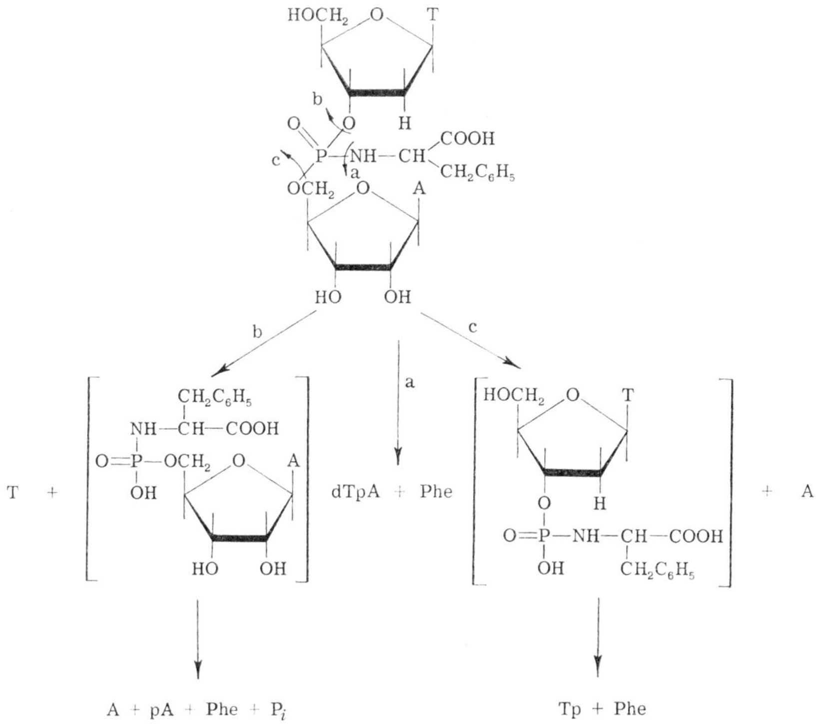

FIG. 13

tively, the structure of the radical in the amino component having almost no effect on the rate of hydrolysis. When the amino component in the amidate is an α-amino acid, the route of cleavage in acid solutions is determined by its carboxyl. If the carboxyl group is free, then, besides the phosphoamide bond cleavage (route "a"), the internucleotide linkage also breaks, the cleavage of the latter prevailing (90%). The internucleotide linkage is hydrolyzed in both directions with the release of the 3′- (route "b") and 5′- (route "c") hydroxyl groups at comparable rates (*72*).

This mechanism of cleavage was proved by the study of the behavior of the intermediate compounds: Ado5′$P \rightarrow N$Phe and dThd3′$P \rightarrow N$Phe. The adenylate derivative is hydrolyzed in acidic solutions to the nucleotide and nucleoside, as might be expected (see Section IV, A, 1), but the thymidylate compound gives only the nucleotide.

These results suggest that the "anomalous" behavior of dThd3′P-($\rightarrow$5′Ado) $\rightarrow N$Phe in acid (as compared to that of dThd3′P($\rightarrow$5′Ado) $\rightarrow$ N peptide compounds) is the result of both the strong inductive influence of the α-carboxyl on the amidate nitrogen and, possibly, of the catalytic

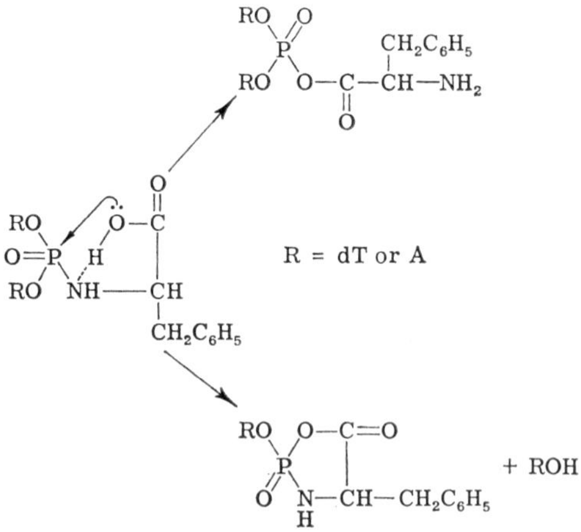

FIG. 14

participation of this carboxyl in the cleavage of the internucleotide center (Fig. 14).

The nucleophilic attack on the phosphorus atom by the carboxyl may lead to the formation of either the mixed dT-A amino acid anhydride (the amino group leaves) or the cyclic anhydride of the mononucleotide-(P-N)-amino acid (the nucleoside is the leaving group). Todd *et al.* (*77*) reported similar conversions of *N*-diphenyl and *N*-di-(*p*-nitrobenzyl)phosphoamino acids.

In alkaline solution, the phosphoamide bond in dThd3'P(→5'Ado) → N amidates is stable, as in mononucleotide derivatives, regardless of the structure of the amino component. However, their internucleotide linkage is much more labile, as compared to dThd3'P(→ 5'Ado) (*75*). The latter is

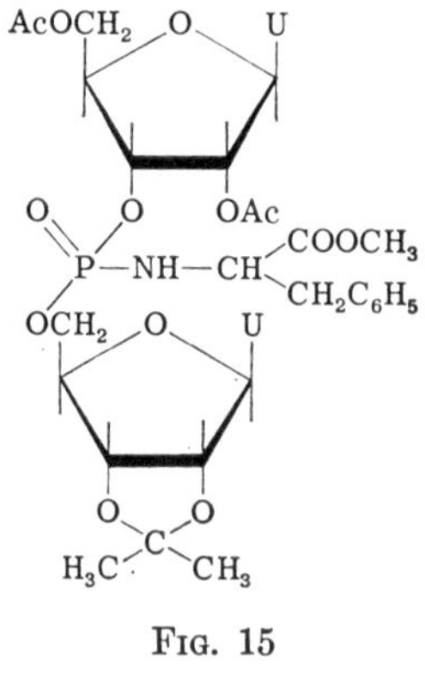

FIG. 15

known not to hydrolyze in 1 *N* NaOH (2 hours, 37°), but in its amidates under these conditions the phosphoester linkages are completely hydrolyzed, the P-O3′ bond cleavage being preferred.

The 2′-hydroxyl fixed near the internucleotide phosphorus (*78*) plays an important role in the course of alkaline hydrolysis of the internucleotide phosphoamide bond (*78*). When this hydroxyl is blocked, for example, by acetylation, the P-N linkage is stable under mild alkaline treatment; this was demonstrated with the phenylalanine amidate of the blocked UpU (Fig. 15).

But removal of the acetyl groups pH 10.5, 37°, 20 hours occasions a simultaneous cleavage of Urd3′*P*(→5′Urd)→*N*Phe. The mechanism of alkaline cleavage (shown in Fig. 16) was suggested by the structure and amounts of the hydrolysis products.

The cleavage of the phosphoamide linkage (route "a") prevails over the "b" type of cleavage.

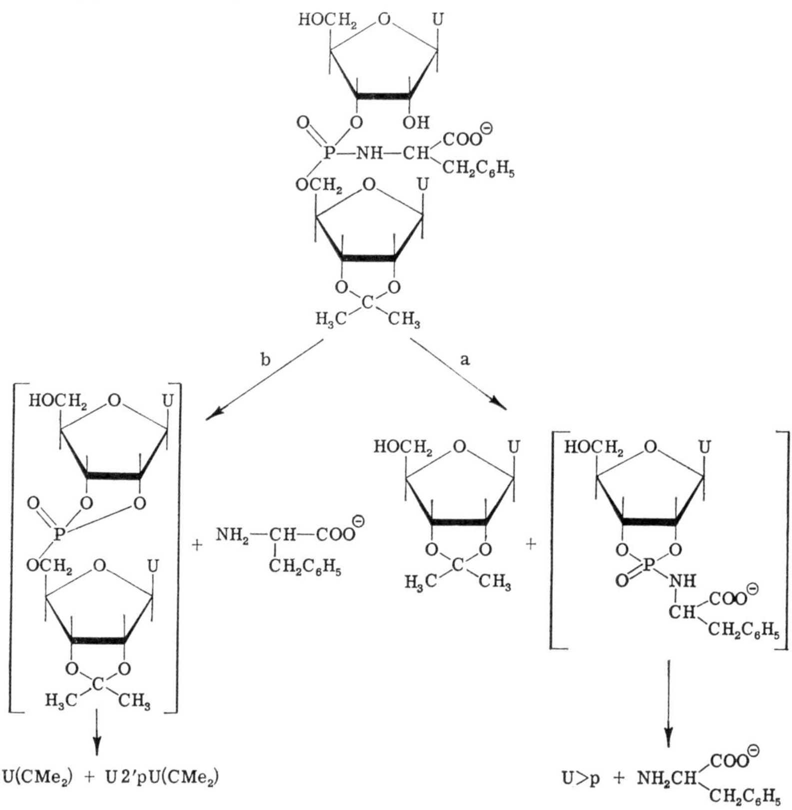

FIG. 16

It is obvious that the alkaline destruction of the phosphoamide center in this compound is not the result only of hydrolysis (i.e., of attack of the internucleotide phosphorus by a water molecule). The main cause of the breakage of the P-N linkage is intramolecular catalysis with the participation of the 2′-hydroxyl group, as compounds containing no 2′-hydroxyl, for example, dThd3′*P*(→5′Ado) → *N* amidates (*72*) and dThd3′*P*-(→5′dThd) → *N*Phe are stable under the similar conditions (50).

The 2′-hydroxyl group is the only proton-donating agent for the leaving amino group. It is possible that this group is also responsible for the nucleophilic attack of the phosphorus atom; however, it may also be suggested that the intermediary role is carried out by the carboxyl group (the mixed anhydride formed thereby is liable to 2′-hydroxyl attack).

The evidence obtained in the course of study of amides of dinucleoside phosphates that are derivatives of both deoxyribo- and ribonucleotides should be taken into account when isolating and determining the structure of native DNA- and RNA-peptides.

If a P-N linkage is formed by an internucleotide phosphorus of DNA (or any polydeoxynucleotide) and a peptide amino group, it will be more stable in acid than the same bond formed by the terminal phosphate residue of a polynucleotide, but the mechanism of cleavage of the phosphoamide center is the same: it is only the P-N bond that is broken. The possibility of intramolecular catalysis should be also considered, for example, with the participation of the carboxyl groups from aspartic and glutamic acids whose proportion in the DNA-peptides is quite appreciable (*11*). If the carboxyl group is brought close to the internucleotide P-N linkage as a result of secondary structure, cleavage at the P-N center is quite possible, the internucleotide linkage being the more likely one to break, causing depolymerization of the polynucleotide chain. A similar effect (DNA depolymeiization) takes place in the course of the alkaline hydrolysis of the DNA-peptides with the nucleotide-peptide bonds formed by the internucleotide phosphorus and the peptide amino group. Regardless of the length of the peptide fragment, the phosphoramidate will cleave on alkaline treatment with breakage of the internucleotide bonds that are stable, in DNA itself, under similar conditions.

RNA-peptides of similar structure must be much more labile in alkali. It is hard to deny the existence of such structures *in vivo*, but they are extremely difficult to isolate, especially when catalytic participation of peptide carboxyl groups is possible. The RNA polynucleotide chain is thus mainly preserved, but the process is accompanied by the isomerization of the internucleotide bond at the points of former contact with the peptide. Partial depolymerization of RNA due to the cleavage of the internucleotide bond in the P-N center is also not impossible. It can be suggested also that the meaning of the process of modification of 2′-hydroxyl in RNA (methylation) in cells is to exclude the 2′-hydroxyl group from the catalytic reactions described above.

B. Nucleotide-(5′→O)-seryl(or threonyl)-peptide Phosphoesters

In the Introduction, it was suggested (*11, 12*) that there may exist nucleotide-peptides in which the fragments are linked by way of the nucleotide phosphorus and the hydroxyl group of a hydroxyamino acid, to give a phosphoric ester bond.

W. F. Berkowitz and A. Bendich (*54*) showed, with deoxythymidylyl-(5′ → *O*)-serine, that in acid solution (0.5 *N* HCl, 160 min, 100°) the nucleotide-peptide linkage is cleaved in two directions to form deoxythymidine and *O*-phosphoserine (route "a") as well as deoxythymidine 5′-phosphate and serine (route "b"). The authors pointed out that under these conditions the initial substance is preserved and that the phosphate-serine linkage is more liable to hydrolyze than is the phosphate-thymidine bond.

In this connection, model nucleotide-(P-O)-peptide esters (compounds of type IV) with the phosphodiester bond between the nucleotide and the N-terminal hydroxyamino acid have been studied. Comparison of nucleotidyl-(5′ → *O*)-serine and nucleotidyl-(5′ → *O*)-seryl(or threonyl)-peptides (*53*) showed that the cleavage of the nucleoside-phosphate linkage and formation of *O*-phosphoserine (route "a") in these compounds depend on the neighboring (α)-carboxyl group. When the α-carboxyl is blocked, (for example, by a peptide linkage), the phosphodiester bond in acid is cleaved only by the "b" route. It is of importance that this acid cleavage of nucleotidyl-(5′ → *O*)-seryl(or threonyl)peptide esters is also extremely slow (*41, 53*). But, nucleotidyl-(5′ → *O*)-serines hydrolyzed at somewhat greater rate along the "a" and "b" routes. The possibility is not excluded that the liability to cleavage and its mode in the compounds with a free α-carboxyl are accounted for by the catalytic participation of this group. The possibility of steric hindrance in the hydrolysis of the peptide derivatives should not be ruled out either, especially if the amino group is also acylated.

Under alkaline conditions, nucleotidyl-(5′ → *O*)-serine and nucleotidyl-(5′ → *O*)-seryl(or threonyl)-peptide esters follow different routes of hydrolysis depending on whether their amino groups are blocked or free (*53*). Extensive β-elimination occurs in the compounds with acylated amino groups (such as a nucleotide-(5′ → *O*)-peptide with a serine located within the peptide chain) to form nucleoside 5′-phosphate. The rate of this elimination decreases with the increase in the length of the peptide fragment; in addition, the threonyl derivatives are more stable than those of serine.

R = Nucleoside
X = OH or amino acid residue

FIG. 17

As was to be expected, the serine derivatives with free amino groups (such as nucleotide-(5′ → *O*)-peptide esters formed by N-terminal serine) do not undergo β-elimination. But in these compounds, the phosphodiester linkage does not remain stable on alkaline hydrolysis: *O* → *N* migration of the nucleotide residue (Fig. 17).

On alkaline hydrolysis, uridylyl-(5′ → *O*)-serine gives uridylyl-(5′ → *N*)-serine, which decomposes under these conditions as was indicated above, (see Section IV, A, 1, c). Hydrolysis of uridylyl-(5′ → *O*)-serylglycine yields alkali-resistant uridylyl-(5′ → *N*)-serylglycine.

The evidence obtained with model nucleotidyl-(5′ → *O*)-serine compounds allows some suggestions about the behavior of the nucleotide-(P-O)-peptide ester bond in native nucleotide-peptides. Such compounds should be sufficiently stable in neutral and acid solution. But strong acid hydrolysis results in a cleavage of the P-O linkage between nucleotide and hydroxyamino acid. *O*-Phosphohydroxyamino acids require for their formation the proximity of the nucleotide-peptide bond and the carboxyl group. But even so, hydrolysis will almost unfailingly break the linkage between the hydroxyamino acid and the nucleotide phosphate; therefore, the rate of accumulation of *O*-phosphoserine and related compounds in the reaction mixture should be insignificant compared with the rate of formation of the terminal phosphate residue of the polynucleotide.

Under alkaline conditions, the properties of polynucleotide-(P-O)-seryl(or threonyl)-peptide esters should depend on the state of the amino group of the hydroxyamino acid. If this group is part of a peptide linkage, the polynucleotide chains in the compounds are easily eliminated by mild alkaline hydrolysis. Such processes are easily detected by accumulation of aminoacrylic acid derivatives in the hydrolyzate. If, in the native DNA-peptide, the nucleotide-peptide P-O linkage involves the hydroxyl of the N-terminal serine (or threonine), then *O* → *N* migration of the nucleotide-peptide linkage will occur in alkali. It is quite possible that even mild alkaline treatment of such a native biopolymer will convert it into a mixed biopolymer with the phosphoamide linkage. In any case, one should be very careful in drawing conclusions about the nature of the nucleotide-peptide linkage in native RNA- and DNA-peptides based on an analysis of the acid and alkaline hydrolysis products. In this connection, it should be noted that the *O*-phosphoserine that Bendich *et al.* (*11, 12*) found in the acid hydrolyzates of DNA-peptides is no unequivocal proof that in the native DNA-peptide the linkage is formed by the serine hydroxyl. Such a DNA-peptide is more likely to have the phosphoamide linkage formed by the N-terminal serine.

C. Specific Cleavage of Nucleotide-peptide Bonds

Nucleophilic substitution at the phosphorus atom in a nucleotide-(P-N)-peptide with participation of the carboxyl and hydroxyl groups indicates that such a reaction can occur in aqueous solution without hydrolysis of the phosphoamide bond. This suggests a method of phosphoamide cleavage by hydroxylamines at pH 4.5–5.0 (*79, 80*) (Fig. 18).

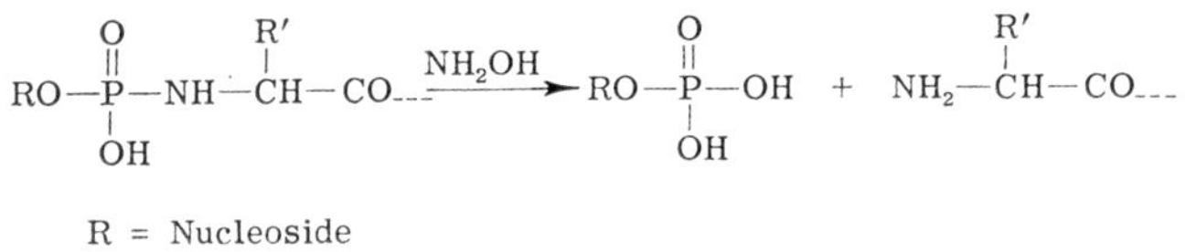

FIG. 18

The phosphoamide linkage is 60–70% cleaved within 10 minutes on treatment with 4 *M* hydroxylamine or *N*-methylhydroxylamine. A much slower (10–15%) rate of cleavage was recorded with *O*-methylhydroxylamine (under the same conditions). Under similar conditions, but without hydroxylamine, the P-N bond in a nucleotide-(P-N)-peptide is stable. As nucleotide-(P-O)-seryl(or threonyl)peptides are not cleaved under the same conditions with the strongest agent—*N*-methylhydroxylamine (*80*)—this method may be considered specific for the cleavage of the phosphoamide linkage. Hydroxylamine cleavage can be combined with a dinitrophenylation procedure (*79*) for the determination of the structure of the N-terminal amino acid participating in the formation of the phosphoamide linkage.

Thus, hydroxylamine may, apparently, be employed not only for the determination of the ester linkage in *O*-aminoacyl-tRNA and the anhydride linkage in aminoacyladenylates, but also for the determination of the phosphoamide bond in natural nucleotide-peptides. In the latter case, the reaction proceeds under mild acid conditions and quantitative determination should be made not of the hydroxamic acid, but of the amino group of the amino acid. There are grounds for believing that this method will be helpful in establishing the nature of the linkage in native DNA- and RNA-peptides where cleavage and migration of the nucleotide-peptide bonds are hardly possible.

D. Concluding Remarks

The general conclusion prompted by the study of the behavior of synthetic nucleotide-(P-N)-peptide phosphoramidates with different kinds of nucleotide-peptide linkage (i.e., compounds exemplified by structures I, II, III in Fig. 1) is that all of them are hydrolyzed by acid. The P-N linkage is usually so hydrolyzed, but sometimes the phosphorus-nucleoside linkage is also very labile. This is especially true for cases where the P-N linkage is close to a hydroxyl or carboxyl group. The phosphoramidate formed by internucleotide phosphorus having α-carboxyl groups nearby are hydrolyzed in acid almost exclusively only at the internucleotide linkage. In alkali, the stability of the phosphoamide linkage depends upon the structure of the compound (the presence of the neighboring proton-donating or nucleophilic groups). Isolated P-N linkages are extremely stable at pH 7.0–14.0, but are unstable in all pH ranges when close to proton-donating groups and nucleophilic groups (a combination of hydroxyl and carboxyl groups or just one hydroxyl group that carries out both these functions). Thus, the idea of the nucleotide-peptides of the phosphoramidate type being stable at alkaline pH values should be ruled out. The results obtained with the model compounds show that some nucleotide-(P-N)-peptide phosphoramidates are extremely labile not only in acid, but also in alkaline media.

Nucleotide-(P-O)-seryl(or threonyl)-peptide esters are more resistant to acid than the respective compounds with the P-N bond. In alkali, the phosphodiester nucleotide-peptide linkage is much more labile than the phosphoamide one; in this case, the linkage is noticeably influenced by neighboring nucleophilic groups (for example, the amino group). Under favorable conditions, $O \rightarrow N$ migration of the nucleotide-peptide linkage is quite possible. All these results throw some light on the catalytic mechanisms with participation of the neighboring groups, which may result in transformations of the nucleotide-peptides at the molecular level.

V. Nucleotide-(P-N)-peptide Phosphoramidates as Models of "Active Nucleotides"

A. Possible Models of Nucleotide Activation by Nucleotidyl Transferases

The evidence concerning the behavior of synthetic nucleotide-peptides (see Section IV) may also be of interest in the study of activation mechanisms of nucleotide residues in various biosynthetic processes. We know that the transfer of nucleotide residues begins with nucleoside 5′-triphosphates and finishes with formation of DNA or RNA of various molecular weights, or with nucleotide anhydrides (nucleotide coenzymes). Almost nothing is known about the stages of this pathway and the specific enzymes that take part in the conversions.

The mode of action of ligases (*81, 82*) may be quite relevant. Formation of the intermediate complex of ligase–adenylate with a covalent bond between AMP and the enzyme (*81*) testifies to the fact that the nucleotide is activated not only by way of the anhydride with the mineral acid, but also through a derivative with the functional group of the enzyme active site. Similar intermediate compounds (enzyme–phosphates) have been found in the transfer of orthophosphate (*24*, *83–88*) and pyrophosphate (*25*, *26*, *89*, *90*). According to the analytical results, serine and histidine play important roles in the activation and binding of phosphate groups (*91-93*).

No conclusions about the mechanisms of the conversion of nucleotides at the molecular level are possible without isolation and structural identification of intermediate enzyme–nucleotide complexes, but such an approach requires information on the possible ways of activation of the phosphate group by the functional groups of the enzymes and the influence of the neighboring functional groups. This kind of knowledge may facilitate the isolation of the enzyme–nucleotide intermediate (which must be extremely labile), the determination of the mode of linkage, and the establishment of the inactivation mechanism in this intermediate. Thus, the task is to choose suitable models and to study their properties.

With this in mind, nucleotide-(P-N)-peptide phosphoramidates deserve special attention. All that we know of their behavior testifies to the difference in the states of the phosphate residue depending on the structure of the nucleotide and peptide. The phosphoamide bond is very stable when there is no protonization of the amidate nitrogen. But under other conditions, when the P-N linkage has neighboring groups capable of donating their proton, the nitrogen linked to the phosphorus is immediately protonized and the electron-donating amidate group turns into an electron acceptor. As a result, the reactivity of the phosphate group increases and it readily reacts (*48*, *49*) not only with strong nucleophilic agents (phosphoric acid and its esters), but also with the hydroxyl group, a very weak nucleophilic agent. Intramolecular catalysis in nucleotide-(P-N)-peptide phosphoramidates, with participation of the two neighboring groups (one a proton-donor, the other the hydroxyl or the carboxyl group), may entail the transfer of the nucleotide phosphorus to the hydroxyl, to form, for example, *O*-phosphoserines, or to the carboxyl to form the mixed anhydride. The latter may undergo many kinds of conversions depending on what nucleophilic group is nearby. Hence it is not very difficult to imagine the migration of the nucleotide residue inside the active center of the enzyme. It is possible that similar phenomena are responsible for "aging" (inactivation) of the ligase–adenylate complex (*81*). Isolation of such complexes should be carried out taking into account other possibilities of migration of the nucleotide residue. Any conclusions about the nature of the covalent bond in the native enzyme–nucleotide intermediates should be made very carefully.

B. Ribonucleoside-5′-phosphoamidase

One of the possible approaches toward the elucidation of the biological functions of the nucleotide amidate derivatives is to study their enzymatic transformations and the properties and mechanisms of action of the enzymes that catalyze these reactions.

Attempts to find an enzyme affecting the phosphoamide bond in nucleotidyl-(5′ → *N*)-amino acids has resulted in discovery of a nucleoside-5′-phosphoamidase (*94*) in some animal tissues and in cell-free extracts of *Escherichia coli*. This enzyme hydrolyzes the phosphoamide bond to form the nucleotide and amino acid. The enzyme has been isolated from rabbit liver (*94*) by fractionation of liver extracts by ammonium sulfate with subsequent chromatography on Sephadex G-100. A pH of 5.0 is the optimum of the enzyme activity; Mg^{2+} and Mn^{2+} have no activating action. Tables II and III contain the relevant information on the action of the partially purified enzyme on various derivatives of phosphoric acid and nucleotidyl-amino acids (or peptides) (*95*).

The data listed in Tables II and III show that a number of mono- and diesters of phosphoric acid usually employed as substrates for phosphomono- and -diesterase are not hydrolyzed by the enzyme. Neither does it affect the phosphoamide bond in simple amides of phosphoric acid:

TABLE II
PHOSPHORIC ACID DERIVATIVES NOT HYDROLYZED BY PHOSPHOAMIDASE[a]

Phosphomonoesters	Phosphodiesters	Phosphoramidates
Glycerol-2-P	Bis-*p*-nitrophenyl-P	Creatine phosphate
D-Glucose-6-P	*p*-Nitrophenyl ester of pT	Phosphoimidazole
5′-AMP		*N*-Phospho-DL-Phe, Me ester
3′-AMP		
5′-UMP	*p*-Nitrophenyl ester of pU	
3′-UMP		*N*-Benzyl hydrogen phospho-DL-Phe
O-Phosphoserine	Methyl ester of pU	
	Benzyl ester of pU	
	d(TpT)	
	dTpU	

[a] The reaction mixture, containing 5 μmoles of substrate, 25 μmoles of acetate buffer (pH 5.0) and 54 μg of enzyme (specific activity 24 units[b]) in a total volume of 0.25 ml, was incubated for 1 hour at 37°.

[b] One unit of ribonucleoside-5′-phosphoamidase activity is defined as the quantity of enzyme necessary to hydrolyze 1 μmole of uridylyl-(5′ → *N*)-DL-phenylalanine, methylester for 10 minutes at 37°.

TABLE III
THE ACTION OF PHOSPHOAMIDASE ON NUCLEOTIDE-(P-N)-PEPTIDES[a]

Substrate	Hydrolysis, 10 min (moles)
Adenylyl-(5′ → *N*)NH_2	3.38
Adenylyl-(5′ → *N*)-DL-phenylalanine	1.20
Adenylyl-(5′ → *N*)-D-phenylalanine	1.65
Adenylyl-(5′ → *N*)-DL-leucine	1.05
Adenylyl-(5′ → *N*)-D-leucine	1.46
Adenylyl-(5′ → *N*)-L-leucine	0.49
Deoxyadenylyl-(5′ → *N*)-DL-phenylalanine	0.34
Uridylyl-(5′ → *N*)NH_2	2.80
Uridylyl-(5′ → *N*)-DL-phenylalanine	0.92
Uridylyl-(5′ → *N*)-D-phenylalanine	1.20
Uridylyl-(5′ → *N*)-DL-serine	0.70
Uridylyl-(5′ → *N*)-DL-phenylalanylglycine	0.12
Uridylyl-(5′ → *N*)-DL-phenylalanylvaline	0
2′:3′-P-uridylyl-(5′ → *N*)-DL-phenylalanine	0
Deoxyuridylyl-(5′ → *N*)-DL-phenylalanine	0
Polyuridylyl-(5′ → *N*)-DL-phenylalanine	0
Guanylyl-(5′ → *N*)NH_2	3.0
Guanylyl-(5′ → *N*)-DL-phenylalanine	1.12
Guanylyl-(5′ → *N*)-DL-phenylalanylglycine	0
Guanylyl-(5′ → *N*)-DL-phenylalanylvaline	0
Guanylyl-(5′ → *N*)-DL-phenylalanylvalylalanine	0
Cytidylyl-(5′ → *N*)NH_2	2.34
Deoxythymidylyl-(5′ → *N*)-DL-phenylalanine	0
Deoxythymidylyl-(3′ → *N*)-DL-phenylalanine	0
Deoxythymidylyl-3′(→ 5′-thymidine) → *N*-DL-phenylalanine	0
Deoxythymidylyl-3′(→ 5′-uridine) → *N*-DL-phenylalanine	0

[a] All amino acid and peptide derivatives of nucleotides were used as methyl esters. Content of the sample: 5 μmoles of substrate, 15 μmoles of acetate buffer, pH 5.0, 49 μg of enzyme (specific activity 20.4 units). Volume = 0.15 ml.

creatine phosphate, phosphoimidazole, and *N*-phospho-DL-phenylalanine. *N*-Benzyl hydrogen phosphophenylalanine used as a substrate differs from nucleotide-(P-N)-amino acids only in one detail: the nucleoside residue is substituted by the residue of another alcohol, e.g., benzyl alcohol. This substitution is sufficient to prevent the compound from being hydrolyzed. This means that the enzyme is specific for only one type of compound—nucleotidyl amidates.

The information in Table III shows that the rate of hydrolysis is also affected by the two parts of the nucleotidyl amino acid—the nucleoside and the amide fragments. If the amino acid is substituted by peptide and the mononucleotide by polynucleotide, the rate of hydrolysis decreases. Compounds in which the amino acid is bound to the internucleotide phosphorus are not hydrolyzed either. It is of interest that the enzyme is capable of hydrolyzing both L- and D-amino acid derivatives of nucleotides. Nucleotide 5′-amidates (ammonium derivatives) are the most readily hydrolyzed compounds.

The study of the relationship between the hydrolysis rate of uridylyl-(5′ → *N*)-amidates that have different structures and basicities showed (see Table IV) the rate of the enzymatic hydrolysis to increase as a rule with the increase in the basicity of the amine (*96*).

At the same time, the relationship between the rate of acid hydrolysis (see the $\tau/2$ values in Table IV) and the basicity is of a more complex nature. For the derivatives of amines having a $pK_a < 10$, the rate of hydrolysis also increases with growing basicity. But amides of strong amines, such as *n*-butylamine, isopropylamine, and cyclohexylamine, are hydrolyzed by acid at a very slow rate, comparable to that of the weak amine derivatives. The decrease in the hydrolysis rate of the P-N bond in the strong amine derivatives is, apparently, due to the decrease in the protonization rate of the amide nitrogen resulting from strong $p_\pi - d_\pi$ overlap. This fact suggests that the mechanism of enzymatic hydrolysis

TABLE IV
ENZYMATIC AND ACID HYDROLYSIS OF URIDYLYL-(5′ → *N*)-PHOSPHORAMIDATES BY PHOSPHOAMIDASE

Substrate	Enzymatic hydrolysis[a] 10 min, (moles)	$\tau/2$ 0.05 *N* HCl 37° (min)	pK_a of amine
UMP-pNHC$_6$H$_4$NO$_2$	0	Stable	1.0
UMP-mNHC$_6$H$_4$Cl	—	165	3.52
UMP-β-naphthylamidate	—	116	4.11
UMP-NHC$_6$H$_5$	0	96	4.58
UMP-pNHC$_6$H$_4$CH$_3$	0.19	66	5.07
UMP-pNHC$_6$H$_4$OCH$_3$	0.19	58	5.29
UMP-NHCH$_2$CF$_3$	1.3	38	5.7
UMP-NHCH(CH$_2$C$_6$H$_5$)COOCH$_3$	1.03	22	7.5
UMP-NHCH(CH$_2$C$_6$H$_5$)COOH	1.41	—	9.13
UMP-NHCH$_2$C$_6$H$_5$	1.60	8	9.34
UMP-NHCH(CH$_3$)CH$_2$C$_6$H$_5$	1.47	42	10.0
UMP-NHCH$_2$CH$_2$CH$_2$CH$_3$	1.61	78	10.43
UMP-NHCH(CH$_3$)$_2$	1.83	127	10.63
UMP-NHC$_6$H$_{11}$	1.73	144	10.64

[a] Content of the sample: 5 μmoles of substrate, 15 μmoles of acetate buffer, pH 5.0, 32 μg of enzyme (specific activity 32 units), total volume 0.15 ml.

of the phosphoamide bond includes a protonization stage. The difference in the rates of enzymatic and acid hydrolysis of the strong amine derivatives may be interpreted as a proof of formation of additional intermediate of enzymatic hydrolysis (phosphorylation of the enzyme by the nucleotide). It is possible that the rate of decomposition of this intermediate (the release of nucleotide) is slower than the rates for the preceding stages—protonization of phosphoramide and phosphorylation of the enzyme. As a result, the rates of formation of nucleotide from nucleotide-amidates formed by amines of different basicity are comparable. It should be also noted that the degree of enzymatic hydrolysis depends on the steric structure of the amine. At almost the same pK values of amines, the derivatives containing less bulky radicals are hydrolyzed at greater rates.

It may be concluded that the amine component does not determine the specificity of the enzyme but affects the rate of the enzymatic reaction: the more bulky its structure, the slower the rate of the reaction.

The specificity of the enzyme action (see Table III) is much more dependent on the structure of the sugar residue, e.g., introduction of an additional 3′-phosphate group or 2′:3′-isopropylidene group (*97*) and substitution of ribose by deoxyribose or β-D-arabinose (*97*) slows the enzymatic hydrolysis. The derivatives of nucleotidyl-(3′ → N)-amino acids are not hydrolyzed.

It is evident that polymer structures are not substrates for nucleoside-5′-phosphoamidase. The hydrolysis reaction found is evidently connected with some processes running at the mononucleotide level or, to be more exact, at the monoribonucleoside 5′-phosphate level.

There arises a question about the biological meaning of this reaction. On the one hand, the enzyme displays high specificity toward the type of linkage and the monoribonucleotide. It is reasonable to conclude that the only contribution of the other moiety of the molecule is to control the behavior of the phosphate residue. This casts doubt on the idea that the enzymatic hydrolysis of the P-N bond has any special significance. This reaction, is most likely the reflection of some biochemical processes occurring at the monoribonucleotide level. Any consideration of the nature of the processes involving participation of such types of structures should be based upon the chemical possibilities of this class of substances. From the abundant information in the literature (*98, 99*) and also from the evidence discussed in Sections IV and V, A, it is clear that the most characteristic feature of phosphoramides and their esters is their ability to participate in phosphorylation reactions. Nucleotide imidazolides have the greatest reactivity, and it is only because of their extremely high reactivity that they are not used in preparative chemistry. Free imidazole is known to catalyze phosphorylation reactions. This compound is the primary phosphate acceptor, and imidazolylphosphates are true phosphorylating agents. This readiness to accept and (or) transfer phosphate residues makes imidazolides good models as carriers of the phosphate or nucleotide residues. Similar transfer systems may occur in nature. Taking into consideration that the enzymes are responsible for the activity of imidazolides by means of intramolecular catalysis with participation of various groups, it may be expected that their role in transphosphorylation processes will be very effective.

The obvious question arises whether imidazole catalysis or that with other amino groups of the enzyme is indispensable in the reactions of nucleotide transfer. Here the function of phosphorylating agents is carried out by nucleoside 5′-triphosphates. At physiological pH values, the hydroxyls of the phosphate groups are ionized. The excess electrons of the negatively charged oxygen, on the one hand, decrease the reactivity of the phosphorus atom (due to conjugation of electrons of oxygen and phosphorus); but on the other hand, they electrostatically prevent the nucleophilic agent carrying the free electron pair from approaching the phosphorus atom. The negative charges of the triphosphate are partially shielded by formation of chelates with metals. But, if the formation of chelate were sufficient for the activation of pyrophosphate, the rate of hydrolysis of the pyrophosphate linkage would greatly increase at mild alkaline pH values and in the presence of metal ions (the nucleophilicity of the hydroxyl ion exceeds by many orders of magnitude that of the water molecule). But this is not the case. It is believed that the enzymes catalyzing nucleoside triphosphate reactions should carry out further chemical activation of the complex of metal and nucleoside triphosphate. Formation of the active phosphoramidate intermediate of phosphate (nucleotide) with the enzyme may be one of the forms of additional activation. The scheme of the reaction mechanism shown in Fig. 19 may be suggested.

The reaction takes place in two stages. The first results in formation of the phosphoramidate intermediate. A histidine residue localized in the active site of the enzyme is the primary acceptor of the nucleotide. The necessary cofactor of this stage of the reaction is the metal ion, which forms a chelate with the β- and γ-phosphorus in the nucleoside 5′-triphosphate and helps the inorganic pyrophosphate to leave as a strong acid anion, facilitating thereby the substitution at the α-phosphorus. The second stage is the transfer of the nucleotide residue to the acceptor. In all probability this reaction is connected with the activation of the P-N linkage carried out by the proton-donating group of the enzyme.

What happens if the enzyme system with the supposed mechanism of action accepts a synthetic nucleotidylamide, which may serve as a model of the enzyme–substrate complex? It is difficult to believe that amino acid

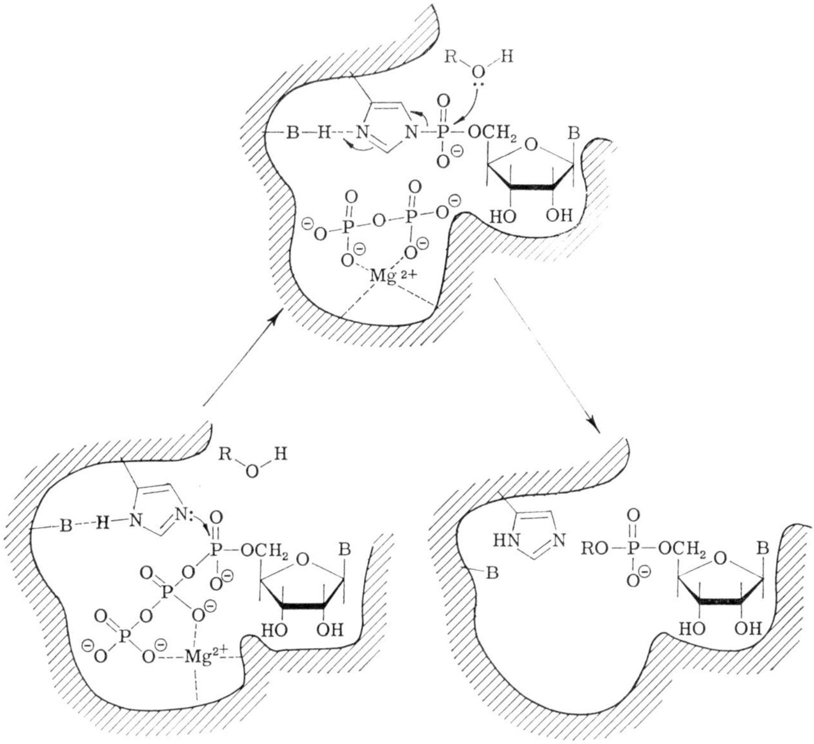

FIG. 19

and, especially, the peptide derivatives of nucleotides can substitute nucleoside triphosphate in the reaction because of, for example, steric hindrance.

But they would be able to approach the active site of the enzyme if the structure of the enzyme were changed, as for example by partial denaturation or dissociation into subunits. The enzyme specificity for the nucleoside in the triphosphate would probably predetermine its specificity for the nucleotide fragment of the nucleotide-amide molecule. As a result of the affinity of the nucleotide for the enzyme, the synthetic phosphoramidate would take a definite place in the active site of the enzyme. If a proton-donating group is in close proximity, the P-N linkage may be activated and the nucleotide residue transported to the enzyme active site. Then the nucleotide must be transferred to the acceptor, but the acceptor is absent, water plays this role, and hydrolysis of nucleotide-amide by means of the so-called specific phosphoamidase occurs. It is clear that the shorter and simpler the amine component in the phosphoramidate, the easier is its way to the active site of the enzyme and the greater is the reaction rate. This is in agreement with the experimental evidence (see Tables III and IV). The nature of the amino group in the nucleotide-amide is clearly of no importance (whether it is an amine or α-amino acid or α-amino peptide). The essential point is the type of linkage and the size of the amine moiety (the ability of this fragment to create steric hindrances).

If we take into consideration the specificity of the isolated phosphoamidase toward amides of the natural ribonucleotides, it may be concluded that it is connected with the metabolism of ribonucleic acids or of nucleotide coenzymes. Ribonucleoside-5′-phosphoamidase is of interest in that it may be used as a model for the study of the mechanism of action of the enzymes involved in nucleotide exchange.

VI. Conclusion

The aim of this review is not only to discuss what is known about the covalent linkage between the components of nucleic acids and proteins, although there is no denying that this evidence will contribute to the question of the nature and manner of linkage of the residual protein associated with DNA or RNA and of the role of this peptide fragment in the functioning of nucleic acids. Rather, the main task has been to draw the attention of molecular biologists to nucleotide-peptide models. Such synthetic nucleotide-peptides may prove very useful in elucidating the mode of action of enzymes of nucleotide exchange from the point of view of the activation of the phosphate residue of the mononucleotide, as well as from that of noncovalent interactions between the nucleotide and the active center of enzymes. Application of physical methods (for example, ORD, circular dichroism) for the study of noncovalent interactions in the molecules of model nucleotide-peptides with closely arranged amino acid radicals and heterocyclic bases may open new perspectives for investigation of the structural arrangement of nucleoproteins in ribosomes and viruses. Besides, synthetic nucleotide-(P-N)-peptides may be also useful for elaborating a model for the enzymatic synthesis of internucleotide linkages, as these compounds have some features of enzyme-substrate complexes. The steric closeness of such synthetic "enzyme-substrate complexes" (attained by using templates) may give rise to formation of new internucleotide linkages. One may hope that this principle may become a basic one for the self-assembly of oligonucleotides to form double-stranded polynucleotides.

Acknowledgments

I am deeply grateful to Academician M. A. Prokofiev, Drs. A. A. Bogdanov, R. K. Ledneva and N. N. Preobrazhenskaya for helpful discussions, suggestions, and criticism.

I wish to express my appreciation to Dr. N. N. Preobrazhenskaya for assistance in the course of preparing the manuscript and Mrs. T. Kheifits for the translation of this review.

References

1. A. A. Bogdanov, *Usp. Sovrem. Biol.* **55**, 321 (1963).
2. R. Bergkvist, *Acta Chem. Scand.* **12**, 364 (1958).
3. A. D. Brown, *Biochem. J.* **71**, 5P (1959).
4. G. Harris and A. Wiseman, *Biochim. Biophys. Acta* **55**, 929 (1962).
5. K. Ö. Persson, *Acta Chem. Scand.* **17**, 2750 (1963).
6. R. G. Hansen and E. Hageman, *Arch. Biochem. Biophys.* **62**, 511 (1956).
7. F. Binkley, *J. Am. Chem. Soc.* **81**, 1257 (1959); **82**, 987 (1960).
8. D. Wilken and R. Hansen, *J. Biol. Chem.* **236**, 1051 (1961).
9. M. F. Khanina, T. V. Wenkstern, and A. A. Baev, *Biokhimiya* **29**, 142 (1964).
10. E. Hase, S. Mihara, H. Otsuka, and H. Tamiya, *Biochim. Biophys. Acta* **32**, 298 (1959).
11. A. Bendich and H. S. Rosenkranz, This series **1**, 219 (1963).
12. A. Bendich, E. Borenfreund, G. C. Korngold, M. Krim, and M. E. Balis, *in* "Acidi Nucleici e Loro Funzione Biologica" (A. Baselli, ed.), p. 214. Istituto Lombardo, *Pavia*, 1964.
13. M. Champagne, A. Mazen, and J. Pouyet, *Biochim. Biophys. Acta* **87**, 682 (1964).
14. M. E. Balis, J. S. Salser, and A. Elder, *Nature* **203**, 1170 (1964).
15. A. A. Bogdanov, E. G. Antonovich, G. V. Terganova, and M. A. Prokofiev, *Dokl. Akad. Nauk SSSR* **150**, 1373 (1963).
16. G. V. Terganova, N. V. Birjukova, P. Kuliev, A. A. Bogdanov, and M. A. Prokofiev, *Dokl. Akad. Nauk SSSR* in press.
17. J. Potter and A. Dounce, *J. Am. Chem. Soc.* **78**, 3078 (1956).
18. A. A. Bogdanov, M. A. Prokofiev, E. G. Antonovich, G. V. Terganova, and V. M. Anisimova, *Biokhimiya* **27**, 266 (1962).
19. A. A. Bogdanov, E. G. Antonovich, G. V. Terganova, and M. A. Prokofiev, *Biokhimiya* **27**, 442 (1962).
20. A. A. Bogdanov, E. G. Antonovich, G. V. Terganova, and M. A. Prokofiev, *Biokhimiya* **27**, 1054 (1962).
21. Yu. F. Drigin, A. A. Bogdanov, and M. A. Prokofiev, *Khim. Prirod. Soed.* **1966**, 218.
22. J. Josse and J. Eigner, *Ann. Rev. Biochem.* **35**, 631 (1966).
23. A. Kornberg, *Advan. Enzymol.* **18**, 191 (1955).
24. R. E. Beyer, *Arch. Biochem. Biophys.* **127**, 43 (1968).
25. R. L. Switzer, *Biochem. Biophys. Res. Commun.* **32**, 320 (1968).
26. H. J. Evans and H. G. Wood, *Federation Proc.* **27**, 588 (1968).
27. B. Weiss and C. C. Richardson, *J. Biol. Chem.* **242**, 4270 (1967).
28. A. Becker, G. Lyn, M. Gefter, and J. Hurwitz, *Proc. Natl. Acad. Sci. U.S.* **58**, 1996 (1967).
29. H. G. Zachau and H. Feldmann, This series **4**, 217 (1965).
30. M. B. Hoagland, *in* "The Nucleic Acids" (E. Chargaff, J. N. Davidson, eds.), Vol. III, p. 349. Academic Press, New York, 1960.
31. T. S. Rjabova, R. N. Glebov, Z. A. Shabarova, and M. A. Prokofiev, *Dokl. Akad. Nauk SSSR* **153**, 363 (1963).
32. T. S. Rjabova, Z. A. Shabarova, and M. A. Prokofiev, *Biokhimiya* **30**, 235 (1965).

33. N. I. Sokolova, N. N. Preobrazhenskaya, Z. A. Shabarova, and M. A. Prokofiev, *Zh. Obshch. Khim* p. 257 (1963).
34. N. N. Preobrazhenskaya, N. I. Sokolova, Z. A. Shabarova, and M. A. Prokofiev, *Khim. Prirod. Sped.* **1965,** 342.
35. O. E. Vorobjov, Z. A. Shabarova, and M. A. Prokofiev, *Zh. Obshch. Khim.* **34,** 359 (1964).
36. O. E. Vorobjov, Z. A. Shabarova, and M. A. Prokofiev, *Dokl. Akad. Nauk SSSR* **158,** 143 (1964).
37. O. E. Vorobjov, Z. A. Shabarova, and M. A. Prokofiev, *Vestn. Mosk. Univ., Ser II: Khim.* **6,** 66 (1964).
38. T. S. Rjabova, Z. A. Shabarova, and M. A. Prokofiev, *Vestn. Mosk. Univ., Ser. II: Khim.* **3,** 89 (1965).
39. V. A. Bakanova, Z. A. Shabarova, and M. A. Prokofiev, *Khim. Prirod. Soed.* **1966,** 35.
40. N. I. Sokolova, R. K. Ledneva, P. P. Purigin, and Z. A. Shabarova, *Khim. Prirod. Soed.* **1967,** 290.
41. E. P. Saveliev, B. A. Juodka, Z. A. Shabarova, and M. A. Prokofiev, *Vestn. Mosk. Univ., Ser. II: Khim.* **5,** 128 (1967).
42. B. A. Juodka, Z. A. Shabarova, and M. A. Prokofiev, *Vestn. Mosk. Univ., Ser. II: Khim.* **5,** 104 (1968).
43. B. A. Juodka, I. V. Obruchnikov, V. K. Nedbye, Z. A. Shabarova, and M. A. Prokofiev, *Biokhimiya* **34,** 647 (1969).
44. E. P. Saveliev, E. S. Gromova, Z. A. Shabarova, and M. A. Prokofiev, *Izv. Akad. Nauk SSSR* in press.
45. N. G. Shinsky, N. N. Preobrazhenskaya, Z. A. Shabarova, and M. A. Prokofiev, *Zh. Obshch. Khim.* in press.
46. Z. A. Shabarova, L. G. Satarova, and M. A. Prokofiev, *Dokl. Akad. Nauk SSSR* **123,** 864 (1958).
47. Z. A. Shabarova, L. G. Andronova, M. Bezdeck, and M. A. Prokofiev, *Dokl. Akad. Nauk SSSR* **130,** 346 (1960).
48. L. G. Andronova, Z. A. Shabarova, T. S. Rjabova, and M. A. Prokofiev, *Zh. Obshch. Khim.* **31,** 3243 (1961).
49. Z. A. Shabarova, T. S. Rjabova, and M. A. Prokofiev, *Dokl. Akad. Nauk SSSR* **136,** 1116 (1961).
50. O. E. Vorobjov, N. I. Sokolova, V. I. Melnikova, Z. A. Shabarova, and M. A. Prokofiev, *Dokl. Akad. Nauk SSSR* **166,** 95 (1966).
51. N. I. Sokolova, V. I. Melnikova, Z. A. Shabarova, and M. A. Prokofiev, *Vestn. Mosk. Univ., Ser. II: Khim.* **5,** 119 (1966).
52. N. I. Sokolova, G. I. Gurova, Z. A. Shabarova, and M. A. Prokofiev, *Vestn. Mosk. Univ., Ser. II: Khim.* **5,** (1969).
53. B. A. Juodka, E. P. Saveliev, Z. A. Shabarova, and M. A. Prokofiev, *Biokhimiya* **33,** 907 (1968).
54. W. F. Berkowitz and A. Bendich, *Biochemistry* **4,** 1979 (1965).
55. N. I. Sokolova, L. G. Gatinskaya, Z. A. Shabarova, and M. A. Prokofiev, *Khim. Nauka i Promy.* **14,** 583 (1969).
56. E. P. Saveliev, Z. A. Shabarova, and M. A. Prokofiev, *Zh. Obshch. Khim.* in press.
57. A. B. Soloviova, N. N. Preobrazhenskaya, N. I. Sokolova, and Z. A. Shabarova, *Zh. Obshch. Khim.* **37,** 431 (1967).
58. O. E. Vorobjov, Dissertation, Mosc. Univ., 1965.

59. N. I. Sokolova, Z. A. Stumbravichute, P. P. Purigin, Z. A. Shabarova, and M. A. Prokofiev, *Dokl. Akad. Nauk SSSR* **174,** 722 (1967).
60. T. S. Rjabova and Z. A. Shabarova, *Vestn. Mosk. Univ., Ser II: Khim.* **1,** 114 (1966).
61. E. P. Saveliev, T. S. Rjabova, Z. A. Shabarova, and I. P. Beletzkaya, *Dokl. Akad. Nauk SSSR* **155,** 1457 (1964).
62. M. Halman, A. Lapidot, and D. Samuel, *J. Chem. Soc.* **1963,** 1299.
63. J. D. Chanley and E. Feageson, *J. Am. Chem. Soc.* **85,** 1181 (1963).
64. A. W. Garrison and Ch. E. Boozer, *J. Am. Chem. Soc.* **90,** 3486 (1968).
65. J. Kraut and L. H. Jensen, *Acta Cryst.* **16,** 79 (1963).
66. E. Shefter and K. N. Trueblood, *Acta Cryst.* **18,** 1067 (1965).
67. M. Sundaralingam, *Acta Cryst.* **21,** 495 (1966).
68. M. Sundaralingam and L. H. Jensen, *J. Mol. Biol.* **13,** 914 (1965).
69. M. Sundaralingam and L. H. Jensen, *J. Mol. Biol.* **13,** 930 (1965).
70. E. Shefter, M. Barlow, R. Sparks, and K. Trueblood, *J. Am. Chem. Soc.* **86,** 1872 (1964).
71. E. P. Saveliev, N. N. Preobrazhenskaya, Z. A. Shabarova, and M. A. Prokofiev, *Khim. Prirod. Soed.* **1967,** 121.
72. N. I. Sokolova, G. I. Gurova, L. G. Gatinskaya, Z. A. Shabarova, and M. A. Prokofiev, *Mol. Biol. (USSR)* **6** (1969).
73. H. Ogura, G. Arata, and Sh. Fujiwara, *J. Mol. Spectr.* **23,** 76 (1967).
74. J. S. Salser and M. E. Balis, *Biochim. Biophys. Acta* **149,** 220 (1967).
75. N. N. Preobrazhenskaya, Z. A. Shabarova, and M. A. Prokofiev, *Dokl. Akad. Nauk SSSR* **174,** 100 (1967).
76. N. G. Shinsky, N. N. Preobrazhenskaya, R. K. Ledneva, Z. A. Shabarova, and M. A. Prokofiev, *Izv. Akad. Nauk SSSR, Sez. Khim.* **1969,** 2707.
77. V. M. Clark, A. R. Macrae, J. F. P. Richter, and Lord Todd, *Tetrahedron, Suppl.* **7,** 307 (1966).
78. O. E. Vorobjov, Z. A. Shabarova, and M. A. Prokofiev, *Dokl. Akad. Nauk SSSR* in press.
79. T. S. Rjabova, Z. A. Shabarova, and M. A. Prokofiev, *Dokl. Akad. Nauk SSSR* **162,** 1068 (1965).
80. B. A. Juodka, V. K. Nedbye, Z. A. Shabarova, and M. A. Prokofiev, *Biokhimiya* **34,** No. 4 (1969).
81. B. Weiss, A. Thompson, and Ch. C. Richardson, *J. Biol. Chem.* **243,** 4556 (1968).
82. B. M. Olivera, Z. W. Hall, and I. R. Lehman, *Proc. Natl. Acad. Sci. U.S.*, **61,** 237 (1968).
83. V. A. Najjar and M. E. Pullman, *Science* **119,** 631 (1954).
84. J. H. Schwartz and F. Lipmann, *Proc. Natl. Acad. Sci. U.S.* **47,** 1996 (1961).
85. R. F. Ramaley, W. A. Bridger, R. W. Moyer, and P. D. Boyer, *J. Biol. Chem.* **42,** 4287 (1967).
86. L. Engstrom, *Arkiv Kemi* **19,** 129 (1962).
87. J. H. Schwartz, *Proc. Natl. Acad. Sci. U.S.* **49,** 871 (1963).
88. C. Lazdunski and M. Lazdunski, *Biochim. Biophys. Acta* **113,** 551 (1966).
89. R. A. Cooper and H. L. Kornberg, *Biochim. Biophys. Acta* **141,** 211 (1967).
90. R. A. Cooper and H. L. Kornberg, *Biochem. J.* **105,** 49C (1967).
91. P. D. Boyer, M. De Luca, K. E. Ebner, D. E. Hultquist, and J. B. Peter, *J. Biol. Chem.* **237,** PC 3306 (1962).
92. L. L. Bieber and P. D. Boyer, *J. Biol. Chem.* **241,** 5375 (1966).

93. J. R. Stevens-Clark, R. A. Conklin, A. Fujimoto, and R. A. Smith, *J. Biol. Chem.* **243,** 4474 (1968).

94. R. K. Ledneva, Z. A. Shabarova, and M. A. Prokofiev, *Dokl. Akad. Nauk SSSR* **157,** 473 (1964).

95. R. K. Ledneva, Z. A. Shabarova, and M. A. Prokofiev, *Dokl. Akad. Nauk SSSR* **172,** 977 (1967).

96. R. K. Ledneva, N. N. Preobrazhenskaya, Z. A. Shabarova, and M. A. Prokofiev, *Mol. Biol.* (*USSR*) in press.

97. S. M. Dudkin, R. K. Ledneva, Z. A. Shabarova, and M. A. Prokofiev, *Dokl. Akad. Nauk SSSR* in press.

98. H. G. Khorana, "Some Recent Developments in the Chemistry of Phosphate Esters of Biological Interest," Wiley, New York, 1962.

99. A. M. Michelson, "The Chemistry of Nucleosides and Nucleotides." Academic Press, New York, 1963.

The Crystal Structures of Purines, Pyrimidines and Their Intermolecular Complexes

DONALD VOET* AND
ALEXANDER RICH

*Department of Biology,
Massachusetts Institute
of Technology,
Cambridge, Massachusetts*

I. X-Ray Crystallography and the Nucleic Acids 183
II. The Experimental X-Ray Data 186
III. What Are the Molecular Structures? 196
A. The Angles between Chemical Bonds and Their Lengths . . 196
B. How Flat Are the Rings? 199
C. Where Are the Protons on the Bases? 203
D. Thymine and Uracil Derivatives Are Similar 205
IV. How Do the Purines and the Pyrimidines Interact? 205
A. Hydrogen Bonding between Bases: Theory and Selective Affinity 205
B. The Geometry of Hydrogen Bonding between Bases: Angles and Distances 213
C. Bases Form Hydrogen-Bonded Pairs with Themselves . . 220
D. Bases Hydrogen-Bond with Other Bases 229
V. Summary and Conclusions 243
Appendix I. Bond Distances 246
Appendix II. Bond Angles 252
Appendix III. Deviations from Planarity 258
References 260

I. X-Ray Crystallography and the Nucleic Acids

An important step in the development of molecular biology was the discovery of the two-stranded helical structure of DNA (*1*). This structure defined in a concrete way the manner in which polynucleotides carry out the information transfer reactions that are a central part of molecular biology. The structure of DNA itself illustrates characteristics of life: it has a highly specific organization and has the ability to replicate (*2*). The replicative process is carried out by means of specific hydrogen bonding or pairing of the planar purine and pyrimidine side chains of the poly-

* Present address: Department of Chemistry, University of Pennsylvania, Philadelphia, Pennsylvania.

nucleotide. An understanding of DNA or of nucleic acid chemistry in general is strongly dependent upon an understanding of the molecular structure and the interactions of nucleotides. This review concerns itself with a survey of the X-ray crystallographic investigations carried out on the biologically important purines and pyrimidines. We direct our attention toward the molecular geometry of these compounds and place special emphasis on the H-bonding potentialities of these molecules. In particular, the survey includes the structures of the crystalline intermolecular complexes formed by these materials. The H-bonding revealed in these studies frequently has a direct relevance to the role of H-bonding in biological systems.

Hydrogen bonding between purines and pyrimidines is widely distributed in biological systems. In addition to that found in double-helical DNA, it is believed that specific H-bonding is also found in the double-helical RNA in viruses. In both of these systems the formation of the double helix is one step in the biological replication of the molecule. Similar specificity is also encountered in the synthesis of RNA by RNA polymerase using DNA as a template. The system is similar to that synthesizing double-helical DNA except for the use of ribonucleotides to synthesize messenger RNA's as well as other RNA molecules.

A most interesting application of specific H-bonding is found in protein synthesis. At the present time it is generally believed that the specificity of codon–anticodon interactions is due to H-bonding between the purines and pyrimidines of messenger RNA and those of transfer RNA. However, the nature of the H-bonding in this system is not known. There are some types of ambiguity in the coding properties of transfer RNA that have been interpreted in terms of altered forms of H-bonding in the third nucleotide of the codon–anticodon complex (*3*).

Nucleotides are also widely used in biological systems as parts of small molecules. In particular, the nucleotide coenzymes play important roles in energy transfer reactions and in a variety of enzymatic processes. The nature of their interactions with the enzymes is not known at the present time, but it is likely that specificity of H-bonding is an important component of these systems.

Even though there have been tremendous advances in the field of molecular biology, we do not know a great deal more about the structure of DNA than was postulated by Watson and Crick in 1953. The reason for this lack of more detailed knowledge of the structure of DNA stems from the inapplicability of high-resolution single-crystal diffraction techniques to the study of the DNA molecule. The significant thing about DNA as well as all naturally occurring polynucleotides is the fact that the ordering of bases along the chain is not periodic. This is a consequence of the fact that they are information-containing structures. Therefore, the diffraction pattern from these molecules cannot provide information concerning the detailed molecular geometry of the purine–pyrimidine interactions. Oriented fibers of DNA have been extensively studied by X-ray diffraction methods (*4–9*). This work leads to direct information concerning the repeating part of the structure, the polynucleotide chain, but only indirectly provides information concerning the bases themselves.

Most of our knowledge concerning the conformation of nucleic acids and the interaction of the bases is based upon studies of model molecules related to nucleotides. The dependence on models is true for DNA as well as for the conformation of double-helical RNA (*10–14*) and of other polynucleotides (*15–17*). The interpretation of the structures of all of these molecules is strongly dependent upon the results of single-crystal analyses of these model substances. Recently a number of workers have crystallized transfer RNA (*18, 19*). It is most likely that the interpretation of the results of the structural studies of these substances will be greatly influenced by model studies. Model-building studies have already been attempted on the structure of transfer RNA as well as on codon–anticodon interactions (*20, 21*). Thus, it is likely that knowledge of the structure and interactions of nucleic acid components will remain indispensable to the nucleic acid chemist as it becomes possible to elucidate the detailed molecular structures of even larger units, such as ribosomes or viruses.

There have been some reviews of purine and pyrimidine structure determinations in the past (*17, 22–29*). However, most of these reviews have been more general in scope, and they have provided a less detailed analysis of purines and pyrimidines than is given here. The reviews that dealt fully with the structures of purines and pyrimidines were those of Pauling and Corey (*22*) and of Spencer (*24, 25*). However, since these reviews were written, the number of purine and pyrimidine structures determined has increased over tenfold and the power and accuracy of X-ray diffraction techniques has also greatly increased.

In this review, we are mainly concerned with the structures and interactions of purines and pyrimidines in the crystalline state. However, because of their close similarity to the pyrimidines, we also discuss the crystal structure of the barbiturates. The structure and conformation of the sugar and the phosphate components of nucleosides, nucleotides, and nucleic acids and their steric relationship to the purines and pyrimidines have been reviewed (*28–31*). We do not cover the structures of polynucleotides and nucleic acids as determined by fiber X-ray diffraction methods, as this subject also has been recently reviewed (*17*). However, we do point out the relevance of single-crystal analysis of purine–pyrimidine interactions to the fiber analysis of polynucleotides.

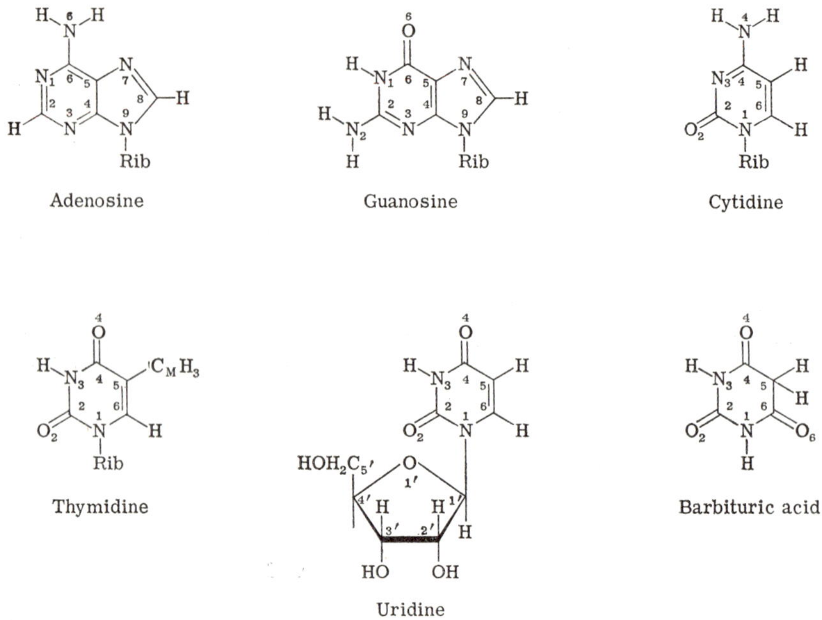

FIG. 1. The chemical structure and standard atom numbering system of adenosine, guanosine, cytidine, thymidine, uridine, and barbituric acid.

We preface this review with some general comments concerning the X-ray diffraction method. In particular we stress the problems of accuracy in structure determinations so that the noncrystallographically oriented reader will be able to assess the extent to which a structure determination is reliable. We present a complete tabulation of the molecular geometry of the reliable structures for which there is three-dimensional diffraction information. In the literature search, the catalogue of purine, pyrimidine and barbiturate structures prepared by Craven and Tamberg (*32*) was invaluable.

The chemical structures for adenosine, guanosine, cytidine, thymidine, uridine, and barbituric acid, together with the approved[1] atom numbering scheme, are shown in Fig. 1.

II. The Experimental X-Ray Data

The data surveyed in this review are the crystal structures listed in Table I. These include 134 structure determinations of various purines, pyrimidines, and barbiturates. Twenty-five of these are H-bonded inter-

[1] International Union of Pure and Applied Chemistry, Definitive Rules B-2. 11, Examples 17 and 24 [*J. Am. Chem. Soc.* **82**, 5545 (1960)].

molecular complexes. Table I also presents information related to the accuracy of the crystal structure determination. For convenience, the material is separated into sections dealing with derivatives of adenine, guanine, additional purines, cytosine, uracil, thymine, barbiturates,[2] and additional pyrimidines. There is some duplication of entries in Table I, since data concerning intermolecular complexes containing two types of derivatives are entered under the heading of each derivative.

In single-crystal X-ray diffraction work, a beam of monochromatic X-rays is scattered from the crystal by diffraction. The intensity of each diffracted beam is measured, but the relative phasing of the different diffracted beams cannot be measured directly. Thus the object of a structure determination can be considered to be the determination of the phase angles, ϕ_{hkl}, of a (mathematically) complex quantity known as the structure factor (F_{hkl}).[3] The structure factor of the crystallographic plane with Miller indices (h,k,l) can be expressed as:

$$\begin{aligned} F_{hkl} &= \sum_{j=1}^{N} f_j \exp[2\pi i(hx_j + ky_j + lz_j)] \\ &= |F_{hkl}| \exp(i\phi_{hkl}) \end{aligned} \tag{1}$$

where there are N atoms in a unit cell with positional coordinates given in fractions of a unit cell edge (x_j,y_j,z_j), f_j is the atomic scattering factor for atom j, and $i = \sqrt{-1}$. The amplitude of the structure factor, $|F_{hkl}|$, is easily determined from the intensity measurements since

$$I_{hkl} = K|F_{hkl}|^2 \tag{2}$$

where I_{hkl} is the measured intensity of the reflection from the (hkl) plane and K is a known quantity related to the geometry of the data collection technique and to the size and composition of the crystal. The phase angle, ϕ_{hkl}, is not measurable experimentally. Nevertheless, it must be determined in order to solve a crystal structure, because the electron density, ρ, at the point (x,y,z) in the unit cell can be expressed as:

[2] The trivial names for the barbiturate derivatives represent the following compounds: alloxan: 5,5-dihydroxybarbituric acid; alloxantin: 5-hydroxy-5-(5′-hydroxy-5′-barbituryl) barbituric acid; dilituric acid: 5-nitrobarbituric acid; violuric acid: 5-hydroxyimino-barbituric acid; dialuric acid: 5-hydroxybarbituric acid; Veronal: 5,5-diethylbarbituric acid; amytal: 5-ethyl-5-isoamylbarbituric acid; phenobarbital: 5-ethyl-5-phenylbarbituric acid.

[3] For a more comprehensive review of principles and practice in X-ray structure determinations, see, for example: K. C. Holmes and D. M. Blow, "The Use of X-Ray Diffraction in the Study of Protein and Nucleic Acid Structures." Wiley (Interscience), New York, 1965; G. H. Stout and L. H. Jensen, "X-Ray Structure Determination." Macmillan, New York, 1968.

TABLE I

Crystallographic Data

Crystal[d]	Space group	Z	Number unique reflections meas.	X-Ray Detector	R	σ (bond dist.) (Å)	σ (bond angle) (°)	H atoms located	Ref.
	A. Adenine Crystallographic Data								
Adenine·HCl[c]	$P2/c$	4	260	F	0.061	0.01	—	No	*(57a,b)*
$Cu(Ade)_2$	$P\bar{1}$	2	2882	C	0.075	—	—	No	*(51a)*
Aristeromycin·HBr	$P2_1$	—	—	—	—	—	—	No	*(51b)*
9MeAde·2HBr	$Pna2_1$	4	247	F	0.11	0.1	6	No	*(45a,c)*
dAdo·H_2O	$P2_1$	2	1400	F	0.078	0.01	1.2	Yes	*(36)*
9MeAde	$P2_1/c$	4	385	F	0.091	0.011	0.9	No	*(189)*
5′AMP	$P2_1$	2	1197	F	0.068	0.013	0.8	Yes	*(43)*
3′AMP·H_2O	$P2_1$	2	1411	F	0.043	0.006	0.3	Yes	*(44)*
Ado-3′:5′-P · $2H_2O$	$P2_12_12_1$	8[b]	800	F + C	0.13	—	—	No	*(49)*
Ado-2′-*P*-5′-Urd	$P2_1$	2	1786	F	0.065	0.012	0.7	Yes	*(50)*
9MeAde·1MeThy	$P2_1/m$	2[a]	1361	F	0.081	0.005	0.2	Yes	*(37a,b)*
9EtAde·1MeUra	$P\bar{1}$	2	2700	F	0.14	0.008	0.6	No	*(38)*
9EtAde·1Me5FUra	$P\bar{1}$	2	1264	F	0.178	—	—	No	*(48)*
9EtAde·1Me5BrUra	$P\bar{1}$	2	3064	F	0.127	0.025	1	No	*(46a,b)*
9MeAde·1Me5BrUra	$P\bar{1}$	2	1120	F	0.104	0.021	—	No	*(47)*
9Et2,6Am_2Pur·$(1MeThy)_2$·H_2O	$P\bar{1}$	2	2963	C	0.100	0.013	0.8	Yes	*(54b)*
9Et8BrAde·1Me5BrUra	$P2_1/c$	4	2720	C	0.086	0.02	1.8	No	*(53a)*
9EtAde·$(1Me5IUra)_2$	$P2_1/c$	4	2856	C	0.092	0.04	1.8	No	*(53b)*
9Et8BrAde·9Et8BrHyp	$P\bar{1}$	2	2148	C	0.091	0.02	1.2	No	*(53c)*
Ado·5BrUrd	$P22_12_1$	4	2511	F	0.138	0.03	2	No	*(39a,b)*
5′Br5′dAdo·Riboflavin·$3H_2O$	$P2_12_12_1$	4	1152	C	0.164	0.05	4	No	*(40)*
9EtAde·5iPr5BrAllylBarb.	$P\bar{1}$	2	1718	C	0.102	0.02	1	No	*(41)*
8Br9EtAde_2·Phenobarbital	$P\bar{1}$	2	3000	C	0.084	0.008	1.0	Yes	*(42)*
9Et2AmPur·1Me5FUra	$P2_1/c$	4	2211	F	0.11	0.01	0.8	Yes	*(52a,b)*
9Et2AmPur·1Me5BrUra	$P2_1/c$	4	2683	C	0.12	0.015	1	No	*(52b)*
9Et2,6Am_2Pur·$(1Me5FUra)_2$	$P\bar{1}$	2	1195	F	0.154	0.04	—	No	*(55)*
9Et2,6Am_2Pur·$(1Me5IUra)_2$	$P2_1/c$	4	2855	C	0.116	0.05	2.3	No	*(54a,b)*

Crystal[d]	Space group	Z	Number unique reflections meas.	X-Ray Detector	R	σ (bond dist.) (Å)	σ (bond angle) (°)	H atoms located	Ref.
	B. Guanine Crystallographic Data								
Guanine	$P2_1/n$	4	467	C	0.101	—	—	No	*(56)*
Guanosine	$P2_1$	4[b]	2967	C	0.036	—	—	Yes	*(56)*
9EtGua·1MeCyt	$P\bar{1}$	2	1976	F	0.112	0.007	0.4	Yes	*(58a,b)*
9EtGua·1Me5FCyt	$P\bar{1}$	2	1900	F	0.129	0.009	0.5	Yes	*(58b)*
dGuo·5BrdCyd	$P22_12_1$	4	1350	F	0.135	0.05	3	No	*(59a,b)*
9MeGua·1Me5BrCyt	$P2_1/c$	4	2200	F	0.20	—	—	No	*(60)*
9MeGua·HBr	$P2_1/c$	4	783	F	0.082	0.03	2	No	*(61)*
Gua·HCl·H_2O	$P2_1/c$	4	1600	F	0.073	0.06	0.3	Yes	*(62a,b)*
Gua·HCl·H_2O[c]	$P2_1/c$	4	340	F	0.17	0.04	—	No	*(63)*
	C. Additional Purine Crystallographic Data								
Purine	$Pna2_1$	4	568	F	0.070	0.006	0.5	Yes	*(64)*
Caffeine	$P2_1/a$	4	602	F	0.146	0.02	1.0	Yes	*(66)*
Caff.·5ClSalicylic Acid	Pbcn	8	2519	C	0.077	0.007	0.3	Yes	*(65)*
Theo.·5ClSalicylic Acid	$P2_1/c$	4	2338	C	0.087	0.008	0.4	Yes	*(67)*
Theophylline	$P2_1$	2	1051	F	0.113	0.01	0.5	Yes	*(68)*
Inosine	$P2_1$	4[b]	1431	C	0.032	—	—	Yes	*(56)*
Ino-5′-P [Na salt]	$C222_1$	8	1420	C	0.10	0.02	1.3	No	*(76)*
Ino-5′-P [Na_2 salt]	$C222_1$	8	2165	F	0.12	—	—	No	*(75)*
Ino-5′-P [Ba salt]	$P2_12_12_1$	8[b]	2819	F	0.12	—	—	No	*(75)*
1,3,7,9-Me_4Uric acid	$P2_1/c$	4	973	F	0.112	0.011	0.7	Yes	*(69)*
1,3,7,9-Me_4Uric (alt. form)	$P2_1/a$	4	—	F	—	—	—	No	*(70)*
1,3,7,9-Me_4Uric·Pyrene	Pc	2	525	F	0.174	0.035	2.2	No	*(71a,b)*
$(1,3,7,9\text{-}Me_4Uric)_2$·3,4BzPyrene	P1	1	740	F	0.185	—	—	No	*(72)*
1,3,7,9-Me_4Uric·Coronene	P1	1	742	F	0.19	—	—	No	*(73)*
Uric acid	$P2_1/a$	4	1030	F	0.066	0.005	0.4	Yes	*(74)*
6SIno	$P2_12_12_1$	8[b]	2729	C	0.067	0.009	0.4	Yes	*(77)*

C. Additional Purine Crystallographic Data (Continued)

9Et8BrHyp·9Et8BrAde	$P\bar{1}$	2	2148	C	0.091	0.02	1.2	No	(*53c*)
9EtHyp·5FUra	$P2_1/c$	4	1464	C	0.055	0.007	0.5	Yes	(*78*)
9βClEt7,8H_2(9*H*)imid-azo[2,1-*i*]PurMethiodide	$P2_1/c$	4	786	F	0.020	—	—	No	(*79*)
7*H*-tetrazolo[5,1-*i*]-Purine·H_2O	$Pna2_1$	4	758	F	0.090	0.009	0.6	Yes	(*80*)
2S6MePurine·H_2O	$P2_1/a$	4	320	F	0.15	—	—	No	(*81*)

D. Cytosine Crystallographic Data

Cytosine	$P2_12_12_1$	4	613	F	0.085	0.003	0.6	Yes	(*82*)
Cytosine·H_2O	$P2_1/c$	4	1280	F	0.11	0.004	0.4	Yes	(*83*)
1MeCytosine	$P\bar{1}$	2	830	F	0.115	—	—	No	(*87*)
Cytidine	$P2_12_12_1$	4	1195	F	0.080	0.007	0.5	Yes	(*86*)
Cyt-5-acetic acid	$P2_1/c$	4	1258	F	0.096	0.006	0.3	Yes	(*88*)
3′CMP (orthorhombic)	$P22_12_1$	4	1196	F	0.045	0.008	0.6	Yes	(*91*)
3′CMP (monoclinic)	$P2_1$	2	1332	F	0.073	0.009	1	Yes	(*92*)
1MeCyt·HBr	Pnma	8	349	F	0.12	0.04	2	No	(*45a,b*)
Cyt·Acridine·H_2O	Pbca	8	1037	C	0.112	—	—	No	(*89*)
bis-Cyt Cu(II) Cl	$P2_1/c$	4	558	F	0.15	—	—	No	(*90*)
1MeCyt·9EtGua	$P\bar{1}$	2	1350	F	0.112	0.007	0.4	Yes	(*58a,b*)
1Me5FCyt·9EtGua	$P\bar{1}$	2	1900	F	0.129	0.009	0.5	Yes	(*58b*)
1Me5BrCyt·9EtGua	$P2_1/c$	4	2200	F	0.20	—	—	No	(*60*)
5BrdCyd·dGuo	$P22_12_1$	4	1350	F	0.135	0.05	3	No	(*59a,b*)
Cyt·5FUra·H_2O	$P\bar{1}$	2	1189	F	0.16	0.01	1	No	(*84*)
1MeCyt·5FUra	Pbca	8	1500	C	0.094	0.01	1	No	(*85*)

E. Uracil and Thymine Crystallographic Data

Uracil	$P2_1/c$	4	1163	C	0.045	0.003	0.2	Yes	(*93a,b*)
1MeUra	Ibam	8[a]	517	F	0.11	0.008	0.6	No	(*94*)
5NO_2Ura·H_2O	$P2_1/c$	4	1210	F	0.062	0.006	0.4	Yes	(*95*)
Thymine·H_2O	$P2_1/c$	4	1068	F	0.078	0.01	0.5	Yes	(*104*)
1MeThy	$P2_1/c$	4	1166	F	0.072	0.004	0.2	Yes	(*105*)
5Et6MeUra	$P\bar{1}$	2	1368	F	0.080	0.01	1	Yes	(*107*)
2,4-S_2Uracil	$P2_1/c$	4	1020	C	0.057	0.009	0.5	Yes	(*102*)
2,4-Se_2Uracil	$P2_1/c$	4	649	F	0.136	—	—	No	(*103*)
5FdUrd	$P2_12_12_1$	4	1187	C	0.100	0.011	0.7	Yes	(*98*)
5BrUrd	$P2_1$	2	1221	F	0.070	0.02	1.0	No	(*97a,b*)

E. Uracil and Thymine Crystallographic Data (Continued)

5BrdUrd	$P2_1$	2	1312	F	0.074	0.02	1.0	Yes	(*97a,b*)
5IdUrd	P1	1	1434	C	0.054	0.028	1.9	No	(*99*)
Thymidine	$P2_12_12_1$	4	1203	C	0.08	—	—	Yes	(*114*)
S-S(5SdUrd)$_2$	$P2_12_12_1$	4	2328	C	0.054	0.008	0.4	Yes	(*115*)
4SUrd	C2	4	1400	C	0.10	—	—	No	(*116*)
S-S(4SUrd)$_2$	$P2_1$	2	—	C	0.079	—	—	Yes	(*117*)
1MeUra·HBr	Pnma	8	226	F	0.17	0.05	2	No	(*118*)
Urd-3′:5′-P	$P2_1$	2	2405	C	0.08	0.03	2	No	(*119*)
Ado-2′-*P*-5′-Urd	$P2_1$	2	1786	F	0.088	0.12	0.7	Yes	(*50*)
Urd-5′-P [Ba salt]	$C222_1$	8	2502	F	0.098	0.03	2	No	(*96*)
dThd-5′-P[Ca salt]	$P2_1$	2	1575	F	0.118	0.023	1.4	Yes	(*106*)
5FOro·H_2O [Rb salt]	Pbca	8	1155	C	0.14	0.05	3.6	No	(*100*)
H_2Ura	$P2_1/c$	4	715	C	0.10	—	—	No	(*101*)
Photodimers of									
1MeThy(*trans-anti*)	$P2_1/c$	2[a]	1048	C	0.05	0.004	0.2	Yes	(*108*)
Thy(*trans-anti*)	$P2_1/c$	2[a]	833	C	0.048	0.003	0.2	Yes	(*109a,b*)
1,3Me_2Thy(*cis-syn*)	$P2_1/c$	4	—	C	0.113	—	—	No	(*110*)
1,3Me_2Thy(*cis-anti*)	Cc	4	1308	C	0.035	0.003	0.2	Yes	(*111*)
Ura(*cis-syn*)	$P2_1/c$	4	2135	C	0.045	0.0025	—	Yes	(*112*)
5OH-6,4′(5′MePyri-mid-2′-one)H_2Thd	$P\bar{1}$	2	1844	F	0.102	—	—	No	(*113*)
1MeThy·9MeAde	$P2_1/m$	2[a]	1361	F	0.081	0.005	0.2	Yes	(*37a,b*)
1MeUra·9EtAde	$P\bar{1}$	2	2700	F	0.14	0.008	0.6	No	(*38*)
1Me5FUra·9EtAde	$P\bar{1}$	2	1264	F	0.178	—	—	No	(*48*)
1Me5BrUra·9EtAde	$P\bar{1}$	2	3064	F	0.127	0.125	1	No	(*46a,b*)
1Me5BrUra·9MeAde	$P\bar{1}$	2	1120	F	0.104	0.021	—	No	(*47*)
1Me5BrUra·9Et8BrAde	$P2_1/c$	4	2720	C	0.086	0.02	1.8	No	(*53a*)
(1Me5IUra)$_2$·9EtAde	$P2_1/c$	4	2856	C	0.092	0.04	1.8	No	(*53b*)
5BrUrd·Ado	$P22_12_1$	4	2511	F	0.138	0.03	2	No	(*39a,b*)
1Me5FUra·9Et2AmPur	$P2_1/c$	4	2211	F	0.11	0.01	0.8	Yes	(*52a,b*)
1Me5BrUra·9Et2AmPur	$P2_1/c$	4	2683	C	0.12	0.015	1	No	(*52b*)
(1Me5FUra)$_2$·9Et2,6Am_2Pur	$P\bar{1}$	2	1195	F	0.154	—	—	No	(*55*)
(1Me5IUra)$_2$·9Et2,6Am_2Pur	$P2_1/c$	4	2855	C	0.116	0.05	2.3	No	(*54a,b*)
5FUra·Cyt·H_2O	$P\bar{1}$	2	1189	F	0.16	0.01	1	No	(*84*)
5FUra·1MeCyt	Pbca	8	1500	C	0.095	0.01	1	No	(*85*)
5FUra·9EtHyp	$P2_1/c$	4	1464	C	0.055	0.007	0.5	Yes	(*78*)
(1MeThy)$_2$·9Et2,6Am_2Pur·H_2O	$P\bar{1}$	2	2963	C	0.100	0.013	0.8	Yes	(*54b*)

TABLE I (*Continued*)

Crystal[d]	Space group	Z	Number unique reflections meas.	X-Ray Detector	R	σ (bond dist.) (Å)	σ (bond angle) (°)	H atoms located	Ref.
			F. Barbiturate Crystallographic Data						
Barbituric acid·$2H_2O$	Pnma	8	556	F	0.14	0.010	0.7	Yes	(*120*)
Barbituric acid	$P2_1/c$	4	770	F	0.102	0.007	0.5	Yes	(*121*)
Barb. [NH_4 salt]	$P2_1/c$	4	1202	F + C	0.150	0.011	0.8	Yes	(*122*)
Alloxan·$3H_2O$	C2/m	4[a]	858	F	0.097	0.006	0.5	Yes	(*123*)
5-Oxobarbituric acid	$P4_12_12$	4[a]	323	F	0.093	0.007	—	Yes	(*124*)
Alloxan	$P\bar{1}$	2	1173	F + C	0.087	0.005	0.3	Yes	(*125*)
Alloxantin·$2H_2O$	$P\bar{1}$	1[a]	200	F	0.034	0.014	1.0	Yes	(*126*)
Dilituric acid	$P2_1/c$	4	1080	F	0.093	0.005	0.3	Yes	(*127*)
Dilituric acid·$3H_2O$	$P2_1/c$	8[b]	1800	F	0.129	0.01	0.3	No	(*128*)
Violuric acid·H_2O	$Cmc2_1$	4[a]	397	C	0.059	0.008	0.6	Yes	(*129*)
Veronal I	$R\bar{3}$	18	1617	C	0.041	0.003	0.2	Yes	(*130*)
Veronal II	C2/c	4[a]	840	C	0.062	0.003	0.3	Yes	(*130*)
Amytal I	C2/c	8	1861	C	0.096	0.004	0.3	Yes	(*130*)
Amytal II	$P2_1/c$	8[b]	4139	C	0.072	0.004	0.3	Yes	(*130*)
5-(6′Br3′Et2′MeBenz-imidazolium)Barb.	$P2_1/c$	8[b]	4804	F	0.173	0.02	2	No	(*131*)
Dialuric·H_2O	$P2_1/n$	4	1250	F	0.080	—	—	Yes	(*132*)
Veronal [K salt]	$P2_12_12_1$	4	1189	F	0.242	0.05	5	No	(*133*)
2-S-Barb.	$P2_1/c$	8[b]	—	—	0.13	—	—	No	(*134*)
Phenobarb. · 8Br9EtAde$_2$	$P\bar{1}$	2	3000	C	0.084	—	—	Yes	(*42*)
5iPr5BrAllylBarb. · 9EtAde	$P\bar{1}$	2	1718	C	0.102	0.02	1	No	(*41*)

Crystal[d]	Space group	Z	Number unique reflections meas.	X-Ray Detector	R	σ (bond dist.) (Å)	σ (bond angle) (°)	H atoms located	Ref.
			G. Additional Pyrimidine Crystallographic Data						
Pyrimidine	$Pna2_1$	4	374	F	0.087	0.008	0.5	Yes	(*135*)
2Am4Me6ClPyr	$P2_1/a$	4	—	F	0.24	0.02	—	No	(*136*)
4Am2,6Cl$_2$Pyr	$P2_1/a$	4	—	F	0.17	0.02	1.2	No	(*137*)
5Br4,6Am$_2$Pyr	$P2_1/a$	4	—	F	0.11	0.04	—	No	(*137*)
2(4′Am5′AmPyrimidyl) 2-penten-4-one	$P2_1/c$	4	1005	F	0.128	0.01	0.8	No	(*138a,b*)
Isocytosine	$P2_1/n$	8[b]	1828	F	0.016	0.003	0.1	Yes	(*139a,b*)
Hydrolyzed Thiamine-PP	$P\bar{1}$	2	2398	F	0.13	0.011	0.6	Yes	(*140*)
Thiamine·HCl	$P2_1/c$	4	3039	C	0.080	0.008	0.5	Yes	(*141*)
Thiamine-PP	$P2_1/c$	4	—	F	0.12	0.015	—	No	(*142*)
Na$_2$(OH)$_4$pyrimido-[5,4-*d*]pyrimidine	$P2_1/c$	2[a]	439	F	0.075	0.012	1.4	Yes	(*143*)
1,3,10-Me$_3$Isoalloxazonium Iodide	$P2_1/n$	4	932	C	0.054	—	—	No	(*144*)
Riboflavin·HBr·H_2O	$P2_12_12_1$	4	2267	F	0.09	0.02	1.5	Yes	(*145*)
Ribofl.·5′Br5′dAdo·$3H_2O$	$P2_12_12_1$	4	1152	C	0.164	0.05	4	No	(*40*)
10MeIsoalloxazine·HBr·$2H_2O$	$P2_1/c$	4	1686	C	0.048	0.007	0.8	Yes	(*146a,b*)
8-Azaguanine·H_2O	$P2_1/c$	4	2952	C	0.053	0.0018	0.12	Yes	(*147a-c*)
Xanthazol·H_2O	$P\bar{1}$	2	4000	F	0.14	0.011	0.8	Yes	(*148*)

[a] There is only half a molecule per asymmetric unit in the unit cell.

[b] There are two independent molecules per asymmetric unit in the unit cell.

[c] This structure was determined in two dimensions only.

[d] The following symbols, most of them recommended by IUPAC and IUB, are used in the Tables and Appendices in this article [cf. *J. Biol. Chem.* **241**, 527 (1966); **241**, 2491 (1966); and prefatory note in this volume, p. ix].

Ade	adenine	Ado	adenosine	P or p	phosphate	Barb.	barbituric acid	Oro	orotic acid
Gua	guanine	Guo	guanosine	Me	Methyl	Caff.	caffeine	Uric	uric acid
Hyp	hypoxanthine	Ino	inosine	Et	ethyl	Theo.	theophylline		
Cyt	cytosine	Cyd	cytidine	iPr	isopropyl	Ribofl.	riboflavin		
Ura	uracil	Urd	uridine	Am	amino	Phenobarb.	phenobarbital		
Thy	thymine	Thd	ribothymidine	Bz	benz-	Pyr	pyrimidine		
		d	deoxy			Pur	purine		

$$\rho(x,y,z) = \sum_{h,k,l} |F_{hkl}| \exp[-2\pi i(hx + ky + lz)] \exp[i\phi_{hkl}] \quad (3)$$

In this equation, the ϕ_{hkl}'s are the only unknown quantities, except for $\rho(x,y,z)$. Therefore, in crystal structure determinations, it is necessary to determine these quantities. The values of the phase angles are usually deduced by means of a variety of different crystallographic techniques, including some that utilize phase information that is implicity contained in the distribution of intensities. A structure is considered to be solved when the phases of its structure factors have been determined, since we can then calculate the electron density map, the peaks of which represent atomic positions.

Many factors influence the accuracy of a crystal structure determination; however, only a few of these are important for most crystals. Most important are the number and nature of the unique atoms whose positions must be determined, the number of unique reflections whose intensities have been measured, and the precision with which these measurements are made. In theory, there are an infinite number of reflections that exist for a perfect single crystal. However, only a finite number of these reflections are accessible for measurement, this number varying inversely with the wavelength of the X-rays. The intensity of reflections falls off quite rapidly with increasing resolution (i.e., shorter spacings) of the reflections. CuKα X-radiation (λ = 1.54 Å), which is used in the majority of crystallographic measurements, is usually short enough to measure all the reflections that are significantly above the level of background radiation. This wavelength is of the same order of magnitude as the length of chemical bonds. The number of reflections accessible to intensity measurement for a particular X-ray wavelength varies directly with the volume of the unit cell of the crystal, which in turn is roughly proportional to the number of atoms contained in the unit cell.

The unit cell is the repeating unit out of which a crystal is built. Most unit cells have internal symmetry that consists of a combination of point and translational symmetries. The nature of this symmetry is expressed by the space-group symbol (*33*), which is given for each derivative in column 2 of Table I. For a given space group, there is a fixed, whole number of equivalent, symmetry-related volumes for each unit cell. These volumes are known as asymmetric units. Very often a single molecule (or a chemically unique unit in the case of an intermolecular complex) occupies a single asymmetric unit of the unit cell. However, this situation need not be the case. If the molecule contains internal symmetry, occasionally this symmetry may be incorporated in the crystal as part of its crystallographic symmetry. Alternately, any whole number of molecules or asymmetric sections of symmetrical molecules may combine to make up an asymmetric unit of a unit cell. The number of molecules or chemically unique units in a unit cell for each derivative is expressed by the quantity Z, which is given in column 3 of Table I. If this quantity differs from the number of asymmetric units per cell for the space group of the derivative, this fact is noted by reference to a footnote of the table. The greater the internal symmetry of a unit cell, the less the number of unique atoms it contains. However, the total number of X-ray reflections from a crystal also has a correspondingly increased symmetry so that there are a decreasing number of unique reflections with increasing symmetry of a unit cell. Therefore, as far as the ultimate accuracy of the structure is concerned, there is no particular advantage in having a unit cell of increased symmetry. The number of unique X-ray reflections are proportional only to the volume of the asymmetric unit of the unit cell. Column 4 of Table I lists the number of unique reflections measured in the structural determinations. For those structures in which the ratio of the number of measured unique reflections to the number of crystallographically unique atoms in the structure is small, the accuracy with which the structure is known is much less than for those structures for which this ratio is large. Indeed, the great accuracy of crystallographic structural determinations is due to the normally large value of this ratio, which greatly overdetermines the parameters describing the structure.

As an illustration of the preceding discussion, let us consider the structure of the complex 1-methylcytosine · 5-fluorouracil (*85*). This structure contains a single complex per asymmetric unit of the unit cell, i.e., it contains 19 crystallographically unique, nonhydrogen atoms. Its cell dimensions are: $a = 18.35$ Å, $b = 11.40$ Å, and $c = 10.26$ Å with $\alpha = \beta = \gamma = 90°$.

The geometry of X-ray diffraction is described by the Bragg equation:

$$2\,d \sin\theta = n\lambda \tag{4}$$

where d is the interplanar spacing between the crystallographic planes under consideration, θ is the angle of reflection of the X-rays from the plane, and n is the order of the reflection. The maximum value of n occurs when $\sin\theta = 1$. Therefore, for CuKα radiation, for planes normal to the a axis ($d = 18.35$), $n_a = 24$. Similarly, $n_b = 15$ and $n_c = 13$. Therefore the total number of reflections accessible to measurement with CuKα radiation is approximately $(4\pi/3)n_a n_b n_c = 17{,}900$ (the three-dimensional array of reflections is ellipsoidal). However, as can be seen in Table I, there are 8 asymmetric units in the unit cell of the crystal, and correspondingly the reflections are divided into 8 symmetry equivalent octants so that the number of unique reflections accessible is $17{,}900/8 = 2237$. The number of reflections actually measured in the structural determination being considered was 1500, which is approximately 70% of the total number possible. Such a fraction is typical of most structural determinations. Most of the unmeasured reflections are usually physically inaccessible for experimental reasons or they have intensities too weak to measure. Also, in many cases, the space group of the crystal causes certain classes of reflections to be extinct.

The ratio of unique experimental measurements (reflections) to unique atoms in this structural determination is approximately 100:1. This large ratio, which is typical of crystallographic investigations, is responsible for the great accuracy of the structural investigation. Thus, 100 measurements are made to obtain the three positional coordinates for one atom as well as information related to its thermal vibrations.

An important exception to the observation that the ultimate accuracy of a structural determination is unaffected by the nature of the space group of the crystal occurs when the space group contains a center of symmetry (a rather common occurrence in crystal structures). Normally the phase angle can have any value, i.e., $-\pi \leq \phi_{hkl} \leq \pi$. However, in crystals whose space groups contain a center of symmetry, this distribution is restricted to the values 0 or π. Therefore, from Eq. (1) it can be seen that for centrosymmetric crystals $F_{hkl} = \pm|F_{hkl}|$. Thus, the presence of a center of symmetry in a crystal structure greatly simplifies the matter of phase determination. Because of these restrictions on the phase of a centrosymmetric crystal, it is not necessary to refine its structure to the same extent as that of an acentric structure in order to determine it to the same degree of accuracy. Among the space-group symbols in Table I, those that are P1, $P2_1$, Pc, $P22_12_1$, C2, $C222_1$, $P2_12_12_1$, and $Pna2_1$ correspond to space groups containing no center of symmetry. The other space-group symbols in Table I correspond to space groups that are centrosymmetric.

Many crystals containing biologically interesting molecules are acentric because such molecules often contain asymmetric carbon atoms. A molecule containing an asymmetric carbon atom cannot be superimposed on its inverted image and, therefore, cannot crystallize in a centrosymmetric crystal. Thus, the presence of an asymmetric carbon atom in a crystal makes the solution of a crystal structure more difficult than it otherwise might be. The space groups of crystals containing such compounds as nucleosides and nucleotides are restricted to being acentric. However, planar purine or pyrimidine derivatives often crystallize with a center of symmetry.

Heavy atoms such as bromine or barium, if present, contribute a relatively large proportion of the total electron density in the structure. Refining the structure thus fixes the position of the heavy atoms to a greater extent than those of the lighter atoms. Therefore, if a heavy-atom structure has been refined to the same extent as a comparable structure containing no heavy atoms, the parameters describing the light atoms in the heavy-atom structure will be known less accurately than similar parameters describing the light-atom structure.

However, it is much easier to determine the crystal structure of a compound containing a heavy atom. From Eq. (1) it can be seen that the phase of a given structure factor would be approximately that calculated if the unit cell contained only the heavy atom since the atomic structure factors (f_j) for heavy atoms are much larger than those of the light atoms and the individual contributions of the light atoms to the structure factor largely cancel each other out. The electron density map [Eq. (3)], made using these heavy-atom phases and the corresponding experimental values of the structure factor amplitudes, often reveals the light-atom positions. Therefore, although the presence of a heavy atom in a structure renders the ultimate accuracy of the light-atom parameters less than they otherwise would be, it is often indispensable in solving the crystal structure.

The intensities of X-ray reflections can be measured by the extent of blackening they cause on a photographic film or by the use of a radiation counter, such as a scintillation counter. Films have been widely used for a number of decades. However, during the past decade the use of diffractometers with radiation counters has become prevalent. Film blackening is usually measured by comparing it by eye to a standard scale. Thus, with films, one might expect a precision of radiation intensity measurement of no better than 10%. This figure for radiation counters is about 5% or less. Therefore, if other things are equal, one should expect more accurate structures using radiation counters rather than using photographic films. Column 5 of Table I lists an "F" if photographic film was used to measure the radiation intensities; "C" if a radiation counter was used.

The commonly used measure of the extent of refinement of a structure is the error residual, R. R is defined by the ratio

$$R = \frac{\sum\limits_{\substack{\text{all}\\ \text{reflections}}} ||F_o| - |F_c||}{\sum\limits_{\substack{\text{all}\\ \text{reflections}}} |F_o|} \tag{5}$$

where $|F_o|$ is the amplitude of the observed structure factor and $|F_c|$ is the amplitude of the calculated structure factor. The closer the calculated structure is to the "true" structure, the closer $|F_c|$ will be to $|F_o|$ and the smaller will be the residual. Of course the closeness of approach of the residual to zero is limited by errors in the measurement of $|F_o|$. For a typical structure in which the reflection intensities are measured by photographic films, a well-refined structure has a residual in the neighborhood of 0.10. For reflection intensities measured by a radiation counter, this quantity should be somewhat less. It should be pointed out that, for an entirely wrong structure that contains a center of symmetry, $R = 0.83$, whereas for an acentric structure it is 0.59 (*34*). Thus, as mentioned above, it is necessary to refine an acentric structure to a greater extent than a comparable centric structure in order to have the same degree of confidence in the parameters describing the structure. A reasonable rule of thumb, for both centric and acentric structures, is that if the residual of a structure is greater than 0.20, the qualitative nature of the postulated structure may be in doubt. The residuals for all the structures in Table I are given in column 6 of the table.

The most reliable measure of the extent to which a structure has been determined is the estimated standard deviations of the positions and thermal vibrations of its atoms. These quantities are measured by the estimated standard deviations of the bond distances and bond angles in the structure. The values of the standard deviations for bonds involving only light atoms are presented in columns 7 and 8 of Table I. For a reasonably well-refined structure, the maximum estimated standard deviations of the bond lengths and the bond angles should be no greater than 0.01 Å and 1°, respectively.

Another rough measure of the confidence one can place in a structure is the ease with which H atoms can be located. Hydrogen, with only one electron per atom, has the lowest electron density of all the elements, and therefore has very little effect on the extent of refinement of a structure. Thus, it is not possible to locate H atoms except in a well-refined structure. If H atoms have been located in a given structure, this fact is listed in column 9 of Table I. However, even well-refined structures have large uncertainties in the positions of H atoms; thus, it is difficult to make meaningful comparisons between these positions in similar structures. Therefore, H atom positions are discussed only in reference to the establishment of protonation sites. More accurate H atom parameters can be found by using the technique of neutron diffraction, which measures the positions of atomic nuclei rather than the distribution of electron density measured by X-ray diffraction. However, neutron diffraction is experimentally much more difficult to carry out than is X-ray diffraction. Thus, there has only been one investigation by neutron diffraction of a compound of the type reviewed here, that of the barbiturate derivative, violuric acid monohydrate (*35*).

III. What Are the Molecular Structures?

A. The Angles between Chemical Bonds and Their Lengths

The solution of an X-ray diffraction investigation is an electron density map, the peaks of which represent the positions of the atomic centers. From these positions, we can obtain the lengths of the chemical bonds and the angles between chemical bonds. Appendices I and II contain in tabular form the bond distances and bond angles, respectively, for the three-dimensional structures of purine and pyrimidine derivatives. These appendices show that the corresponding distances and angles in similar derivatives differ significantly among themselves compared with the estimated standard deviations of these quantities. Thus, these differences are real and are probably due to the different environments of the molecules in the crystal lattice. However, the similarities are great enough so that we have averaged the corresponding distances and angles, using the more reliable data. The criterion of reliability was that the estimated standard deviations of the bond lengths and the bond angles as given in Table I be less than 0.05 Å and 3°, respectively. Because of the reality of the variation of a given parameter among related structures, these parameters were given unit weight in the averaging process rather than a value related to their variance. The slight discrepancies between the average values and the quantities necessary to ensure both ring closure and coplanarity of the external bonds with the rings are due to the absence of these constraints in the averaging process. For the distances and angles averaged, a standard deviation from the average, σ, was calculated in the usual manner (*149*).

$$\sigma = \sqrt{\sum_{i=1}^{n} (x_i - x_{\text{av}})^2/(n-1)} \tag{6}$$

where the x_i are the quantities averaged, x_{av} is the average value of the quantities x_i, and n is the number of such quantities. This estimated standard deviation gives some measure of the spread of values found among similar molecules for the particular parameter under consideration. Both the average and the estimated standard deviation of the bond distances and bond angles, respectively, are listed in Appendices I and II. The average values of the bond distances and bond angles of the various bases are shown diagrammatically in Fig. 2. These quantities compare quite well with similar averages as determined by Donohue (*188*), with the standard values set for them by Pauling and Corey (*22*) and by Spencer (*24*) using empirical methods and also agree with the bond lengths predicted using molecular orbital methods (*150, 151*).[4] Indeed, the spread of all these earlier estimates is of the same order of magnitude as the standard deviation

[4] See article by Pullman and Pullman in Volume 9 of this series.

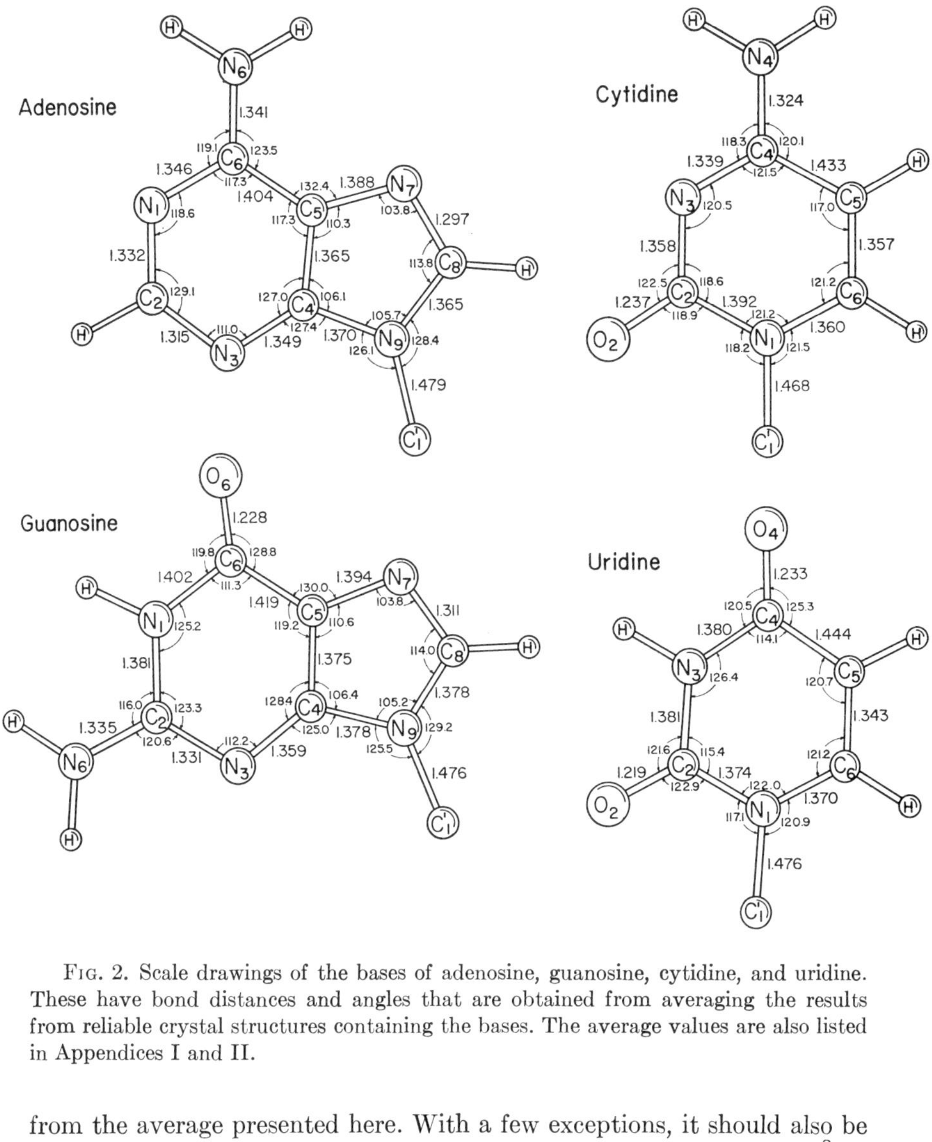

FIG. 2. Scale drawings of the bases of adenosine, guanosine, cytidine, and uridine. These have bond distances and angles that are obtained from averaging the results from reliable crystal structures containing the bases. The average values are also listed in Appendices I and II.

from the average presented here. With a few exceptions, it should also be noted that the values of σ in Appendices I and II do not exceed 0.03 Å or 2°, respectively.

Sundaralingam and Jensen (*29*) and also Singh (*152*) have pointed out that the internal angle at an N atom that is a member of a planar six-membered heterocyclic ring is, on the average, about 10° larger for those N atoms with an extra-annular attachment than for those without such a bond. The data concerning such angles in Appendix II corroborate this conclusion and suggest that it may also be true for five-membered heterocyclic rings, but with an angle change of only 2° or 3°. It can also be seen in Appendix II that the internal ring angles adjacent to the N atom in six-membered rings, i.e., the N′-C-C angle and the N′-C-N angle, each decrease about 5° with extra-annular attachment to their common N atom (N′). With a change of one angle of a planar six-membered ring, the sum of the other five internal angles must change in the opposite sense by an equal amount in order to preserve the planarity of the ring. It is not necessary that these compensating changes occur only at the angles adjacent to the one at which the initial change took place. However, this appears to be the case with the structures described in Appendices I and II.

Figure 3 contains histograms that illustrate the variation of bond lengths for selected bond types found in derivatives of adenine, guanine, cytosine, uracil, and the barbiturates. The bond type illustrated in Fig. 3a is conjugated in the purines and unconjugated in the pyrimidines (see Fig. 1). Therefore this bond should be shorter in the purines than in the pyrimidines (*153*). Figure 3a shows that this is indeed the case. Similarly, the bond type illustrated in Fig. 3b is conjugated in derivatives of adenine and cytosine and unconjugated in derivatives of guanine and uracil. The bond distance distribution in Fig. 3b supports this hypothesis. The C8-N9 bond in the purines is on the average about 0.06 Å longer than the N7-C8 bond (Figs. 3c, 3d). Thus the N7-C8 bond in adenine and guanine derivatives has more double-bond character than C8-N9, as would be expected from their major resonance forms.

The bond distance distributions of the C4-C5 bond in adenine and guanine derivatives and the C5-C6 bond in cytosine and uracil derivatives are illustrated in Fig. 3e. This bond is on the average somewhat shorter in the pyrimidines than in the purines, implying that there is more double-bond character in the former molecules. A similar conclusion was reached from molecular orbital calculations (*150*, *151*).

Narrower distributions of bond lengths are seen in the glycosyl bonds of purines and pyrimidines (Fig. 3f) and in the carbonyl groups attached to the bases (Fig. 3g). Since these bond distances do not change significantly with the type of base, it is likely that these bonds are somewhat isolated electronically from the rest of the molecule. This is of considerable significance in the case of the glycosyl bond, as the bond length is the most important constraint in determining the possible angular positions of the purine or pyrimidine ring relative to the sugar ring (*31*). The relative invariance of the glycosyl bond length suggests that it is unlikely that some portions of a nucleic acid molecule would have more freedom for base rotation than other portions.

The important conclusion from this section is that there is a limited variability in the geometry of the purines and pyrimidines in different environments. Although the differences are small, they are nonetheless well substantiated. This suggests that these differences may also occur in the nucleic acids, where they might be responsible for differences in the reactivity of certain bases in unique environments.

B. How Flat Are the Rings?

From the positions of a set of atomic centers in an electron density map, we can determine the degree of planarity of the purine and pyrimidine rings by calculating the least-squares plane of the set. Deviations from this plane are listed in Appendix III for a selection of purine, pyrimidine, and barbiturate derivatives. The signs of these deviations have relative meaning only within that group of atoms forming a particular derivative.

We expect aromatic rings to be planar (*153*). The greater the deviation of a conjugated system from planarity, the smaller the amount of resonance energy stabilizing the molecular system (*154*). Because the amount of this stabilizing resonance energy decreases rapidly with increasing deviation of a conjugated system from planarity, the deviations of such a system from strict planarity should be small. Appendix III shows that this is the case with purine, pyrimidine, and barbiturate derivatives. In a few derivatives, the space group of the crystal imposes strict planarity on the molecule. In most of the structures listed in Appendix III, the deviations of the atoms from strict planarity are within the limits of the experimental accuracy even though the molecule is not restricted to being planar by the symmetry of the unit cell of the crystal.

The atoms that appear to deviate the most from planarity are those that are substituents on the rings (Appendix III). We expect the first atom of every substituent on a conjugated ring to lie in the plane of the ring because the bonding orbitals of the ring atoms lie in the plane of the ring. However, if a ring atom is slightly out of planarity, it would have little effect on the plane of the ring but it would be magnified by the lever arm of the substituent bond distance, thereby causing the substituent atom to deviate significantly from that plane. It is also reasonable to expect that the substituent atoms would show a greater response to a given distorting force than the ring atoms since the former atoms are held in the plane of the ring by one bond whereas the latter atoms are held in that plane by two bonds.

The only apparent exception to the foregoing remarks are the rather large deviations from planarity of atom C5 of the barbiturate derivatives: barbituric acid (*121*), alloxan trihydrate (*123*), alloxan (*125*), alloxantin dihydrate (*126*), and phenobarbital (*42*). These deviations, which are 0.198, 0.214, 0.23, 0.35 and 0.365 Å, respectively, are at least an order of magnitude larger than other deviations from planarity. It should be noticed, however, that the C5 atom in barbiturate derivatives is a saturated C atom

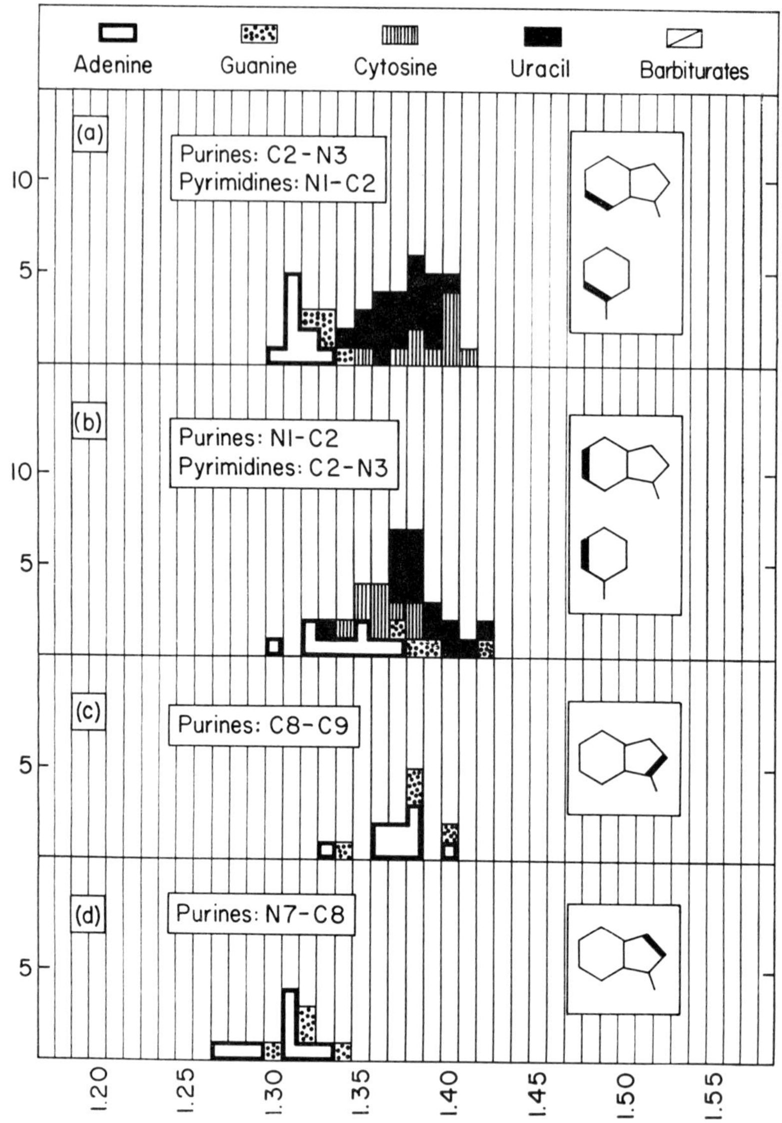

FIG. 3. The distribution of covalent bond distances in derivatives of adenine, guanine, cytosine, uracil, and the barbiturates. The abscissa is the bond distance in Ångstrom units. The ordinate is the number of bonds in a given range of length. The nature of the base is indicated by the key at the beginning of the figure.

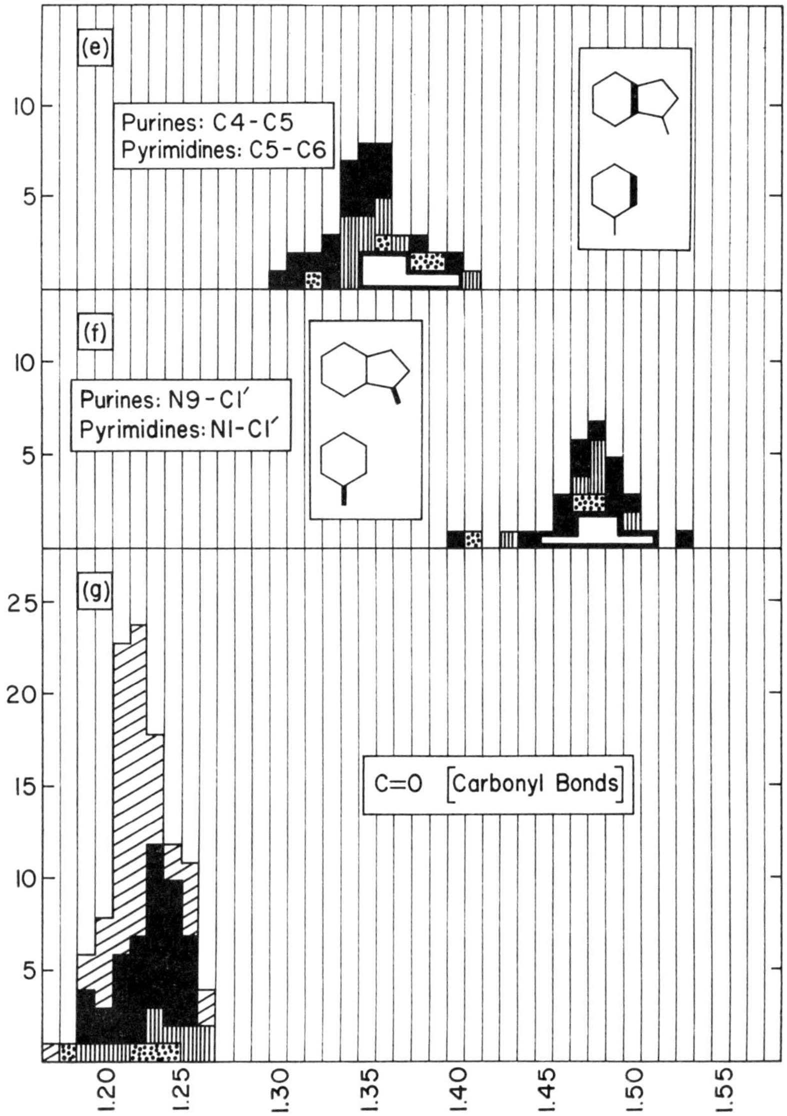

FIG. 3 (*Continued*)

and hence cannot be conjugated with the ring. Nevertheless, C5 could be coplanar with the rest of the ring if the bond angle at C5 is somewhat larger than the normal tetrahedral angle of 109°. This is observed in most barbiturates. However, we might also expect that for some cases atom C5 would deviate from planarity to relieve angular strain at C5. The C4-C5-C6

bond angles in Appendix II for those barbiturate derivatives in which atom C5 deviates from the plane of the ring are significantly smaller (4° on the average) than the average value of this angle. However, it should be pointed out that there are other barbiturate derivatives, amytal I, IIA, and IIB (*130*) and Veronal I and II (*130*), in which atom C5 is essentially coplanar with the ring and in which the C4-C5-C6 angle is also small. In these latter derivatives it is likely that the strain in the C4-C5-C6 angle is distributed about the ring rather than atom C5 deviating from the plane of the ring. It appears that the mode of strain relief of the angle C4-C5-C6 in barbiturate derivatives is mainly controlled by the nature and interactions of the substituents attached to C5. However, not enough data are available to correlate the nature of the substituents with the mechanism for strain relief. Nonetheless, it appears that all those barbiturate derivatives disubstituted at the 5 position with aliphatic groups have planar rings and relieve the strain at ring angle C4-C5-C6 by distributing it about the other ring angles.

The purine, pyrimidine, and barbiturate structures for which planarity data are available reveal little other systematic manner in which the atoms of these structures deviate from planarity. Macintyre has attempted to explain the deviations from planarity of some purine and pyrimidine structures by postulating that the hydridization of the bonds formed by the ring N atoms is partially sp^3 in character (*155*). In many of these structures, however, it appears that the distortions of the molecules occur in response to crowding in their crystal structures. Thus, most of the small deviations from planarity reported for purine and pyrimidine structures can be attributed to the distorting action of the packing forces present in the crystal on the normally planar molecules present. Of course, dihydrouracil, which is nonaromatic in character, is found in the half-chair conformation (*101*).

It should be mentioned that the experimentally observed thickness of conjugated ring systems that are held stacked upon one another only by van der Waals forces is very close to 3.4 Å. In the structures discussed in this paper, the aromatic rings appear to stack together in such a manner. Thus, there is no evidence that these stacked rings interact through the formation of a charge-transfer complex, but rather it appears that the stacking association is stabilized by electrostatic and London dispersion forces as determined by DeVoe and Tinoco (*156*). In a similar hypothesis, Shefter (*157*) has postulated that the fundamental force stabilizing the stacked complexes between xanthines and aromatic molecules (*66*, *67*, *71–73*) is "polarization bonding," as described by Wallwork (*158*), of the α-β unsaturated ketone portion of the xanthine with the π-electron system of the aromatic molecules.

We conclude that the purines and pyrimidines are largely planar structures capable of forming stabilized complexes through stacking. However, in response to a unique environment, the bases can become slightly nonplanar, especially the substituents attached to the ring. Since these latter are important in H-bond formation, this might affect their interactions with other bases.

C. Where Are the Protons on the Bases?

Although the molecular formulas of the purine and pyrimidine bases are normally written as is shown in Fig. 1, it was not always clear that those tautomeric forms for the molecules were the predominant ones. Indeed, it was assumed by Watson and Crick with advice from Donohue that the tautomeric forms shown in Fig. 1 would be those in DNA. There are four reasonable tautomeric forms for adenine, eleven for guanine, three for cytosine, and three for uracil. In addition, there are ten reasonable tautomeric forms for the barbiturates. However, most of these possibilities make only minor contributions.

Experimental determination of the tautomeric form of a molecule by X-ray diffraction techniques is quite straightforward. Even in those structures in which the positions of the hydrogen atoms have not been directly determined, it is often possible to infer these positions by a consideration of the bond distances and angles of the structures. Thus, for example, one would expect the C-O bond length to be on the order of 1.22 Å for a double bond on a ring but nearer 1.43 Å for a single bond (*159*). As Singh has pointed out (*152*), the internal ring angle at a nitrogen atom of a six-membered heterocyclic ring is about 125° if the N atom is covalently bound to an extra-annular H atom and near 117° if the N atom has no extra-annular attachment. These differences can be easily distinguished, even by a poorly refined crystallographic study. The stable tautomeric forms for the neutral purines and pyrimidines are those illustrated in Fig. 1. All the derivatives unsubstituted on the glycosyl N atom (N1 in cytosine and uracil and N9 in adenine and guanine) are protonated in this position. The other types of neutral purine and pyrimidine derivatives whose structures have been determined have also been found to have the keto and amino forms, respectively, rather than the enol and imido forms. However, the structure of isocytosine (*139a,b*) is quite curious in that the two molecules that occupy nonequivalent sites in the asymmetric unit of the unit cell have different tautomeric forms. In one of the molecules, the ring proton is bound to N1 of the pyrimidine ring whereas in the other molecule it is bound to N3. These configurations allow the two types of molecules to complex with each other, thereby lowering the energy of the crystal. Since the energies of the two tautomers are probably not very different, the simultaneous presence of the two distinct tautomeric forms is accounted for.

There remains the problem of determining the position to which the added protons are bound in protonated derivatives of purines and pyrimidines. The solution to this problem is often not easy to determine *a priori*, except for those structures studied by X-ray crystallography. Monoprotonated adenine derivatives such as adenine hydrochloride (*57b*), 5′-AMP (*43*), 3′-AMP (*44*), and adenosine 2′-, uridine 5′-phosphate (Ado-2′-P-5′-Urd)[5]·$4H_2O$ (*50*) all show that adenine is protonated at the N1 position. This position is also protonated when polyriboadenylic acid forms a double-stranded helix at low pH (*173*). It should be pointed out that crystallographic studies on nucleotides in the acid form indicate that these molecules are zwitterions with the phosphate group negatively charged and the base positively charged. Only the adenine base is charged in Ado-2′-P-5′-Urd·$4H_2O$ (*50*). Adenine dihydrobromide (*45a,c*) is the only doubly protonated adenine derivative whose structure has been studied. It is protonated on N7 as well as N1.

The structures of two protonated guanine derivatives have been solved in three dimensions, guanine hydrochloride dihydrate (*62a,b*) and 9-ethylguanine hydrobromide (*61*). They are both protonated on N7.[6] The structure of guanine hydrochloride monohydrate (*63*) was determined in 1950 using only two-dimensional data. It gave ambiguous results concerning the position of protonation. The bond distances and angles for this structure of guanine hydrochloride monohydrate compare very favorably with those of the above two protonated guanine derivatives. By Singh's rule (*152*), N3 appears not to be protonated, while N7 is in the correct position for hydrogen bonding to O6 of a neighboring guanine molecule. Thus, we can reasonably conclude that guanine hydrochloride monohydrate is also protonated at N7.[6]

The three charged derivatives of cytosine, 3′-CMP (orthorhombic form) (*91*), 3′-CMP (monoclinic form) (*92*), and 1-methylcytosine hydrobromide (*45a,b*) are all protonated at N3. Cytosine-5-acetic acid (*88*) is of great interest because of the manner in which N3 of that molecule is protonated. At any time, half of the molecules in this position are randomly protonated; thus a given molecule can be considered to be hemiprotonated at N3. The same system of hemiprotonated cytidine residues is found in the double-stranded helix of polyribocytidylic acid (*176, 177*). The only protonated uracil derivative in Table I is 1-methyluracil hydrobromide (*118*), which is protonated on O4.

The structures of three thiamine derivatives have been determined,

[5] Symbolism recommended by IUPAC-IUB Commission on Biochemical Nomenclature [cf. *J. Biol. Chem.* **241,** 527 (1966)-Section 5.3]. See Preface of this volume.

[6] See article by Shapiro in Volume 8 of this series.

hydrolyzed thiamine pyrophosphate (*140*), thiamine hydrochloride (*141*), and thiamine pyrophosphate (*142*). All these molecules are protonated on the N1 position, which is the pyrimidine ring nitrogen atom opposite the amine group in these derivatives. The first and third of these molecules are zwitterions, as would be expected for the acid forms of purine and pyrimidine derivatives that contain phosphate groups.

Of the structures reported in this paper, the only anionic ones are ammonium barbiturate (*122*) and potassium Veronal (*133*). In the former, C5 loses the proton, whereas in the latter structure, because C5 is here doubly substituted, N1 loses the proton. Otherwise, the tautomeric forms of these molecules remain unchanged.

X-ray studies have thus made it possible to obtain exact information concerning protonation positions in the crystalline state, and most of the results are in agreement with solution studies. Reliable knowledge of protonation is of great value to the nucleic acid chemist who frequently uses protonation or deprotonation as a means of modifying these materials.

D. Thymine and Uracil Derivatives Are Similar

It is of interest to determine the conformational differences, if any, between thymine and uracil in order to determine to what extent they are responsible for the differences between RNA and DNA. Appendices I and II reveal no significant differences between the average bond distances and bond angles of the thymine derivatives and the corresponding quantities for the uracil derivatives. Therefore it must be concluded that any differences between thymine and uracil bases are due to the steric effects of the thymine methyl group itself and to its influence on the electronic structure of the uracil ring. These differences are expressed chemically by the fact that the ionizable proton on N1 of uridine has a pK of 9.4, while the corresponding proton of thymidine has a higher pK, 9.9. However, a change such as this would be reflected only in very small differences in the structure of these molecules. Thus, it is reasonable to assume that most of the differences between RNA and DNA are due to the additional hydroxyl group on the C2′ atom of ribose.[7]

IV. How Do the Purines and the Pyrimidines Interact?

A. Hydrogen Bonding between Bases: Theory and Selective Affinity

Hydrogen bonds are the result of an associative interaction between an acidic, H-bearing atom, such as an amine N atom (called the donor atom), and an electronegative, basic group, such as a carbonyl O atom (the ac-

[7] See article by Yang and Samejima in Volume 9 of this series.

ceptor atom). The attractive forces are roughly in the direction of the covalent bond binding the hydrogen atom to the acidic group. Hydrogen bonds are quite weak in comparison to covalent bonds; the former have a binding energy of only a few kilocalories per mole, whereas the latter have binding energies near 100 kcal per mole.

There has been little agreement among theorists concerning the origin and nature of the attractive forces of H-bonds. Four basic theoretical approaches have been used recently to try to explain H-bonding interactions: (a) pure electrostatic theories, (b) valence-bond theories, (c) charge-transfer theories, and (d) molecular-orbital theories.[4] Electrostatic theories generally assume that the H atom can form only one covalent bond, i.e., the one with the donor atom, and attempt to explain the H-bonding in terms of classical electrostatic interactions between the electrons and nearby nuclei (*153*). Valence-bond theories explain the origin of the H-bond in terms of a weak covalent bond between the hydrogen and the acceptor (*160a,b*). It is usually assumed that the covalent bond makes part of the contribution, but in addition there is an electrostatic component. The latter component is larger than the former in an H-bond. In the formalism of the charge-transfer theory of H-bonding, there is a partial transfer of an electron from a bonding orbital of the acceptor atom to an antibonding orbital of the donor atom. This charge-transfer results in a weakening of the covalent bond between the donor atom and the hydrogen atom, and it produces an attractive interaction between the acceptor atom and the H atom. In this theory, the specific requirement of an H atom in H-bonding is explained as deriving from the small magnitude of the short-range repulsion term between the hydrogen and the acceptor, which is a consequence of the small size of the H atom (*161*). In the molecular orbital treatment of H-bonding,[4] the H-bonded complex, A-H · · · B, is considered to be a single molecule. In complexes of the type discussed here, where there are two H-bonded aromatic molecules, the H-bond facilitates conjugation between the two molecules forming the complex. The subsequent increased delocalization of electrons then leads to an increased resonance energy of the molecular system, which in turn is responsible for the attractive interaction observed in H-bonds between aromatic molecules (*162*, *163*).

As is the case with most present day chemical theories, each of the above theories of H-bonding appears to rationalize satisfactorily certain observed aspects of H-bonding but offers little explanation of others. Each theory has its strong points and weak points. It is obvious, however, that much more theoretical work is necessary before the nature of the H-bond is understood. The major difficulty, of course, is the fact that H-bond energies are only a few kilocalories per mole compared to the much larger covalent bond energies. Since all chemical theories are approximations rather than direct solutions of the Schrödinger equation, a number of different stratagems are adopted in each of the methods discussed above. Most of the theoretical treatments make direct assumptions about the magnitudes of various terms in the quantum mechanical treatment. However, the magnitudes of the energies implicit in these assumptions are frequently equal to, if not greater than, the magnitude of the H-bond energy. Therefore the theoretical treatments are of limited value. However, it should be noted that semiempirical calculations carried over a series of compounds may provide some useful comparative information. A recent review of the status of the various theories of H-bonding has been written by Bratŏz (*161*).

X-Ray investigations are powerful in that they yield direct structural information that allows us to determine not only the geometry of the molecules but also their interactions, as seen, for example, in the formation of H-bonded base pairs among the nucleic acid bases. However, diffraction investigations are limited in this respect because they do not provide a great deal of information concerning the energetics of these interactions. Some information is obtained from the measurements of H-bond lengths since shorter H-bonds are considered to be stronger (*201a*, *201b*). However, these bond lengths do not fully reflect the energetics of the interaction because of constraints inherent in a crystal structure. In order to form a crystal, the molecules must arrange themselves in such a way as to fill space. Thus, packing considerations are of great importance. The total lattice energy of the crystalline array is often as significant in a crystal as the energetics of the interactions between associating molecules. This is a major limitation of purely structural studies.

Most of the interactions between nucleic acid molecules occur in solution. Accordingly, solution studies provide a perspective for interpreting the results of X-ray diffraction investigations. Although there have been many solution studies of polynucleotide interactions, it has been more difficult to obtain quantitative information about monomeric purine and pyrimidine interactions. In an aqueous solution, purines and pyrimidines do not form significant numbers of intermolecular H-bonds due to the successful competition of the water (which has a concentration of 55 moles per liter) for their H-bonding sites. Rather, they tend to stack (*190–192*) since unsaturated π-electron systems in a medium of high dielectric constant are stabilized in such a configuration (*156*). In order to overcome this experimental difficulty, studies have been carried out on the interactions of monomeric purine and pyrimidine derivatives in nonaqueous solutions in which the bases are not stacked. The use of nonaqueous solutions has substantial advantages since in them it is possible directly to visualize the —NH and —NH_2 stretching vibrations in the infrared region of the spectrum. These vibrations shift in frequency when the —NH and —NH_2 groups form H-bonds. From quantitative measurements of these shifts, one can get association constants as well as enthalpy and entropy changes for the reaction. Studies of this type were first carried out in chloroform solution using chloroform-soluble derivatives of adenine and uracil (*193*). It was found that 1-cyclohexyluracil will dimerize with itself at concentrations near 0.1 *M*. The same is true for 9-ethyladenine at an equal concentration. In 0.02 *M* solutions very little self-hydrogen-bonding was detected. However, when 0.02 *M* solutions of 9-ethyladenine and 1-cyclohexyluracil were mixed, the infrared absorption spectrum showed that the H-bonding between these derivatives was extensive.

A quantitative study of this reaction showed that at 25°C the association constant of 1-cyclohexyluracil with itself was 6.1 M^{-1} and that 9-ethyladenine had a self-association constant of 3.1 M^{-1} (*194*). At the same temperature the H-bonded dimer of 9-ethyladenine and 1-cyclohexyluracil had an association constant of 100 M^{-1}. This increase in the association constant was clearly reflected in measurements of the enthalpy for these three reactions. Thus, the enthalpies of dimerization are −4.3 kcal/mole for 1-cyclohexyluracil and −4.0 kcal/mole for 9-ethyladenine, whereas the H-bonded association between 9-ethyl-adenine and 1-cyclohexyluracil has an enthalpy of −6.2 kcal/mole. As would be anticipated, the entropy changes for all three dimerization reactions are about the same, approximately −11 entropy units. These results clearly indicate that there is a stronger association between 9-ethyladenine and 1-cyclohexyluracil than in the self-association of these molecules. This effect has been seen qualitatively with other uracil and adenine derivatives, using both infrared and nuclear magnetic resonance techniques (*185*, *195*, *196*).

The difference in the measured association constants is not reflected geometrically in the structure of the H-bonded dimers. As discussed below (Section IV, C), H-bonded pairs of uracil derivatives are frequently found. They are held together by two H bonds with normal angles and distances. The same is true of H-bonded pairs of adenine derivatives. Study of the molecular structures of H-bonded pairs containing one uracil and one adenine residue (Section IV, D) shows that they are also held together with two H-bonds of normal length. Thus, the basis for the preferential H-bonding between uracil and adenine residues is not reflected in simple geometrical considerations, but rather must reflect differences in the distribution of electrons that makes the binding more energetic in the mixed dimers than in the homologous dimers.

Similar experiments have also been carried out in nonaqueous solutions with guanine and cytosine derivatives (*183–185*). The guanine derivatives strongly self-associate even in dilute solution. However, when cytosine residues are present in the solution, an H-bonded dimer containing one guanine and one cytosine derivative is formed with a higher association constant. Furthermore, when one measures interactions between all the base pairs there is very little association found between adenine and guanine or cytosine derivatives or between uracil and guanine or cytosine derivatives. Thus the H-bonding appears to be almost entirely selective, with stronger associations found between adenine and uracil (or thymine) derivatives and between guanine and cytosine derivatives.

The selective nature of the H-bonding of individual derivatives has been termed electronic complementarity (*183*, *197*). This phrase implies that the electronic distribution within the adenine · uracil and guanine · cytosine pairs is such as to favor the formation of these H-bonded dimers in solution. This phenomenon is a reflection of the electronic distributions within the molecules and thus should be contrasted sharply with the geometrical complementarity that is recognized to be an important feature of the double-helical nucleic acid structure. It is possible to make a regular two-stranded helix for DNA or RNA only if the Watson-Crick pairs are used. If adenine were to pair with bases other than thymine, or thymine with guanine or cytosine, the resultant structure could not be organized into a regular helix. Most discussions of the structure of double-helical nucleic acids stress the geometrical nature of the complementary interactions. However, another important aspect of this interaction is electronic in character. It has a complementarity that reinforces these geometrical arrangements. Both these factors are undoubtedly of importance in stabilizing the structure of the nucleic acids.

Selectivity of H-bonding of monomeric nucleic acid derivatives is seen also in an aqueous solution. Thus, if derivatives of adenosine are fixed covalently to a column material over which cytidine and uridine are slowly passed, there is a selective retardation of uridine compared to cytidine in its elution from the column (*198*). When guanine is attached to the column, the opposite effect occurs, and cytosine is retarded. This is explained by the complementary H-bonding of monomeric nucleosides that is also observed in nonaqueous solutions.

The effect of substituents on the H-bonding association between uracil and adenine derivatives as determined by infrared studies is illustrated diagrammatically in Fig. 4. Figure 4a shows the association constant of 9-ethyladenine with a variety of uracil derivatives, and Fig. 4b shows the association constants of 1-cyclohexyluracil with various adenine derivatives. Both the intermolecular-association constants and the self-association constants are listed. In Fig. 4a it can be seen that methylation of N3 of the uracil derivatives eliminates all self-association. This would be expected since the resulting molecule has no protons for H-bonding. The association

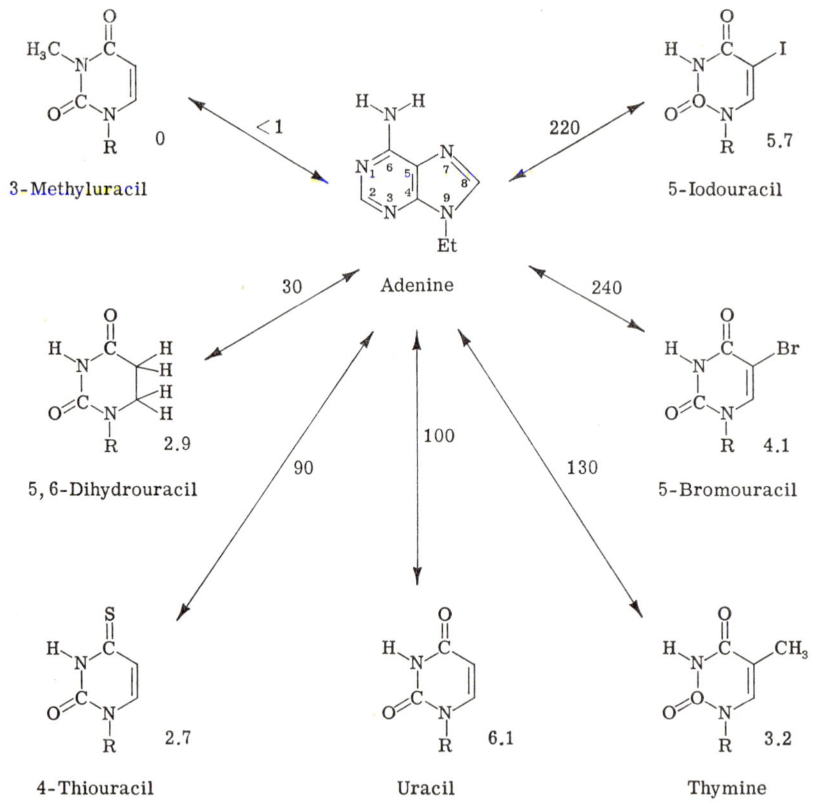

FIG. 4a.

FIG. 4. Association constants (liter/mole) between various substituted 1-cyclohexyluracil derivatives and 9-ethyladenine derivatives in deuterochloroform solution at 25°C. Figures near the double-headed arrows are the association constants between adenine and uracil derivatives, and the numbers shown below the structural formulas are the self-association constants. (a) Association of 9-ethyladenine with uracil derivatives. (b) Association of 1-cyclohexyluracil with adenine derivatives. Taken from Kyogoku *et al.* (*197*).

constant of the 3-methyluracil derivative with 9-ethyladenine is less than one. Since 9-ethyladenine has two protons on its amino group that are capable of forming H-bonds but is sterically constrained to form only one such bond to each 3-methyluracil derivative, this clearly indicates that an H-bonded complex is stable only if it has two H-bonds and can form a cyclic H-bonded dimer. Such an effect is expected since at 25°C the energy of a single H-bond is on the order of thermal energy. Thus a single H-bonded intermolecular complex could not be stable under these conditions. Loss

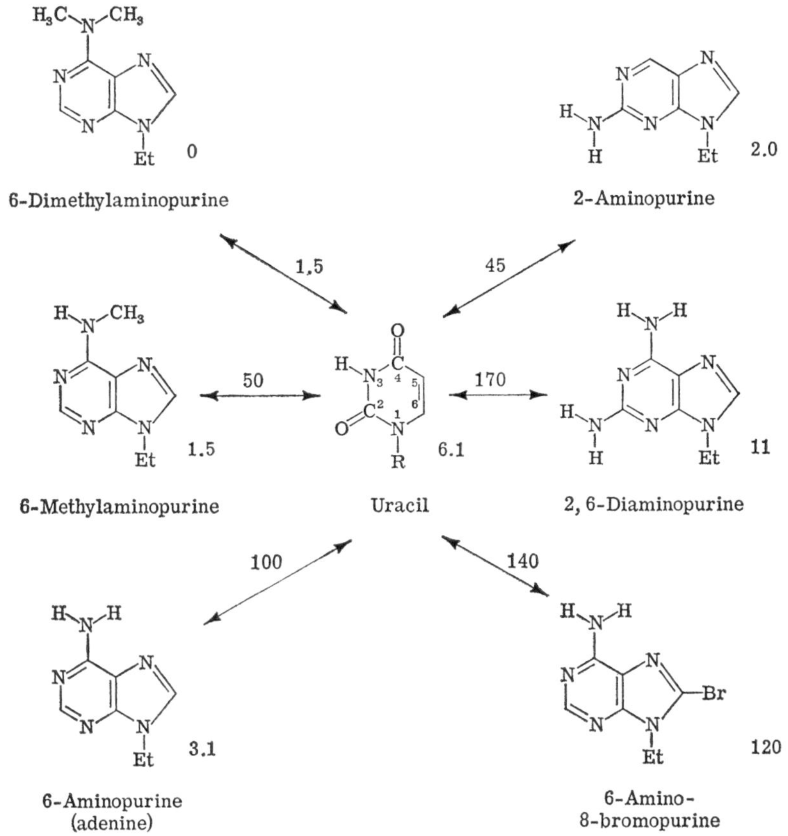

FIG. 4b.

of aromaticity as seen in the 5,6-dihydrouracil derivative also results in a decrease of affinity for H-bonding; the association constant for this derivative with 9-ethyladenine is 30 M^{-1} as compared to 100 M^{-1} with the uracil derivative. One component of this weaker interaction may result from the fact that dihydrouracil has a pK near 11, whereas uracil itself has a pK near 9. The strength of H-bonding is modified by the acidity of the proton involved in H-bonding. Thus, a less acidic proton produces weaker H-bonding while a more acidic proton produces strong H-bonding. This effect may be responsible in part for the increase in the association constant of the 5-bromouracil derivative with 9-ethyladenine to 240 M^{-1}. 5-Bromouracil itself has a pK of 7.8. However, it should be noted that the self-association constants for all uracil derivatives are small, regardless of the nature of their substituents.

Changes in the association constant due to modifications of the adenine ring are shown in Fig. 4b. The dimethylamino derivative has a self-association constant of zero and an association constant of 1.5 with uracil derivatives. This also is undoubtedly due to the inability of the intermolecular complex to form a cyclic dimer. Monomethylating the adenine amino group decreases the association constant by one-half. This might be expected if the contribution made by forming H-bonds to the N1 site were equal to that of the binding to the imidazole site, N7. A similar effect is also seen when an amino group is found only on the C2 position of adenine as in the 2-amino purine derivative. In this position, the uracil residue can H-bond to the amino group and N1, but it no longer can bond to the imidazole nitrogen N7. Loss of this second binding site is reflected in a halving of the association constant. This also suggests that the affinity to the N7 and N1 sites are similar as the N3 position is prevented from forming an H-bond by the steric interference of the adenine imidazole ring with a uracil carbonyl group. Finally, in the 2,6-diaminopurine derivative, the association constant has a value of 170 M^{-1}, an increase that is undoubtedly due to the fact that three H-bonds can form to the uracil derivative. In these derivatives, even though there are substantial changes in the electronic configuration of the molecule and changes in the affinity of H-bonding, the molecules nonetheless retain the property of electronic complementarity since they do not form H-bonds with derivatives of guanine or cytosine (*197*).

Another example of a modification of the uracil derivative is seen when a 6-carbonyl group is present. The association constants of a number of barbituric acid (6-oxyuracil) derivatives have been measured in chloroform solution (*199*). These derivatives form H-bonds selectively with adenine derivatives and with no other bases. However, the association constants of the resultant complexes are an order of magnitude stronger than the association constants found between adenine and uracil derivatives. For example, phenobarbital, which has phenyl and ethyl substituents on C5, has a self-association constant of 8.1 M^{-1}. However, the complex of phenobarbital with 9-ethyladenine has an association constant of 1200 M^{-1}. This large association constant may account for the fact that several crystalline intermolecular complexes of barbiturates with adenine derivatives have been identified (*41, 42*). The enhanced affinity of barbiturates for adenine derivatives may be explained partially by the lower pK value of the barbiturate protons. Phenobarbital has a pK of 7.3. This results in stronger H-bonds with adenine derivatives. It is likely that the enhanced affinity of barbiturates for adenine derivatives is related to their biological activity.

This discussion of the electronic complementarity and the affinity of various purine and pyrimidine derivatives prefaces the presentation of the structural information embodied in the H-bonded base pairs. Many of the base pairs described here utilize a number of the substituents shown in Fig. 4. The solution studies are thus a necessary chemical adjunct to the purely structure X-ray diffraction studies.

B. The Geometry of Hydrogen Bonding between Bases: Angles and Distances

In a crystal, a hydrogen bond, AH · · · B, is believed to exist if the A · · · B distance is somewhat less than the sum of the covalent radii of atoms A and H and the van der Waals radii of atoms H and B. This sum for the grouping N-H · · · O and N-H · · · N is near 3.4 Å (*153*), and therefore the upper limit for which N-H · · · O and N-H · · · N contacts can be considered to be H-bonds is about 3.2 Å for the distance between the donor atom and the acceptor atom. Tables II and III summarize the available information concerning H-bond distances and angles for complexes between pairs of purines, pyrimidines, and their mixtures. These include N-H · · · O (Table II) and N-H · · · N hydrogen bonds (Table III). Donor and acceptor atoms are identified by their number, and in the case where two different types of bases are bonded, the letter before the number identifies the base.

The configuration in which the H-bond occurs is also identified in Tables II and III. Here the word "cyclic" implies that the H-bond is a member of a system of covalent bonds and H-bonds forming a closed ring. If such a complex contains two molecules, it is called a cyclic dimer. The majority of cyclic dimers contain a six-membered ring (not counting H atoms) formed by a complex between molecules and containing two H-bonds. An example of such a complex is the pairing between adenine and thymine in DNA (*1*). If the H-bonded ring contains other than six non-hydrogen atoms, this number is noted in parentheses beside the word "cyclic." When the H-bonded ring binds together four molecules, this is labeled a "tetramer." In some cases, a cyclic dimer is held together by three H-bonds, and thus the system is built out of two fused six-membered rings. Here the H-bonds form three parallel bonds, one of which is common to both six-membered rings. An example of such a cyclic dimer is seen in the Pauling and Corey modification of the Watson-Crick complex between cytosine and guanine (*22*). This is labeled "triple" in Tables II and III. When a cyclic H-bonded system contains a center of symmetry (centric), only half of the H-bonds are crystallographically unique, and only those are listed. Most of the complexes between adenine and thymine found in single crystals are formed through cyclic, seven-membered H-bonded rings involving the imidazole nitrogen, N7, of adenine as indicated in Tables II and III. An isolated single H-bond is labeled "single."

TABLE II

NH · · · O HYDROGEN BOND DISTANCES AND ANGLES

Derivative[a]	NH	O	Type[b]	d(Å)	α(°)
A. Structures Containing One Type of Base					
Guanine·HCl·H_2O	7	4	Cyclic (8), centric	2.62	—
Theophylline	7	6	Cyclic (8), centric	2.76	—
Theophylline·5-Chlorosalicylic acid	7	6	Cyclic (8), centric	2.72	—
Uric acid	1	2	Cyclic, centric	2.83	—
	7	8	Cyclic, centric	2.79	—
	3	8	Single	2.73	—
	9	6	Single	2.81	—
Cytosine	4	2	Cyclic	3.03	—
	4	2	Cyclic (4), tetramer	2.98	—
Cytosine·H_2O	4	2	Cyclic	2.99	—
Cytosine-5-acetic acid	4	2	Cyclic, triple, cent.	2.79	120
	4	2	Cyclic (4), tetramer	2.82	136
1-Methylcytosine	4	2	Cyclic (8), tetramer	2.90	—
Cytosine·Acridine·H_2O	4	2	Cyclic	2.93	—
Uracil	1	4	Single	2.86	—
	3	4	Cyclic, centric	2.86	—
1-Methyluracil	3	4	Cyclic, centric	2.83	—
5-Nitrouracil	3	4	Cyclic, centric	2.85	—
5-Fluoroorotate[Rb salt]·H_2O	1	4	Cyclic	2.85	—
Thymine·H_2O	1	2	Cyclic, centric	2.84	111
	3	4	Cyclic, centric	2.84	113
1-Methylthymine	3	4	Cyclic, centric	2.83	—
5-Ethyl-6-methyluracil	1	2	Cyclic, centric	2.78	—
	3	4	Cyclic, centric	2.82	—
Photodimer of thymine	1	4	Single	2.83	—
(*trans-anti* isomer)	3	2	Single	2.86	—
Barbituric acid·$2H_2O$	1	4	Single	2.84	—
Barbituric acid	1	4	Single	2.80	—
	3	2	Cyclic, centric	2.90	—
Ammonium barbiturate	1	4	Single	2.86	—
	3	2	Cyclic	2.80	—
Alloxan·$3H_2O$	3	4	Cyclic, centric	2.88	—
A. Structures Containing One Type of Base (Continued)					
Alloxan	1	4	Single	2.88	—
	3	2	Cyclic, centric	3.04	—
Alloxantin·$2H_2O$	1	4	Single	2.82	—
Dilituric acid	1	4	Single	2.79	—
	3	2	Cyclic, centric	2.85	—
Dilituric acid·$3H_2O$	A1	B2	Cyclic	2.83	—
	B1	A2	Cyclic	2.83	—
	A3	B4	Cyclic	2.86	—
	B3	A4	Cyclic	2.85	—
Violuric acid·H_2O	A1	B2	Cyclic	2.99	—
	B1	A2	Cyclic	3.06	—
Veronal I	1	4	Single	2.87	—
	3	2	Cyclic, centric	2.89	—
Veronal II	1, 3	4, 6	Cyclic, centric	2.87	—
Amytal I	1	2	Cyclic, centric	2.92	—
	3	4	Cyclic, centric	2.89	—
Amytal II	A1	B2	Cyclic	2.91	—
	A3	B4	Cyclic	2.92	—
5-(6′-Bromo-3′-ethyl-2′-methyl-	A1	A2	Cyclic, centric	2.88	—
benzimidazolium)-barbiturate	A1	B2	Cyclic	2.94	—
	B3	A4	Cyclic	2.84	—
Dialuric acid·H_2O	1	2	Cyclic	2.77	—
	3	4	Cyclic, centric	2.80	—
Potassium Veronal	1	4	Single	2.98	—
Isocytosine	A1	B4	Single	2.73	121
	A2	B4	Single	2.82	139
	A2	B4	Cyclic, triple	2.90	117
	B2	A4	Cyclic, triple	2.86	120
8-Azaguanine·H_2O	2	6	Cyclic (7)	2.96	—
B. Structures Containing Two Types of Bases					
9MeAde·1MeThy	A6	T2	Single	2.88	132
	A6	T4	Cyclic (7), imidazole	2.85	133

B. Structures Containing Two Types of Bases (Continued)

9EtAde·1MeUra	A6	U2	Single	2.96	130
Ado·5BrUrd	A6	U4	Cyclic (7), imidazole	2.98	130
	A6	U4	Cyc. (7), rev. imid.	3.10	126
(8Br9EtAde)$_2$·Phenobarbital	A$_1$6	B4	Cyclic	3.19	112
	A$_2$6	B2	Cyclic	2.97	113
9EtAde·1Me5BrUra	A6	U2	Single	2.96	130
	A6	U4	Cyc. (7), rev. imid.	2.98	130
9MeAde·1Me5BrUra	A6	U4	Cyclic (7), imidazole	2.98	—
9EtAde·1Me5FUra	A6	U2	Single	2.97	—
	A6	U4	Cyclic (7), imidazole	2.95	—
9Et8BrAde·1Me5BrUra	A6	U2	Cyclic	3.06	113
9EtAde·(1Me5IUra)$_2$	A6	U$_1$4	Cyclic (7), imidazole	3.06	—
	A6	U$_2$2	Cyclic	3.16	—
9Et8BrAde·9Et8BrHyp	A6	HX6	Cyclic (7), imidazole	2.99	128
9Et2AmPur·1Me5FUra	P2	U2	Cyclic	2.88	122
9Et2AmPur·1Me5BrUra	P2	U2	Cyclic	2.97	116
9Et2,6Am$_2$Pur·(1Me5IUra)$_2$	P2	U$_1$4	Cyclic, triple	2.87	121
	P6	U$_1$2	Cyclic, triple	2.92	122
	P6	U$_2$4	Cyclic (7), imidazole	2.98	129
9Et2,6Am$_2$Pur·(1Me5FUra)$_2$	P2	U$_1$4	Cyclic, triple	2.89	—
	P6	U$_1$2	Cyclic, triple	2.81	—
	P6	U$_2$2	Cyc. (7), rev. imid.	2.87	—
9Et2,6Am$_2$Pur·(1MeThy)$_2$·H_2O	P2	T$_1$2	Cyclic, triple	2.89	123
	P6	T$_1$4	Cyclic, triple	2.99	119
	P6	T$_2$2	Cyc. (7), rev. imid.	2.99	128

B. Structures Containing Two Types of Bases (Continued)

5′Br5′dAdo·Riboflavin·H_2O	A6	R2	Cyc. (7), rev. imid.	2.82	129
9EtAde·5iPr5BrAllylBarb.	A6	B4	Cyclic	3.34	108
	A6	B6	Cyclic (7), imidazole	3.12	128
9EtGua·1MeCyt	G2	C2	Cyclic, triple	2.81	123
9EtGua·1Me5FUra	C4	G6	Cyclic, triple	2.94	116
	G2	C2	Cyclic, triple	2.82	122
	C4	G6	Cyclic, triple	2.96	128
dGuo·5BrdCyd	G2	C2	Cyclic, triple	2.78	117
	C4	G6	Cyclic, triple	2.83	118
9EtGua·1Me5BrCyt	G2	C2	Cyclic, triple	2.91	—
	C4	G6	Cyclic, triple	2.86	—
	C4	C2	Cyclic (4), tetramer	2.92	—
	G2	G6	Cyclic (4), tetramer	3.18	—
9EtHyp·5FUra	H1	U2	Single	2.80	—
	U3	U4	Cyclic, centric	2.81	—
Cyt·5FUra·H_2O	C4	U2	Cyclic	3.03	117
	C4	U2	Cyc. (4), tet., cent.	2.95	145
	U2	U4	Cyclic, centric	2.80	115
1MeCyt·5FUra	C1	C2	Cyclic, centric	2.85	116
	C4	U2	Cyclic	3.07	—
	C4	U2	Cyc. (4), tet., cent.	2.89	—
	U2	U4	Cyclic, centric	2.82	—

[a] Symbols are defined under Table I.

[b] If the H-bonded ring contains other than six nonhydrogen atoms, this number is noted in parentheses after the word "cyclic." Cyc. = cyclic; rev. imid. = reversed imidazole; tet. = tetramer; cent. = centric.

TABLE III
NH · · · N HYDROGEN BOND DISTANCES AND ANGLES

Derivative[a]	NH	N	Type[a]	d(Å)	α(°)
A. Structures Containing One Type of Base					
Adenine·HCl	6	7	Cyclic (8), centric	2.98	132
Deoxyadenosine·H_2O	6	1	Cyclic (7)	2.88	—
	6	7	Cyclic (7)	3.03	—
9-Methyladenine	6	1	Cyclic (7)	2.96	—
	6	7	Cyclic (7)	3.06	—
Guanine·HCl·H_2O	2	3	Cyclic, centric	3.08	—
Guanine·HCl·$2H_2O$	2	3	Cyclic, centric	3.04	118
Purine	7	9	Single	2.85	—
Cytosine	1	3	Cyclic	2.84	—
Cytosine·H_2O	1	3	Cyclic	2.95	—
1-Methylcytosine	4	3	Cyclic, centric	3.04	—
Cytosine-5-acetic acid	3	3	Cyclic, triple, cent.	2.82	—
Cytosine·Acridine·H_2O	1	3	Cyclic	2.79	—
2Am4Me6ClPyr	2	1	Cyclic, centric	3.21	—
	2	3	Cyclic, centric	3.37	—
4Am2,6Cl_2Pyr	4	3	Cyclic, centric	3.09	116
5Br4,6Am_2Pyr	4	1	Single	2.96	143
	4	3	Cyclic, centric	3.07	122
2-(4′$NH_2$5′Pyrimidyl)-2-penten-4-one	4	3	Cyclic, centric	3.07	—
Isocytosine	A2	B1	Cyclic, centric	2.98	116
	B3	A3	Cyclic, triple	2.91	120
B. Structures Containing Two Types of Bases					
9MeAde·1MeThy	T3	A7	Cyclic (7), imidazole	2.92	119
9EtAde·1MeUra	U3	N7	Cyclic (7), imidazole	2.83	122
Ado·5BrUrd	U3	N7	Cyclic (7), rev. imid.	2.80	119
(8Br9EtAde)$_2$·Phenobarbital	$A_1$6	B1	Cyclic	2.78	124
	$A_2$6	B3	Cyclic	2.80	—
	$A_1$6	$A_1$7	Cyclic (8), centric	3.09	—
	$A_2$6	$A_2$7	Cyclic (8), centric	3.02	138
9EtAde·1Me5BrUra	U3	A7	Cyc. (7), rev. imid.	2.80	120
9MeAde·1Me5BrUra	U3	A7	Cyclic (7), imidazole	2.86	120
	A6	A3	Cyclic, centric	3.00	—
B. Structures Containing Two Types of Bases (Continued)					
9EtAde·1Me5FUra	U3	A7	Cyclic (7), imidazole	2.78	—
9EtAde·1Me5BrUra	U3	A1	Cyclic	2.85	114
	A6	A7	Cyclic (8), centric	3.03	132
9EtAde·(1Me5IUra)$_2$	$U_1$3	A7	Cyclic (7), imidazole	2.82	—
	$U_2$3	A1	Cyclic	2.80	—
9Et8BrAde·9Et8BrHyp	HX1	A7	Cyclic (7), imidazole	2.94	118
	A6	A1	Cyclic, centric	3.01	119
9Et2AmPur·1Me5FUra	U3	P1	Cyclic	2.94	122
	P2	P3	Cyclic, centric	3.04	122
9Et2AmPur·1Me5BrUra	U3	P1	Cyclic	2.80	121
	P2	P3	Cyclic, centric	3.11	116
9Et2,6Am_2Pur·(1Me5IUra)$_2$	$U_1$3	P1	Cyclic, triple	2.91	116
	$U_2$3	P7	Cyclic (7), imidazole	2.83	122
	P2	P3	Cyclic, centric	3.04	122
9Et2,6Am_2Pur·(1Me5FUra)$_2$	$U_1$3	P3	Cyclic, triple	2.88	—
	$U_2$3	P7	Cyc. (7), rev. imid.	2.90	—
	P2	P3	Cyclic, centric	3.09	—
9Et2,6Am_2Pur·(1MeThy)$_2$·H_2O	$T_1$3	P1	Cyclic, triple	2.99	116
	$T_2$3	P7	Cyc. (7), rev. imid.	2.88	116
	P2	P3	Cyclic, centric	3.07	121
5′Br5′dAdo·Riboflavin·$3H_2O$	R3	A7	Cyclic (7), rev. imid.	2.84	115
9EtAde·5iPr5BrAllylBarb.	B1	A7	Cyclic (7), imidazole	2.81	126
	B3	A1	Cyclic	2.79	117
9EtGua·1MeCyt	G1	C3	Cyclic, triple	2.92	120
	C2	C3	Cyclic, centric	3.01	120
	C4	G7	Single	3.11	122
9EtGua·1Me5FCyt	G1	C3	Cyclic, triple	2.94	122
	C2	C3	Cyclic, centric	2.99	121
	C4	G7	Single	2.99	128
dGuo·5BrdCyd	G1	C3	Cyclic, triple	2.92	124
9EtGua·1Me5BrCyt	G1	C3	Cyclic, triple	2.95	—
9EtHyp·5FUra	U1	H7	Single	2.73	—
Cyt·5FUra·H_2O	U1	C3	Cyclic	2.83	—
1MeCyt·5FUra	U1	C3	Cyclic	2.75	—

[a] Symbols are defined under Table I and Table II.

In addition to the H-bond distance, d, between the donor and the acceptor atoms, the angle, α, of the H-bond is listed where it has been determined. The angle α is measured between the line connecting donor and acceptor atoms A and B (which is also used to measure d) and the line between donor atom A and an adjacent atom in the same molecule. In cyclic systems, α is taken as an internal ring angle. In systems with a single H-bond, α is the smallest of the two possible values for this angle. Similarly, for the central H-bond of cyclic systems containing three H-bonds, α is taken as the smallest angle of the two possible choices.

Tables II and III are each divided into two sections listing information concerning H-bonds between like molecules and between unlike molecules. A comparison of these sections reveals that there are no significant differences between H-bonds connecting identical molecules and those connecting different molecules.

Figure 5 contains histograms of the distribution of the values of d for

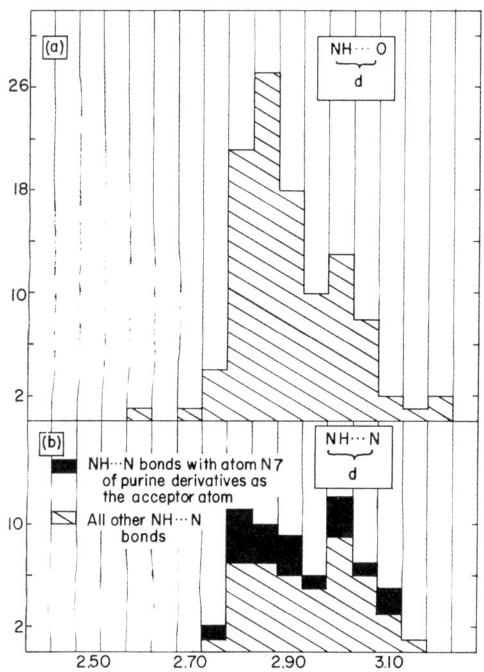

FIG. 5. The distribution of intermolecular H-bond distances involving purines and pyrimidines. The distance d is given for: (a) NH · · · O H-bonds; (b) NH · · · N H-bonds. Filled boxes represent H-bonds in which the purine derivative atom N7 is the acceptor atom. Diagonally shaded boxes represent all other types of H-bonds. The abscissa of the figures is the H-bond distance, d, in Ångstrom units. The ordinate is the number of H-bonds in a given range of length.

N-H · · · O H-bonds (Fig. 5a) and for N-H · · · N H-bonds (Fig. 5b). Figures 6a and 6b show the two distributions for α. The H-bond lengths in Fig. 5b are given different shadings depending on whether or not the acceptor is the N7 atom of a purine. The spread of the distribution of H-bond lengths and angles in Figs. 5 and 6 is approximately four times the spread of the covalent bond distances and angles listed in Appendices I and II and in Fig. 3. This result might be expected, considering the relative strengths of interaction of covalent bonds as compared to those of H-bonds. Tables II and III show that there is no apparent segregation of H-bond lengths or angles according to whether a system contains one, two, or three H-bonds between two molecules. The formation of a second (or third) H-bond in a complex apparently does not produce a decrease in the length of the original one. Since bond length is usually correlated inversely with the strength of the bond, this suggests that incremental H-bonds make only additive rather than cooperative increments to the strength of an H-bonded complex. The H-bond length distribution for N-H · · · N H-bonds tends to be doubly peaked. As seen in Fig. 5b, the reason for this apparent bimodal distribution is partially due to the fact that N-H · · · N distances involving as acceptor atoms the N7 of purines have an average value of 2.90 Å whereas the center of the distribution of all other N-H · · · N distances appears to be at about 2.95 Å. The shortening of N-H · · · N H-bonds involving the N7 of purines has been previously noted (*46a*). It is due to a somewhat larger negative charge on N7 in the

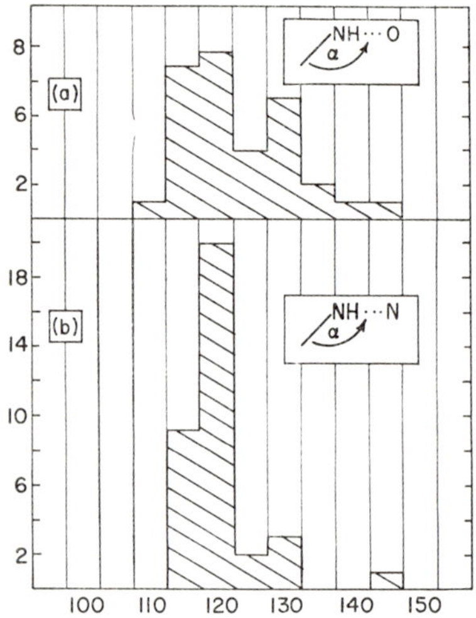

FIG. 6. The distribution of intermolecular H-bond angles, α, for: (a) NH · · · O H-bonds; (b) NH · · · N H-bonds. The abscissa of the figures is the H-bond angle, α, in degrees. The ordinate is the number of H-bond angles in a given angular range.

electronic structures of purines, as discussed by Marsh (*164*). The center of the distribution of N-H · · · O H-bond distances appears to be about 2.85 Å, somewhat shorter than the corresponding quantity for N-H · · · N H-bonds of any type. This result by itself does not establish that the N-H · · · O H-bond is stronger than the N-H · · · N H-bond, since O has a smaller covalent radius than N. However, from a comparison of the electronegativity differences between the donor and the acceptor atoms in these H-bonds (*153*), we would expect that N-H · · · O H-bonds will be somewhat stronger than N-H · · · N bonds. The distribution of H-bond angles for both types of H-bonds appears to be quite similar; both distributions show a peak at about 120° and are skewed somewhat at larger angles.

The results given above add more information to the reviews of H-bonds by Donohue (*165*) and Fuller (*166*) and are also in agreement with the predictions made by Pauling and Corey (*22*) and by Spencer (*25*) for the lengths of H-bonds in the two Watson-Crick pairs in DNA. No attempt is made here to correlate H-bond distances with H-bond angles. However, previous attempts by Donohue (*165*, *168*) and Fuller (*166*) to make such a correlation have shown that the relationship between H-bond lengths and H-bond angles in crystalline materials is apparently a random one.

In the H-bonded structures in which H atoms have been located, the line joining a donor atom to an H atom taking part in H-bonding rarely lies closer than 5° to the line joining the donor atom to the acceptor atom. Donohue (*168*) has discussed this situation and concluded that it is a quite normal phenomenon in crystal structures containing H-bonded complexes.

Of the structures included here, those of caffeine (*66*), caffeine · 5-chlorosalicylic acid (*65*), theophylline · 5-chlorosalicylic acid (*67*), theophylline (*68*), 1,3,7,9-tetramethyluric acid (*69*), 3′-CMP (orthorhombic form) (*91*), uracil (*93a*), 1-methylthymine (*105*), and 5-[1(2′-deoxy-α-D-ribofuranosyl)uracilyl] disulfide (*115*) have been reported to have short CH · · · O contacts. Sutor (*169*) has suggested that these distances may indicate the existence of a CH · · · O H-bond. However, Ramachandran *et al.* (*167*) have suggested that the previously accepted van der Waals contact distance of 2.6 Å for O · · · H suggested by Pauling (*153*) be revised to 2.4 Å. Donohue (*168*) has shown that the CH · · · O contacts that still are considered to be short in light of this revision are due to errors in structural analysis. He therefore concluded that the CH · · · O H-bond probably does not exist in these molecules. This conclusion, however, should not be construed to mean that the CH · · · O bond cannot exist under any circumstances, as there is evidence for its existence in some situations in which the carbon atom becomes positively charged (*170*).

In the hydrates discussed here, there are only a few cases in which the water molecules do not function as H-bond donors, acceptors, or both. It is also frequently observed that the hydroxyl groups of ribose in nucleosides and nucleotides function as H-bond donors, acceptors, or both. They form H-bonds to other ribose hydroxyl groups, to water molecules, or to the electronegative atoms in the purines or pyrimidines. The phosphate group of nucleotides also takes part in such H-bonding. The halogen anion in chloride or bromide salts often acts as an H-bond acceptor. These types of H-bonds are not discussed further here. Hamilton and Ibers (*170*) have recently written a monograph on H-bonding in crystals, and the subject of crystalline hydrates has been recently reviewed by Baur (*171*). A comprehensive treatise on the effects, detection, and properties of H-bonds has been written by Pimentel and McClellan (*172*). In 1952, Donohue pointed out (*165*) that "only in very exceptional cases does a hydrogen atom bonded to nitrogen or oxygen occupy a position such that hydrogen bond formation is impossible." The papers discussed in this review reinforce this statement.

C. Bases Form Hydrogen-Bonded Pairs with Themselves

Most of the crystal structures reviewed here contain only one purine or pyrimidine derivative. Since all the nucleic acid purines and pyrimidines have at least one proton donor and one proton acceptor, they are all capable of forming H-bonded pairs with themselves. This occurs quite frequently and is a reflection of the fact that most H-bonds are formed within crystal structures. Of the 108 crystal structures listed in Table I that contain an individual purine or pyrimidine derivative, 43 form self-pairs in which the purine or pyrimidine is bonded to itself by at least two hydrogen bonds. The diversity of these self-hydrogen-bonded pairs is described here.

Table IV summarizes the literature on pairing arrangements between bases that involve at least two H-bonds. This includes both pairings involving the same base and those involving different bases, as indicated in two separate columns of Table IV. The type of self-pairing geometries are illustrated in the drawings of Figs. 7–14. The numbers listed in the second column of Table IV refer to these figures. In these figures, a dotted line indicates an H-bond and a small dot shows a center of symmetry. The material is presented for derivatives of the various bases, with the figure legends listing the reference numbers for the different crystal structures with that particular pairing. Self-pairing of bases is frequent, both in crystals containing a single base derivative as well as those containing more than one derivative. However, when bulkier substituents, such as sugar residues, are attached to the bases, self-pairing tends to be somewhat less frequent, probably for steric reasons.

Figure 7a shows the pairing made between adenine residues in adenine

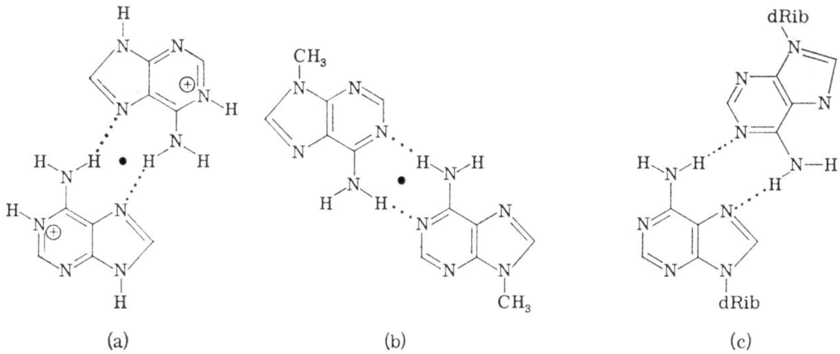

FIG. 7. The pairing between adenine residues as found, for example, in the crystal structures of (a) adenine hydrochloride (*57a,b, 53a, 42*); (b) The complex 9-methyladenine · 1-methyl-5-bromouracil (*47, 52a,b, 53c*); (c) Deoxyadenosine monohydrate (*36*). If more than one complex uses this type of pairing, the additional references are given above.

hydrochloride (*57a,b*), in the complex 9-ethyl-8-bromoadenine · 1-methyl-5-bromouracil (*53a*) and in the complex phenobarbital · (8-bromo-9-ethyladenine)$_2$ (*42*). A projection of the crystal structure of the latter complex is shown in Fig. 18 below. This type of adenine · adenine H-bonding is of interest as it has been shown by Rich *et al.* (*173*) to bind together opposing adenine residues in the protonated form of double helical polyadenylic acid. The same configuration has also been shown to exist (*174*) in concentrated acidic gels of polyadenylic acid. An alternative pairing scheme between adenine residues is found in the structure of the complexes 9-methyladenine · 1-methyl-5-bromouracil (*47*), in the complexes of 9-ethyl-2-aminopurine with 1-methyl-5-fluorouracil (*52a,b*) and with 1-methyl-5-bromouracil (*52b*) and in the complex 9-ethyl-8-bromoadenine · 9-ethyl-8-bromohypoxanthine (*53c*). This involves the amino group and N1 of adenine as is shown in Fig. 7b. Figure 7c shows an additional adenine · adenine pairing geometry that may be considered a hybrid of the two previous arrangements, since N1 is used in one adenine molecule and the N7 is used in the other to form H-bonds with the amino group. This pairing mode is found in the structure of deoxyadenosine monohydrate (*36*). Pullman *et al.* (*175*) calculated the relative stabilities of these various types of self-pairs using molecular orbital methods. They concluded that the pairs illustrated in Figs. 7b and 7c are of equal stability and that, for neutral molecules, the pair illustrated in Fig. 7a is of lower stability.

Figure 8a illustrates the pairing between guanine molecules in guanine hydrochloride monohydrate (*63*). This uses the additional proton added to N7 of the ring. Similar pairing arrangements are found in the crystal structures of theophylline · 5-chlorosalicylic acid (*67*) and theophylline (*68*).

TABLE IV
Pairing Modes between Bases Involving at Least Two Hydrogen Bonds

A. Adenine-Containing Crystals

Crystal[d]	Between adenine derivatives (Fig.)	Between unlike molecules (Fig.)	Dihedral angle between unlike bases (°)
Adenine·HCl	7a	—	—
Deoxyadenosine·H_2O	7c	—	—
9-Methyladenine	7c	—	—
9-Methyladenine·1-Methylthymine	—	15b	0[a]
9-Ethyladenine·1-Methyluracil	—	15b	4.4
9-Ethyladenine·1-Methyl-5-fluorouracil	—	15b	—
9-Ethyladenine·1-Methyl-5-bromouracil	—	15c	4
9-Methyladenine·1-Methyl-5-bromouracil	7b	15b	6.3
Adenosine·5-Bromouridine	—	15c	5.8
9-Ethyl-8-bromoadenosine·1-Methyl-5-bromouracil	7a	15i	6.4
9-Ethyladenine·(1-Methyl-5-iodouracil)$_2$	—	15i, 15b	9.4, 15.7[b]
(8-Bromo-9-ethyladenine)$_2$·Phenobarbital	7a; I:I, II:II	15j	17.5, 16.3[b]
5′Br-5′-deoxyadenosine·Riboflavin·3H_2O	—	15c	—
9EtAde·5iPr5BrAllylbarb.	—	15k	10.5[e]
9Et2AmPur·1Me5FUra	8b	15g	3.8
9Et2AmPur·1Me5BrUra	8b	15l	7.0
9Et2,6Am$_2$Pur·(1Me5FUra)$_2$	8b	15b, 15e	—, —[b]
9Et2,6Am$_2$Pur·(1Me5IUra)	8b	15b, 15d	18.2, 5.4[b]
9Et2,6Am$_2$Pur·(1MeThy)$_2$·H_2O	8b	15c, 15f	7.6, 5.8[b]
9Et8BrAde·9Et8BrHyp	7b	22b	5.3

B. Guanine-Containing Crystals

Crystal	Between guanine derivatives (Fig.)	Between unlike molecules (Fig.)	Dihedral angle between unlike bases (°)
Guanine	8c	—	—
Guanosine	8c	—	—
9-Ethylguanine·1-Methylcytosine	8b	19a	6.5
9-Ethylguanine·1-Methyl-5-fluorocytosine	8b	19a	5.4
Deoxyguanosine·5-Bromodeoxycytidine	—	19a	3.4
9-Ethylguanine·1-Methyl-5-bromocytosine	—	19b	6.5
Guanine·HCl·2H_2O	8b	—	—
Guanine·HCl·H_2O	8a, 8b	—	—

C. Additional Purine Crystals

Crystal	Between like molecules	Between unlike molecules	
Caffeine	8a	—	—
Theophylline	8a	—	—
Theophylline·5-Chlorosalicylic acid	8a	—	—
Uric acid	9	—	—
9Et8BrHyp·9Et8BrAde	—	22b	5.3

D. Cytosine-Containing Crystals

Crystal	Between cytosine derivatives (Fig.)	Between unlike molecules (Fig.)	
Cytosine	10a	—	—
Cytosine·H_2O	10b	—	—
1-Methylcytosine	10d	—	—
Cytosine-5-acetic acid	10e	—	—
Cytosine·Acridine·H_2O	10b	—	—
1-Methylcytosine·9-Ethylguanine	—	19a	6.5
1-Methyl-5-fluorocytosine·9-Ethylguanine	—	19a	5.4
1-Methyl-5-bromocytosine·9-Ethylguanine	—	19a	6.5
5-Bromodeoxycytidine·Deoxyguanosine	—	19b	3.4
Cytosine·5-Fluorouracil·H_2O	10c	22a	—
1-Methylcytosine·5-Fluorouracil	—	22a	—

E. Uracil-Containing Crystals

	Between uracil derivatives (Fig.)	Between unlike molecules (Fig.)	Dihedral angle between unlike bases (°)
Uracil	12a	—	—
1-Methyluracil	12a	—	—
5-Nitrouracil·H_2O	12a	—	—
Thymine·H_2O	12b, 12c	—	—
1-Methylthymine	12a	—	—
5-Ethyl-6-methyluracil	12a, 12b	—	—
2,4-Dithiouracil	12a, 12b	—	—
2,4-Diselenouracil	12a, 12b	—	—
Photodimer of 1MeThy(*trans-anti*)	12a	—	—
1-Methylthymine·9-Methyladenine	—	15b	0[a]
1-Methyluracil·9-Ethyladenine	—	15b	4.4
1Me5FUra·9EtAde	—	15b	—
1Me5BrUra·9EtAde	—	15c	4
1Me5BrUra·9MeAde	—	15b	6.3
5-Bromouridine·Adenosine	—	15c	5.8
1Me5BrUra·9Et8BrAde	—	15i	6.4
$(1Me5IUra)_2$·9EtAde	—	15i, 15b	9.4, 15.7[b]
1Me5FUra·9Et2AmPur	—	15g	3.8
1Me5BrUra·9Et2AmPur	—	15h	7.0
$(1Me5FUra)_2$·9Et2,6Am_2Pur	—	15b, 15e	—, —[b]
$(1Me5IUra)_2$·9Et2,6Am_2Pur	—	15b, 15d	18.2, 5.4[b]
$(1MeThy)_2$·9Et2,6Am_2Pur·H_2O	—	15c, 15f	7.6, 5.8[b]
5-Fluorouracil·Cytosine·H_2O	12a	22a	—
5-Fluorouracil·1-Methylcytosine	12a	22a	—
5-Fluorouracil·9-Ethylhypoxanthine	12a	—	—

F. Barbiturate-Containing Crystals

	Between barbiturate derivatives (Fig.)	Between unlike molecules (Fig.)	Dihedral angle between unlike bases (°)
Barbituric acid	13a	—	—
Ammonium barbiturate	13a	—	—
Alloxan·$3H_2O$	13b	—	—
Alloxan	13a	—	—
Dilituric acid·$3H_2O$	13a, 13b	—	—
Violuric acid·H_2O	13a	—	—
Veronal I	13a	—	—
Veronal II	13b	—	—
Amytal I	13a, 13b	—	—
Amytal II	13a, 13b	—	—
5-(6′Br3′Et2′MeBenzimidazolium) barbiturate	13a; A:A; 13b, A:B	—	—
Dialuric acid·H_2O	13a, 13b	—	—
Dilituric acid	13a	—	—
Phenobarbital·$(8\text{-Bromo-9-ethyladenine})_2$	—	15j	17.5, 16.3[b]
5iPr5BrAllylBarb.·9EtAde	—	15k	10.5[c]

G. Additional Pyrimidine Crystals

	Between like molecules (Fig.)	Between unlike molecules (Fig.)	
4-Amino-2,6-dichloropyrimidine	14a	—	—
5-Bromo-4,6-diaminopyrimidine	14a	—	—
2-(4′$NH_2$5′NH_2pyrimidyl)-1-penten-4-one	14a	—	—
Isocytosine	14b	—	—
Riboflavin·5′Br5′dAdo·$3H_2O$	—	—	17[c]

[a] Planar from considerations of symmetry.
[b] The dihedral angles correspond, in the same order, to the pairing configurations in column 3.
[c] Both dihedral angles are identical by symmetry.
[d] For symbols, see footnote to Table I.

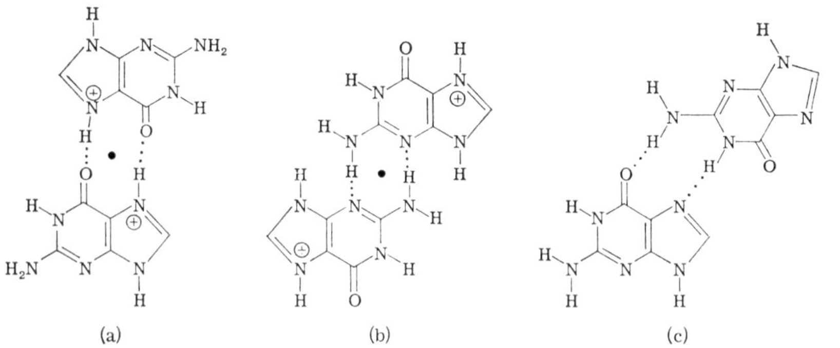

FIG. 8. The pairing modes between guanine residues as found, for example, in the crystal structure of (a) guanine hydrochloride monohydrate (*63, 67, 68*); (b) guanine hydrochloride monohydrate (second mode) (*63, 62a,b, 58a,b, 52a,b, 54a,b, 55*); (c) guanine (*56*).

Figure 8b illustrates an alternate way in which guanine molecules pair. Guanine hydrochloride monohydrate simultaneously pairs in the modes shown in Figs. 8a and 8b. This double pairing gives rise to a structure composed of infinite ribbons of H-bonded guanine molecules. Such extended complexes are commonly observed in purine and pyrimidine structures. A number of other purine derivatives form self-pairs using N3 as shown in Fig. 8b. These include guanine hydrochloride dihydrate (*62a,b*), the guanine residues in the complexes 9-ethylguanine · 1-methylcytosine (*58a,b*) and 9-ethylguanine · 1-methyl-5-fluorocytosine (*58b*), and the purine residues in five structures containing 9-ethyl-2-aminopurine or 9-ethyl-2,6-diaminopurine (*52a,b; 54a,b; 55*). The proximity of an amino group on C2 of these purine derivatives to the ring atom N3 quite readily allows the formation of H-bonded cyclic dimers. This feature is probably of importance in explaining the strong self-association seen in guanine oligonucleotides or even in guanylic acid residues. Figure 8c illustrates the pairing mode between guanine residues in crystals both of guanine and of guanosine (*56*).

Uric acid has three carbonyl groups, two of which are used in forming H-bonds. Figure 9 shows the pairing arrangement found in the structure of uric acid (*74*). It forms infinite ribbons of H-bonded molecules passing through the crystal.

The pyrimidines are smaller than the purines and have fewer electronegative atoms capable of acting as acceptors in H-bond formation. Nonetheless, these molecules form base-pairs as readily as the purine derivatives. The H-bonding arrangements of cytosine derivatives is shown in Fig. 10. In cytosine itself (*82*), the molecule forms a tetramer held together to form an infinite H-bonded network (Fig. 10a). Figure 10b illustrates the pairing found in cytosine monohydrate (*83*). This is identical to that found between two anhydrous cytosine molecules (Fig. 10a). However, the H-bonding

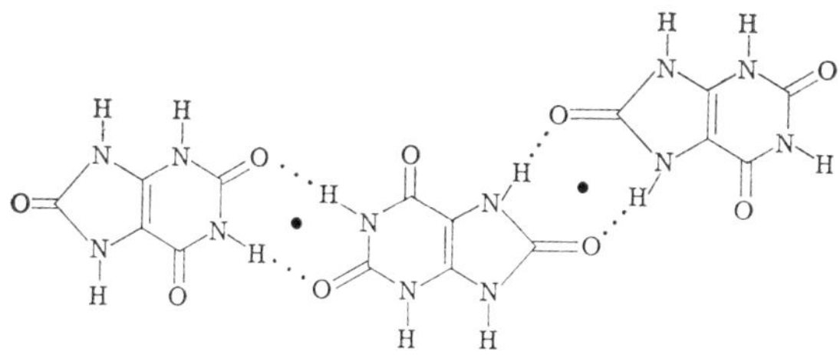

FIG. 9. The type of H-bonding found in the crystal structure of uric acid (*74*).

system of the tetramer of cytosine does not form in the monohydrate because of the H-bonding of the water molecule in the latter. Cytosine residues in the crystalline complex cytosine · acridine monohydrate (*89*) also form pairs by the mode illustrated in Fig. 10b, thereby forming infinite chains of H-bonded cytosine molecules. Figure 10c illustrates the pairing con-

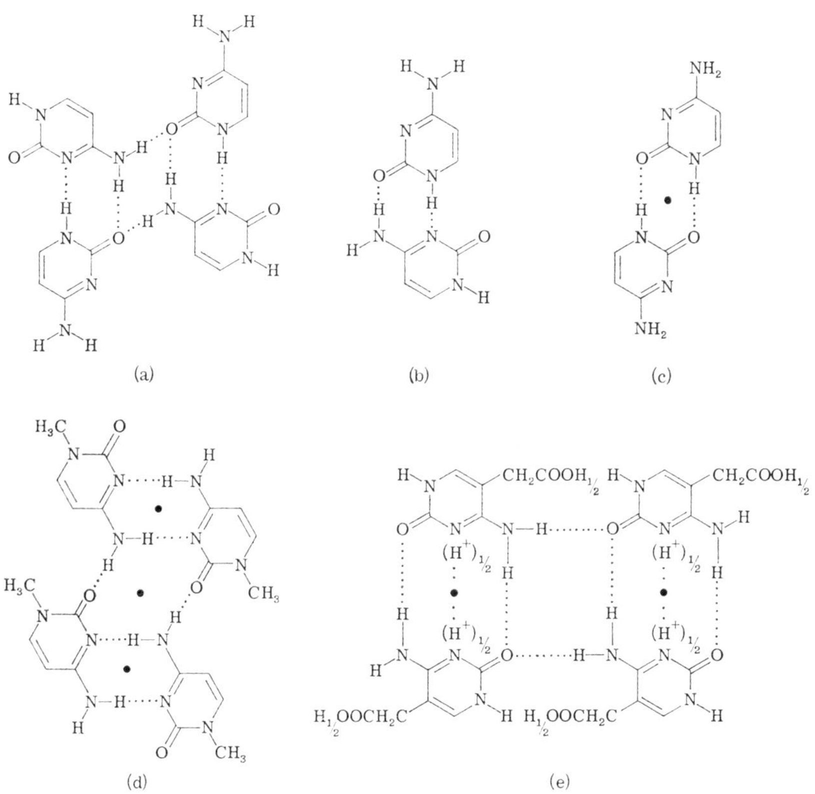

FIG. 10. The types of H-bonding found between cytosine residues as, for example, in the crystal structures of (a) cytosine (*82*); (b) cytosine monohydrate (*83, 89*); (c) the complex cytosine · 5-fluorouracil monohydrate (*84*); (d) 1-methylcytosine (*87*); (e) cytosine-5-acetic acid (*88*).

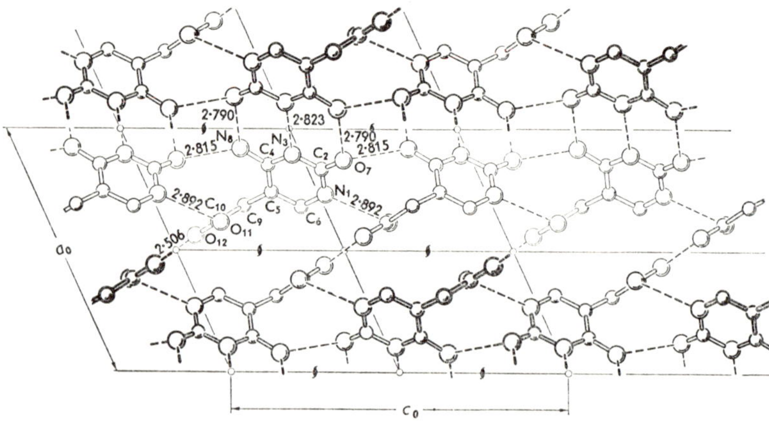

FIG. 11. The crystal structure of cytosine-5-acetic acid as viewed down the *b* axis. The H-bonds are represented by dashed lines. Their lengths are given in Ångstrom units. Taken from Marsh *et al.* (*88*).

figuration found between cytosine residues in the structure of the complex cytosine · 5-fluorouracil monohydrate (*84*). This structure is shown in more detail in Fig. 23 below. Another type of cytosine · cytosine pairing is found in 1-methylcytosine (*87*) (Fig. 10d). The structure consists of an infinite double ribbon of H-bonded molecules. The structure of 1-methylcytosine is quite unusual, however, in that the two molecules forming the H-bonded pair lie in parallel planes displaced from one another by 1.5 Å. The pairing in the structure is made possible by a rotation of the cytosine amino group out of the cytosine plane. It is possible that this unusual structure optimizes its packing energy. Figure 10e illustrates the structure of cytosine-5-acetic acid (*88*), which has the unusual combination of tautomeric forms discussed above (Section III, B). A projection of the crystal structure of this compound is shown in Fig. 11. Hemiprotonation of atom N3 allows the triply H-bonded pair to form in a manner identical to that of the Pauling and Corey modification of the Watson-Crick guanine · cytosine base-pair. This same system of H-bonds is believed to hold together opposing cytosine residues in the double helical form of polycytidylic acid, as Langridge and Rich (*176*) have shown in X-ray diffraction studies. Hartman and Rich (*177*) found the same pairing in solution studies of helical polycytidylic acid.

The types of H-bonded pairing found in uracil derivatives are somewhat simpler than those found in cytosine, perhaps because uracil has fewer protons for H-bonding. Of the twelve uracil derivatives containing uracil · uracil pairs, eleven form the H-bonded arrangement shown in Fig. 12a. These include uracil (*93a,b*), 1-methyluracil (*94*), 5-nitrouracil monohydrate (*95*), 2,4-dithiouracil (*102*), 2,4-diselenouracil (*103*), 1-methylthymine (*105*), 5-ethyl-6-methyluracil (*107*), the *trans-anti* photodimer of 1-methylthymine (*108*) and the complexes 5-fluorouracil · cytosine monohydrate (*84*), 5-fluorouracil · 1-methylcytosine (*85*), and 5-fluorouracil · 9-ethylhypoxanthine (*78*). The twelfth uracil derivative forming self-pairs is thymine monohydrate (*104*), one of the two self-pairing uracil structures containing a water molecule that takes part in H-bonding. Here the thymine simultaneously forms self-pairs of the type shown in Fig. 12b and Fig. 12c to form an infinite chain of thymine molecules. In this structure, 04 is H-bonded to the water molecule, thus preventing the formation of the structure found in Fig. 12a. 2,4-Dithiouracil (*102*), 2,4-diselenouracil (*103*), and 5-ethyl-6-methyluracil (*107*) also form a second self-pair of the type illustrated in Fig. 12b. The large preponderance of uracil derivatives that pair in the form shown in Fig. 12a suggests that the association constant for this geometry may be much larger than that of other uracil · uracil interactions. Pullman *et al.* (*175*) have come to a similar conclusion based on theoretical grounds. The projection of the structure of the complex cytosine · 5-fluorouracil monohydrate, which contains this predominant uracil · uracil pair, is shown in Fig. 23 below.

The *trans-anti* photodimer of thymine (*109a,b*) is the only example among the purine, pyrimidine, and barbiturate structures of a molecule with the chemical capability of forming H-bonded cyclic dimers with neighboring molecules that forms only single H-bonds with them. The reasons for this anomaly are not apparent, especially in light of the existence of an H-bonded cyclic dimer in the structure of a closely related molecule, that of the *trans-anti* photodimer of 1-methylthymine (*108*).

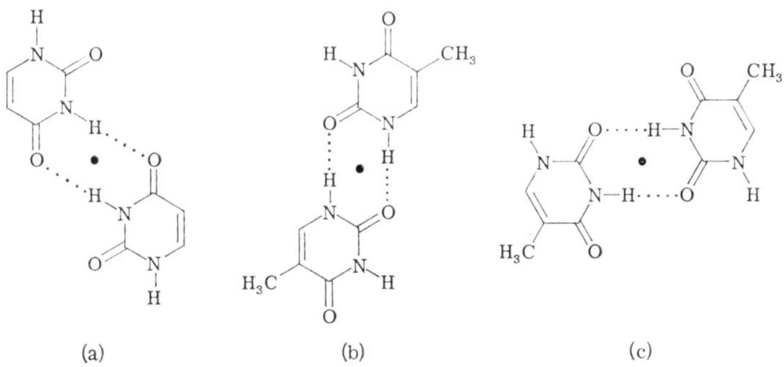

FIG. 12. The types of H-bonding between uracil residues as found, for example, in the crystal structures of (a) uracil (*93a,b*, *94*, *95*, *102*, *103*, *105*, *107*, *108*, *84*, *85*, *78*); (b) thymine monohydrate (*104*, *102*, *103*, *107*); (c) thymine monohydrate (alternate mode) (*104*). The number of references in (a) indicate that that type of H-bonding has been found in 11 different crystal structures.

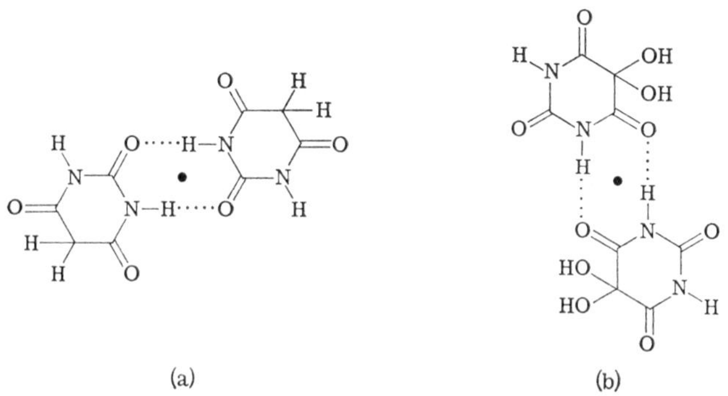

FIG. 13. Hydrogen bonding between barbiturate residues as found, for example, in the crystal structures of (a) barbituric acid (*120, 122, 125, 127–132*); (b) alloxan trihydrate (*123, 128, 130–132*).

There are only two types of pairing interactions found in structures of barbiturate derivatives (Figs. 13a and 13b). The pairing shown in Fig. 13b is analogous to the uracil pairing of Fig. 12a, whereas Fig. 13a illustrates a pairing analogous to both the uracil pairing shown in Fig. 12b and that in Fig. 12c. However, there are eleven structures in which barbiturate derivatives self-pair as shown in Fig. 13a, and only seven in which the barbiturate derivatives self-pair by the scheme shown in Fig. 13b. Thus the addition of the carbonyl oxygen to the C6 of uracil probably changes the order of the association constants of the various possible self-pairs.

Figure 14a shows the pairing found in 4-amino-2,6-dichloropyrimidine (*137*), 5-bromo-4,6-diaminopyrimidine (*137*), and 2-(4′-amino-5′-amino-

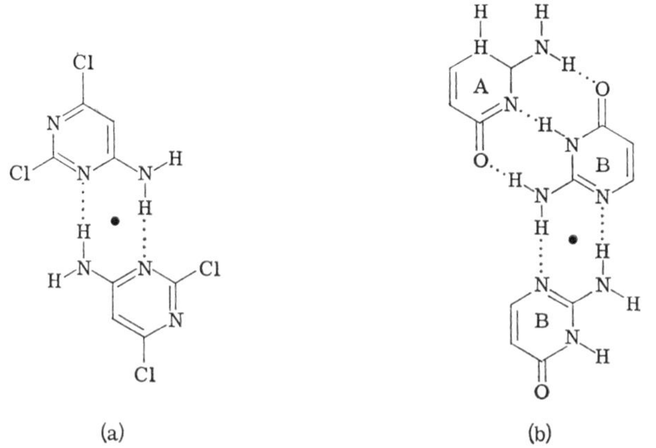

FIG. 14. The H-bonding found between pyrimidine derivatives as seen, for example, in the crystal structures of (a) 4-amino-2,6-dichloropyrimidine (*137, 138a,b*); (b) isocytosine (*139a,b*).

pyrimidyl)-2-penten-4-one (*138a,b*). This pairing is similar to that found in the self-pairing of 1-methylcytosine (Fig. 10d). Figure 14b shows the self-pairing scheme of isocytosine (*139a,b*). As mentioned above (Section III, B), this structure is remarkable in that it is another example in which two tautomers of the same molecule are found in a crystal structure. The H-bonding arrangement between these two pairs is identical to that of the Pauling and Corey modification of the Watson-Crick guanine · cytosine base pair.

Donohue (*178a*) and Donohue and Trueblood (*178b*) have enumerated all the self-pairs that have been found in neutral derivatives of adenine, guanine, cytosine, and uracil that are blocked at the glycosyl nitrogen before most of these derivatives had been discovered. However, there are a number of self-pairs of derivatives of guanine, cytosine, and uracil whose possible existence Donohue has predicted but that have not yet been observed experimentally.

It has been previously observed that a structure rarely fails to take advantage of its full potential to form H-bonds. An interesting exception to this principle is found in the structure of 5-oxobarbituric acid (*124*). This molecule with two N-H groups and 4 carbonyl groups would normally be expected to form four N-H · · · O H-bonds with other such molecules. However, no H-bonds are found in its structure. Instead, the closest intermolecular contacts in the structure of 5-oxobarbituric acid are the C5 · · · O6 and C6 · · · O6 distances, which are 2.79 and 2.97 Å, respectively. These contacts are shorter than the normal C · · · O van der Waals contact of 3.1 Å (*153*). Thus, it is possible that the main forces responsible for the intermolecular interactions in the crystal of 5-oxobarbituric acid may be electrostatic interactions between dipolar carbonyl groups. Perhaps the existence of this rare form of intermolecular interaction in the crystal of 5-oxobarbituric acid is due to the high concentration of carbonyl groups in the molecule of this substance.

D. Bases Hydrogen-Bond with Other Bases

The most striking property of the purine and pyrimidine bases and that which confers on them biological specificity is their ability to form specific, complementary base pairs. The variety of crystals containing more than one type of base is illustrated in the 25 crystal structures listed in Table I that contain intermolecular H-bonded complexes. Twenty-one of these contain H-bonded pairs of guanine and cytosine derivatives or either pairs or triplets containing H-bonded adenine and uracil derivatives. These pairs are therefore complementary in the sense described in the solution studies of H-bonding discussed above (Section IV, A). Many types of mixtures have been made, but only these cocrystallized. The third column of Table

IV lists all the available data on H-bonded pairs between unlike bases involving at least two H bonds. For the sake of convenience, this information is entered in Table IV in the sections concerning both derivatives that take part in the pair. The numbers listed in the third column of Table IV refer to the line drawings in Figs. 15, 19 and 22, which show the pairing configurations.

Adenine has two potential pairs of acceptor and donor sites for H-bonding to uracil; N1 and the amino group, or the imidazole nitrogen, N7 and the amino group. The existence of these pairs of sites was first illustrated clearly by the formation of helical complexes involving one strand of polyadenylic acid with two strands of polyuridylic acid (*179*). These sites

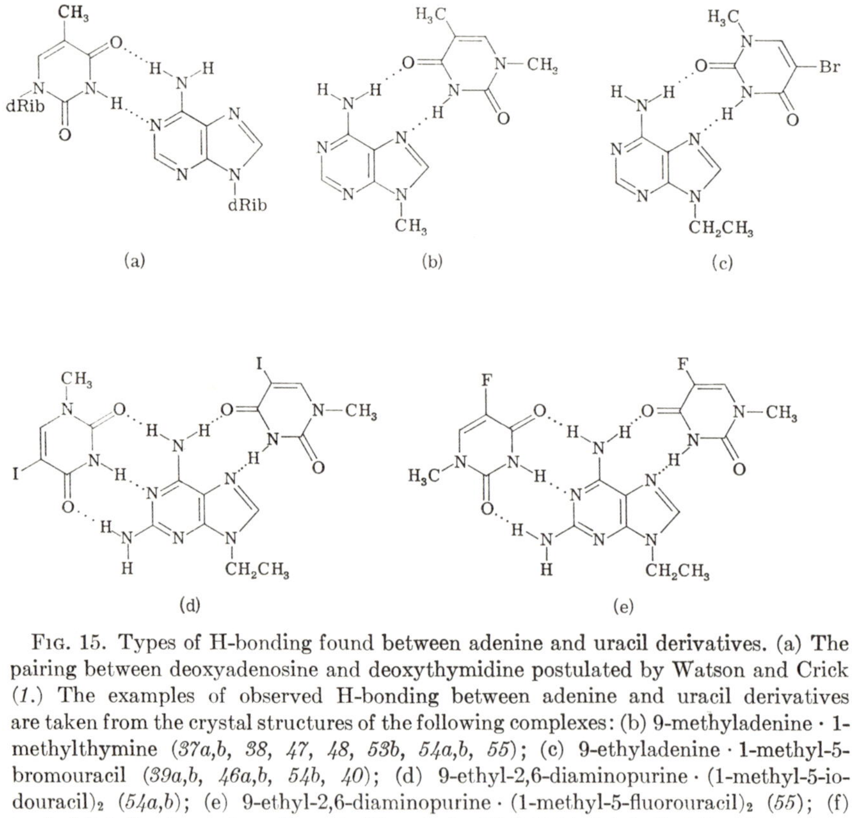

FIG. 15. Types of H-bonding found between adenine and uracil derivatives. (a) The pairing between deoxyadenosine and deoxythymidine postulated by Watson and Crick (*1*.) The examples of observed H-bonding between adenine and uracil derivatives are taken from the crystal structures of the following complexes: (b) 9-methyladenine · 1-methylthymine (*37a,b, 38, 47, 48, 53b, 54a,b, 55*); (c) 9-ethyladenine · 1-methyl-5-bromouracil (*39a,b, 46a,b, 54b, 40*); (d) 9-ethyl-2,6-diaminopurine · (1-methyl-5-iodouracil)$_2$ (*54a,b*); (e) 9-ethyl-2,6-diaminopurine · (1-methyl-5-fluorouracil)$_2$ (*55*); (f) 9-ethyl-2,6-diaminopurine · (1-methylthymine)$_2$ · H_2O (*54b*); (g) 9-ethyl-2-aminopurine · 1-methyl-5-fluorouracil (*52a,b*); (h) 9-ethyl-2-aminopurine·1-methyl-5-bromouracil (*52b*); (i) 8-bromo-9-ethyladenine·1-methyl-5-bromouracil (53*a,b*); (j) phenobarbital·(8-bromo-9-ethyladenine)$_2$ (*42*); (k) 9-ethyladenine · 5-isopropyl-5-bromallylbarbituric acid (*41*). Where the same type of pairing is found in more than one crystal structure, the additional references are cited.

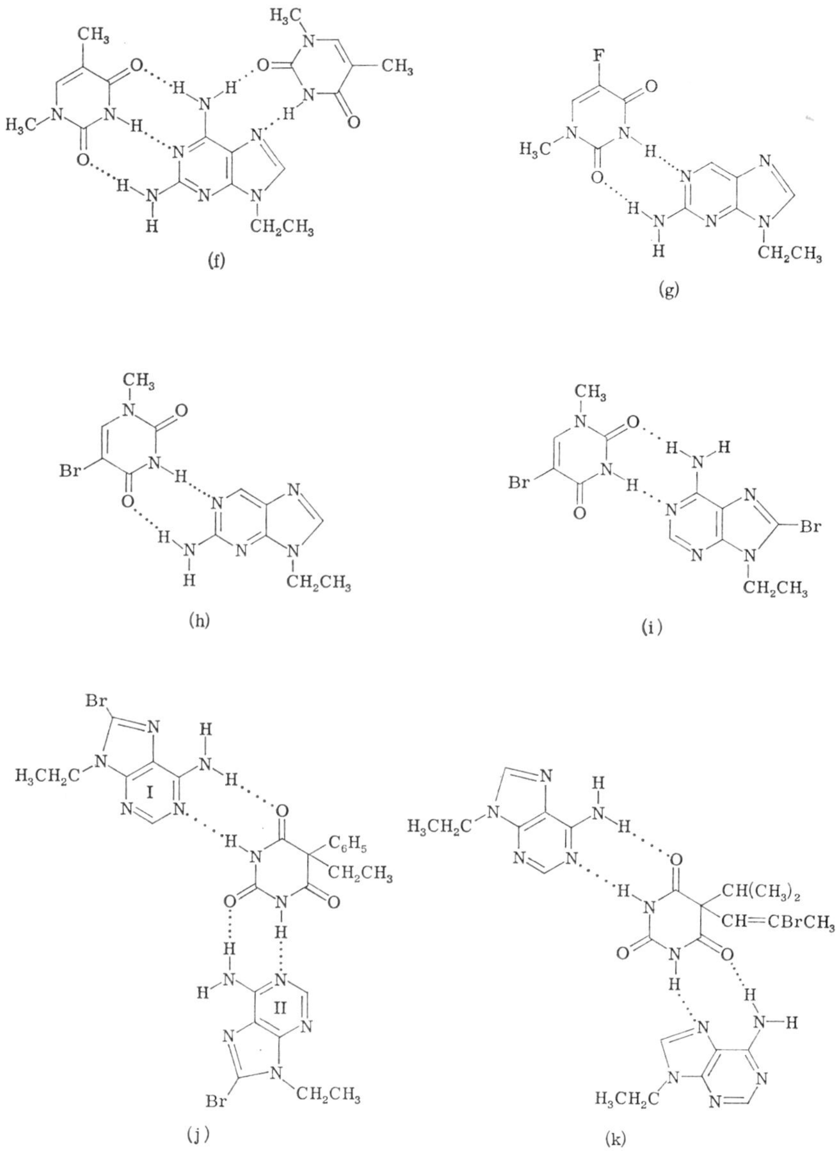

FIG. 15 *(Continued)*

have both been used for pairing in the various crystal structure studies of H-bonded adenine · uracil pairs, although a majority of the 1 : 1 complexes involve H-bonding through N7. The derivatives of adenine can in principle engage in four different types of H-bonding with uracil derivatives. Thus, there can be H-bonding using adenine N1 in bonding to uracil NH-3 and either O4 (the Watson-Crick pair) or O2 of uracil in bonding to the adenine amino group. Two additional pairing configurations can occur when bonding uracil NH-3 to the adenine N7, using either uracil O2 or O4 in bonding to the adenine amino group. All four of these potential types of H-bonding have now been formed. Several complexes have been found involving three bases, two uracil derivatives and one of adenine. In the complexes 9-ethyl-2,6-diaminopurine · (1-methyl-5-fluorouracil)$_2$ (*55*), and 9-ethyl-2,6-diaminopurine . (1-methylthymine)$_2$. H_2O (*54b*), NH-3 of one uracil residue is bonded to the adenine N1 and its carbonyl O4 is bonded to the adenine amino group. At the present time these are the only examples of Watson-Crick pairing between an adenine and a uracil derivative.

Figure 15a is a drawing of the base pair formed between deoxyadenosine and deoxythymidine that was postulated by Watson and Crick to occur in DNA (*1*). It uses N1 of adenine. This pair has never been experimentally observed by itself in crystal structures of complexes containing normal adenine and uracil derivatives. When Hoogsteen (*37a,b*) determine the structure of the complex 9-methyladenine · 1-methylthymine, he discovered that the adenine and the thymine molecule paired in the mode illustrated

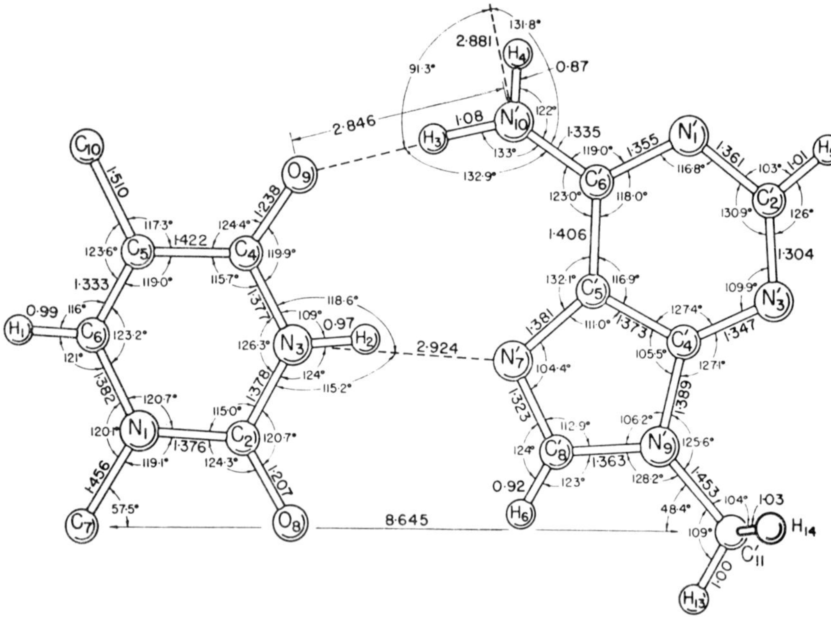

FIG. 16. The molecular dimensions of the intermolecular complex 9-methyladenine · 1-methylthymine. Hydrogen bonds are indicated by dashed lines. Bond distances are given in Ångstrom units. Taken from Hoogsteen (*37b*).

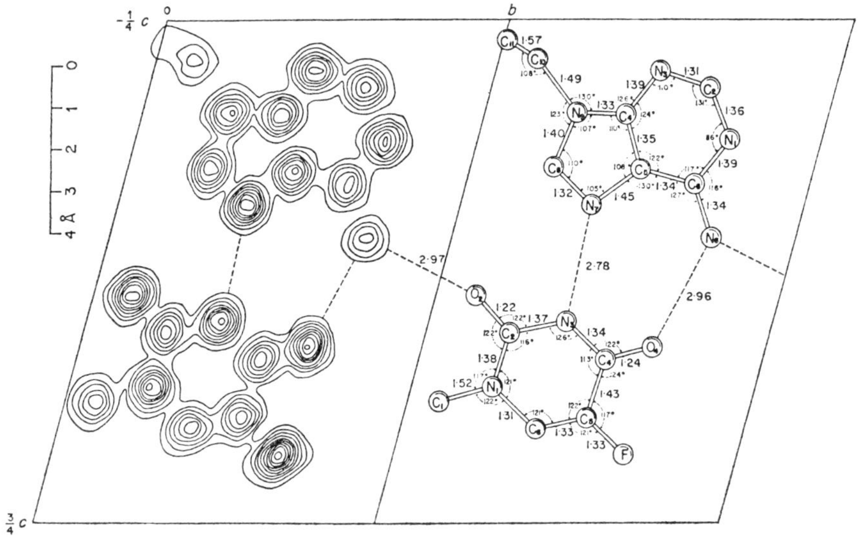

FIG. 17. The molecular arrangement in the crystal structure of the intermolecular complex 9-ethyladenine · 1-methyl-5-fluorouracil. The left side of the figure is a section at $a = 8/30$ of the electron density in the unit cell. Contours are drawn at intervals of approximately 1 electron/Å, starting with the contour $\rho = 1$. The zero contour and negative contours have been omitted. Because of the slight tilt off the plane of the base pair with respect to the *b-c* plane, some atoms are better resolved in other sections. The right half of the figure depicts the adjacent unit cell with bond lengths, given in Ångstrom units, and bond angles indicated. Hydrogen bonds are indicated by dashed lines. Taken from Tomita *et al.* (*48*).

in Fig. 15b. This was the first crystal structure showing this pairing, although it had been previously postulated to occur in the triple-stranded 1:2 complex of polyadenylic acid and polyuridylic acid (*179*). This imidazole structure or bonding (which is sometimes called the Hoogsteen structure) involves N7 of the adenine imidazole ring as a H-bond acceptor atom. It has also been observed in the following six complexes: 9-ethyladenine · 1-methyluracil (*38*), 9-methyladenine · 1-methyl-5-bromouracil (*47*), 9-ethyladenine · 5-fluorouracil (*48*), 9-ethyladenine · (1-methyl-5-iodouracil)$_2$ (*53b*), 9-ethyl-2,6-diaminopurine · (1-methyl-5-iodouracil)$_2$ (*54a,b*) and 9-ethyl-2,6-diaminopurine · (1-methyl-5-fluorouracil)$_2$ (*55*). Figures 16 and 17 show projections of the crystal structures of the complexes 9-methyladenine · 1-methylthymine and 9-ethyladenine · 1-methyl-5-fluorouracil. These figures illustrate the nature of the imidazole bonding. A closely related type of adenine · uracil pair, similar to the imidazole pair but using uracil O2 instead of O4 as the H-bond acceptor, has been observed in the complex adenosine · 5-bromouridine (*39a,b*) as well as in the complexes 9-ethyladenine · 1-methyl-5-bromouracil (*46a,b*), and 9-ethyl-2,6-diaminopurine · (1-methylthymine)$_2$ (*54b*). This reversed imidazole structure is illustrated in Fig. 15c. One can also consider the pairing mode of 5′-bromo-5′-deoxyadenosine with riboflavin (*40*) to be that of the reversed imidazole type if we assume that the pyrimidine ring of isoalloxazine is analogous to uracil. It is of interest that both the imidazole bonding configurations can coexist in the same crystal structure. It is estimated that 6% of the uracil molecules in the complex 9-ethyladenine · 1-methyl-5-bromouracil (*46a,b*) are disordered so that they have the imidazole structure, while 94% of the uracil derivatives have the reversed imidazole bonding. This disorder does not significantly distort the crystal structure because 1-methyl-5-bromouracil has an axis of pseudosymmetry through atoms N3 and C6 due to the fact that a methyl group and a bromine atom are sterically equivalent. Katz *et al.* (*46a,b*) suggested that the complexes adenosine · 5-bromouridine and 9-ethyladenine · 1-methyl-5-bromouracil might form the reversed imidazole structure rather than the imidazole structure because of the inductive effect of the electronegative bromine atom through the pyrimidine ring that tends to make uracil atom O2 more electronegative than uracil atom O4. However, this is unlikely, since Baklagina *et al.* (*47*) subsequently found that a similar complex, 9-methyladenine · 1-methyl-5-bromouracil, exclusively forms the imidazole structure. Molecular orbital calculations[4] (*175*) suggest that the stabilities of some of the possible adenine · thymine pairing schemes are in the following decreasing order: the imidazole structure, the reversed imidazole structure, and finally, the Watson-Crick structure. However, the stabilities of the imidazole and the reversed imidazole structures are almost the same. As mentioned above (Section IV, B), N7 is more electronegative than N1 in purines so that it would tend to form stronger H-bonds.

All three of the structures that contain 9-ethyl-2,6-diaminopurine form H-bonded pairs with two uracil derivatives, each of which is crystallographically independent. The complex 9-ethyl-2,6-diaminopurine · (1-methyl-5-iodouracil)$_2$ (*54a,b*) is illustrated in Fig. 15d. It pairs with three H-bonds to the uracil residue bonding to adenine N1. Here uracil carbonyl O2 H-bonds to the adenine amino group. The uracil residue bonding to adenine N1 in the complex 9-ethyl-2,6-diaminopurine · (1-methyl-5-fluorouracil)$_2$ (*55*) is similar but uses uracil O4 to bind to the adenine amino group as shown in Fig. 15e. Thus it is analogous to the Watson-Crick pair, except that it is further stabilized by an extra H-bond joining together O2 on the uracil derivative with the amino group bonded to C2 of the purine derivative. In both of these foregoing complexes the second uracil derivative is bound in the imidazole configuration. In the complex 9-ethyl-2,6-diaminopurine · (1-methylthymine)$_2$ · H_2O (*54b*) (Fig. 15f), one uracil derivative is in the "Watson-Crick" configuration while the other uracil derivative is in the reversed imidazole configuration. Molecular orbital calculations (*180*) suggest that the pairing scheme involving a Watson-Crick pair and a reversed imidazole pair is the most stable pairing configuration for a trimer consisting of 2,6-diaminopurine and uracil in 1:2 stoichiometry. However, the stability of the pairing configuration containing a Watson-Crick pair and an imidazole pair is only slightly less. Figures 15g and h illustrate the pairing configurations that 9-ethyl-2-aminopurine forms with uracil residues in its complexes with 1-methyl-5-fluorouracil (*52a,b*) and with 1-methyl-5-bromouracil (*52b*). These pairing arrangements use the purine N1, which is also used in the Watson-Crick pair between adenine and uracil.

Model building studies with 8-bromoadenine derivatives show that the bromine atom sterically prevents the formation of the imidazole or the reversed imidazole configurations of adenine · uracil complexes. Such a complex would then be forced to assume either the Watson-Crick configuration or the closely related reversed Watson-Crick configuration, in which uracil atom O2 rather than O4 is the H-bond acceptor. The complex 9-ethyl-8-bromoadenine · 1-methyl-5-bromouracil (*53a*), which is illustrated in Fig. 15i, is in the latter configuration. The second pair in the complex 9-ethyladenine · (1-methyl-5-iodouracil)$_2$ (*53b*) is also in this configuration. This latter complex is the only case of a structure that contains normal adenine and uracil derivatives in 1:2 stoichiometry.

An interesting variation of adenine · uracil pairing is seen in the barbiturate intermolecular complexes. Barbiturates may be regarded as analogs of pseudouridine in that they have substituents attached to atom C5 while pseudouridine has ribose attached to the same atom. The additional proton on N1 of pseudouridine makes a second H-bonding site (N1 and carbonyl O2) that can be used for pairing in addition to the site involving N3 and carbonyl O4 that is used in the Watson-Crick pairing. Thus, in principle, a derivative of pseudouridine should be capable of forming a complex involving two derivatives of adenine. An example of this is seen in the complex phenobarbital · (8-bromo-9-ethyladenine)$_2$ (*42*). A projection of the structure of the complex is found in Fig. 18. It can be seen that the barbiturate is H-bonded to one adenine on the Watson-Crick site (N3 and O4) and to another adenine on the additional site (N1 and O2). It is possible that these two sites for H-bonding in pseudouridine could be used in maintaining the tertiary structure of transfer RNA.

In the complex phenobarbital · (8-bromo-9-ethyladenine)$_2$ (*42*) (Fig. 15j), a single phenobarbital molecule is associated with two crystallographically independent 8-bromo-9-ethyladenine molecules. However, there are a number of other crystalline complexes containing both adenine deriva-

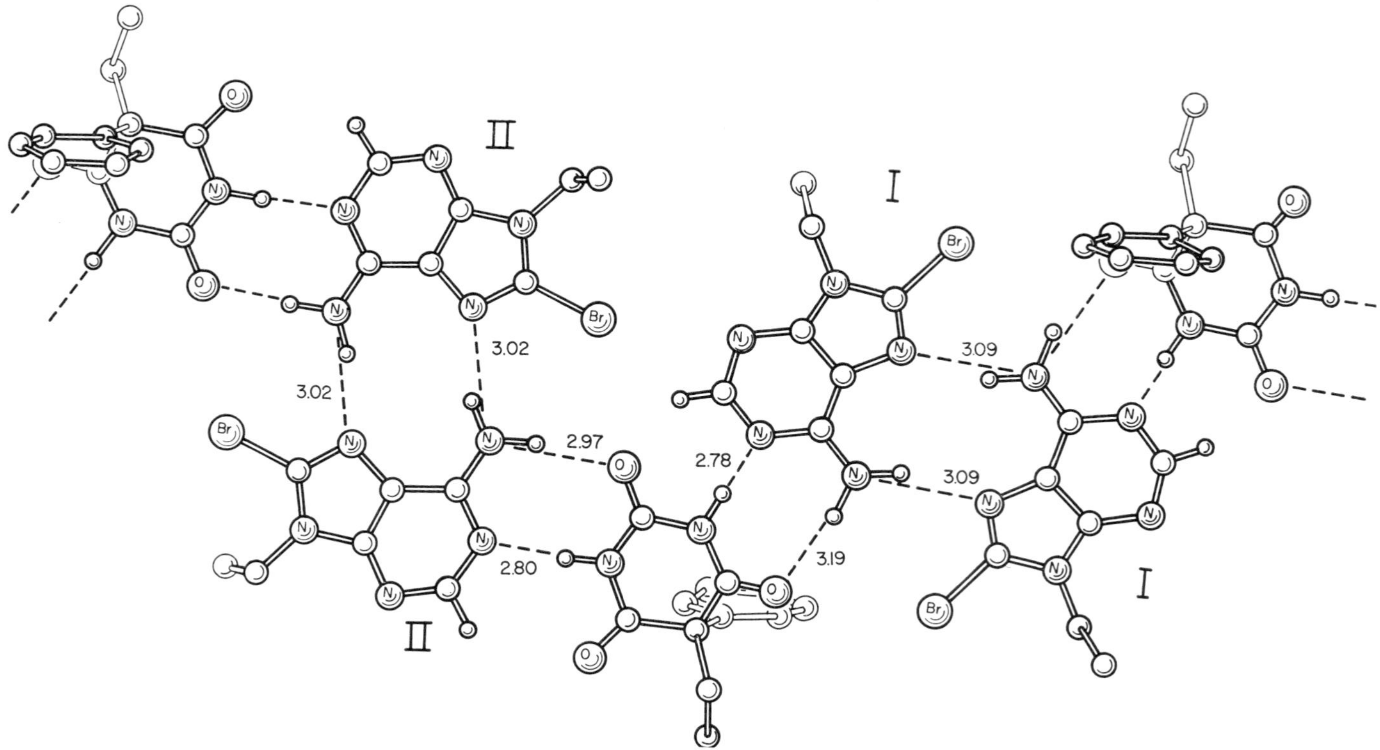

FIG. 18. View of the molecular plane in the intermolecular complex phenobarbital · (8-bromo-9-ethyladenine)$_2$. The H-bonds are represented by dashed lines. Their lengths are given in Ångstrom units. The two crystallographically independent adenine derivatives in the complex are labeled I and II. Taken from Kim and Rich (*42*).

tives and barbiturate derivatives that crystallize in 1:1 stoichiometry (*181*). Figure 15k illustrates the modes of pairing that allow the formation of the infinite H-bonded chains of alternating adenine and barbiturate residues found in the crystalline complex 9-ethyladenine · 5-isopropyl-5-bromallyl-barbituric acid (*41*). Assuming the analogy between the barbiturate and the uracil rings it can be seen that these pairing modes consist of a Watson-Crick pair and an imidazole pair.

There is considerable interest in finding adenine · uracil complexes that have the Watson-Crick structure. There has been much work directed toward establishing the existence of this pairing configuration in double-helical DNA and RNA. X-Ray diffraction studies on fibers of these materials combined with model-building investigations have shown that it is likely that only Watson-Crick pairs exist in double-helical DNA (*4–7*) and RNA (*11–15*). Specifically, the study of low resolution Fourier maps of fibers of lithium DNA led Arnott *et al.* (*5*) and Marvin *et al.* (*7*) to assert that the existence of imidazole pairs in the lithium DNA structure was unlikely. Similar conclusions have been reached by Fuller *et al.* (*6*) concerning fibers of sodium DNA. However, the level of discrimination of these types of analyses have been called into serious question by Donohue (*202a,b*). Nevertheless, the Watson-Crick structure still seems to be the most likely one for DNA. The Watson-Crick structure has also been postulated to occur in the double-stranded 1:1 complex of polyadenylic acid with polyuridylic acid by Sasisekharan and Sigler (*182*). Thus it appears that although the Watson-Crick configuration for an adenine · uracil base-pair may be less stable than some other pairing configurations, steric and other local environmental factors may be important in restricting adenine and uracil (or thymine) residues on opposing strands of double-helical polynucleotides to form only Watson-Crick pairs. However, it is not clear whether similar constraints are important in the structure of transfer RNA or in codon–anticodon interactions.

In contrast to the great diversity of adenine · uracil complexes, the guanine · cytosine crystal structures are all quite similar. This difference in behavior is undoubtedly a direct consequence of the three H-bonds between the latter residues that confer added stability to the pair, both in crystal structures and in solution. The structures of four intermolecular complexes of a guanine and a cytosine derivative have been reported. They are 9-ethylguanine · 1-methylcytosine (*58a,b*), 9-ethylguanine · 1-methyl-5-fluorocytosine (*58b*), deoxyguanosine · 5-bromodeoxycytidine (*59a,b*), and 9-ethylguanine · 1-methyl-5-bromocytosine (60). In all of these, the pairing between guanine and cytosine is the Pauling and Corey modification of the Watson-Crick structure (*22*). This triply H-bonded pairing scheme is shown in Fig. 19a. The structure of the complex 9-ethylguanine · 1-methyl-

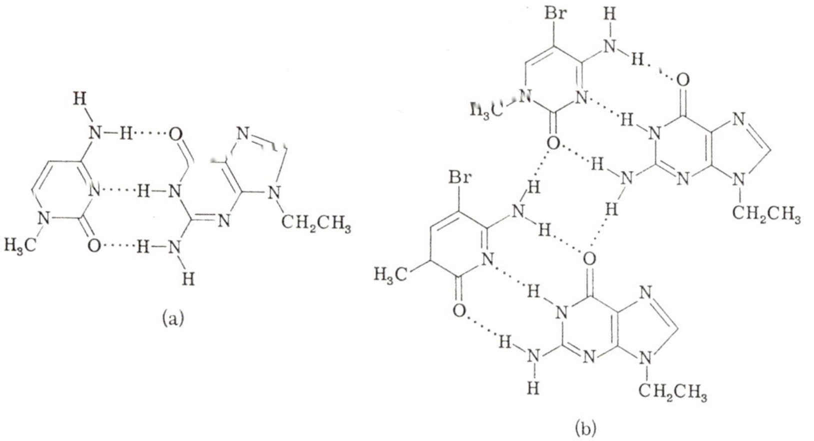

FIG. 19. The types of pairing between guanine and cytosine derivatives as found, for example, in the crystal structures of the complexes (a) 9-ethylguanine · 1-methylcytosine (*58a,b, 59a,b*); (b) 9-ethylguanine · 1-methyl-5-bromocytosine (*60*).

5-bromocytosine is of interest since it also forms a H-bonded tetramer, as is shown in Fig. 19b. This H-bonding configuration causes the structure of this complex to form an infinite double ribbon of H-bonded guanine and cytosine residues. Figure 20 shows a projection of the crystal structure of the complex of deoxyguanosine · 5-bromodeoxycytidine. This figure illustrates how the pairing configuration participates in the crystal structure of the complex.

The existence of "electronic complementarity" in solution is the most plausible reason for the finding that very few crystalline complexes contain other base pairs than those of the type adenine · uracil or guanine · cytosine. There are, however, four interesting exceptions. Three of these involve complexes of 5-fluorouracil, a molecule that has a proton rather than some other substituent attached to atom N1, its glycosyl nitrogen atom. In 9-ethylhypoxanthine · 5-fluorouracil (*78*), there is no cyclic dimer bonding the two different types of molecules. Instead, 9-ethylhypoxanthine and 5-fluorouracil are associated through single H-bonds. The projection of the crystal structure of this complex is shown in Fig. 21. The structures of the complexes cytosine · 5-fluorouracil monohydrate (*84*) and 1-methylcytosine · 5-fluorouracil (*85*) are quite similar. The pairing configuration in both these crystals is illustrated in Fig. 22a. The projection of the structure of cytosine · 5-fluorouracil monohydrate is shown in Fig. 23. These pairs are unusual in that the protonated N1 of 5-fluorouracil is involved in the H-bonding system. The proton on N1 is not present in nucleotides or nucleosides. It is likely that this proton has an effect in modifying the electronic complementarity of the complexing molecules

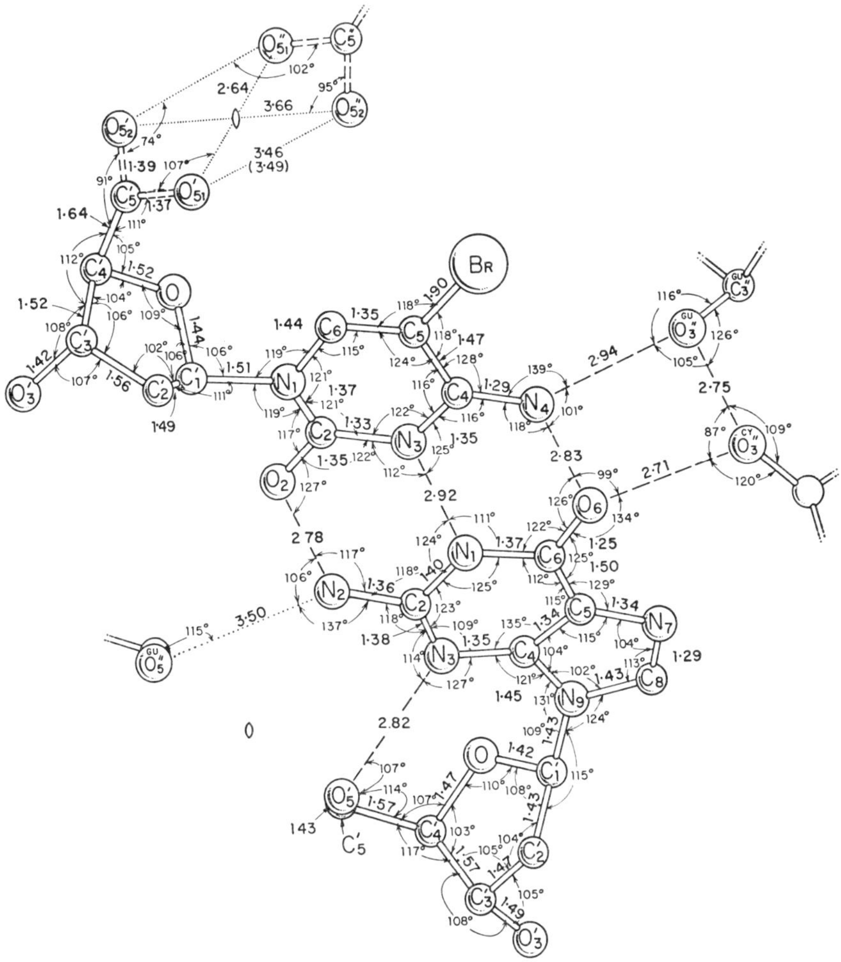

FIG. 20. The view along the *a* axis of the crystal structure of the intermolecular complex deoxyguanosine · 5-bromodeoxycytidine. Atoms belonging to neighboring molecules are labeled GU for deoxyguanosine and CY for 5-bromodeoxycytidine. Hydrogen bonds are indicated with dashed lines. Dotted lines indicate the contact between atom N2 of guanine and neighboring atom O5′ of deoxyguanosine (labeled GU O5″) and also the various contacts for two positions of atom O5′ of 5-bromodeoxycytidine across the twofold axis. The conformations of the two deoxyribose residues are apparent in the projection. Bond lengths are given in Ångstrom units. Taken from Haschemeyer and Sobell (*59b*).

since 1-methyl-5-fluorouracil forms an imidazole complex with 9-ethyladenine (*48*) but does not form complexes with cytosine derivatives.

The remaining unusual base pair involves adenine and hypoxanthine derivatives. Chloroform-soluble derivatives of adenine and hypoxanthine

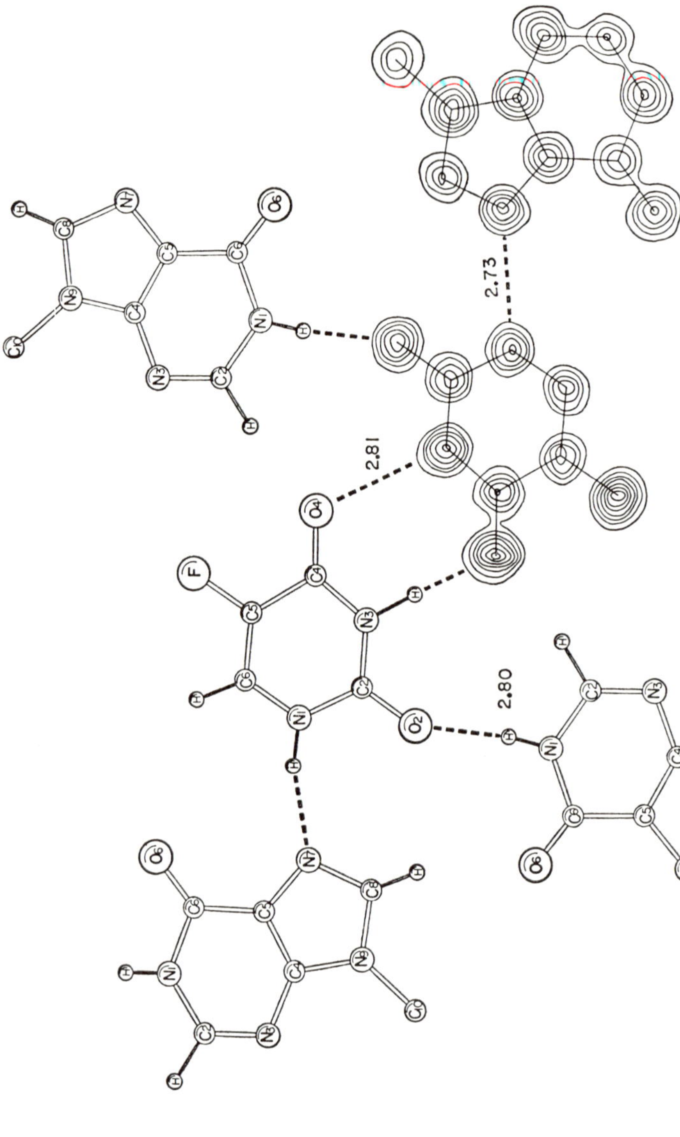

FIG. 21. The electron density map in the plane of the molecular sheet of the 5-fluorouracil · 9-ethylhypoxanthine structure. The contours occur at arbitrary levels of electron density in this section. The terminal carbon on the ethyl group of 9-ethylhypoxanthine does not lie in the plane. The H-bond lengths are indicated in Ångstrom units. A center of symmetry lies between the paired 5-fluorouracil molecules. Taken from Kim and Rich (*78*).

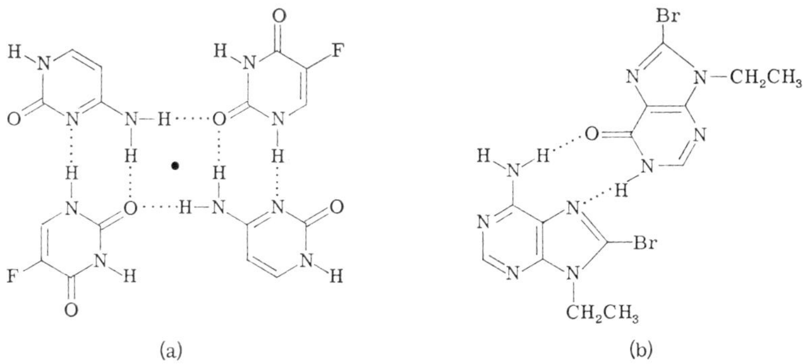

FIG. 22. (a) The H-bonded pairing found with cytosine and 5-fluorouracil (*84, 85*). (b) The pairing between the adenine and hypoxanthine components found in the complex 8-bromo-9-ethyladenine · 8-bromo-9-ethylhypoxanthine (*53c*).

will form H-bonded dimers in solution (*186*) as well as in the solid state (*53c*). The water-soluble polynucleotides, polyinosinic acid and polyadenylic acid, can form two-stranded helices that probably use the same type of H-bonding (*200*). The structure of the crystalline complex 8-bromo-9-ethyladenine · 8-bromo-9-ethylhypoxanthine has been determined (*53c*) (Fig. 22b). This complex is of biological significance because hypoxanthine is known to occur at the third position of the anticodon site of certain species of transfer-RNA. It has been shown by Söll *et al.* (*187a,b*) that these transfer-RNA molecules respond to synthetic messenger RNA sequences containing codons whose third position contains either adenine, cytosine, or uracil. In accounting for such degeneracies in the genetic code, Crick (*3*) has postulated that the third position in codon–anticodon pairing has a certain amount of "wobble" that would permit the formation of an adenine · hypoxanthine base-pair. However, the adenine · hypoxanthine base-pairing configuration postulated by Crick differs from that illustrated in Fig. 22b in that the H-bond acceptor group of the adenine is atom N1 rather than atom N7, a difference that is analogous to that between the Watson-Crick and the imidazole pairing configurations in adenine · uracil complexes. Sakore and Sobell (*53c*) have suggested that an undistorted adenine · hypoxanthine pairing of the sort found in the crystalline complex would be possible at the codon–anticodon site if the adenine ring was in the *syn* rather than the usual *anti* conformation relative to its covalently bound ribose group.

Table IV also lists the known dihedral angles between the least squares planes of the bases forming intermolecular complexes. This information is not presented for pairs between identical bases because two identical planar molecules are often complexed about a center of symmetry and are therefore

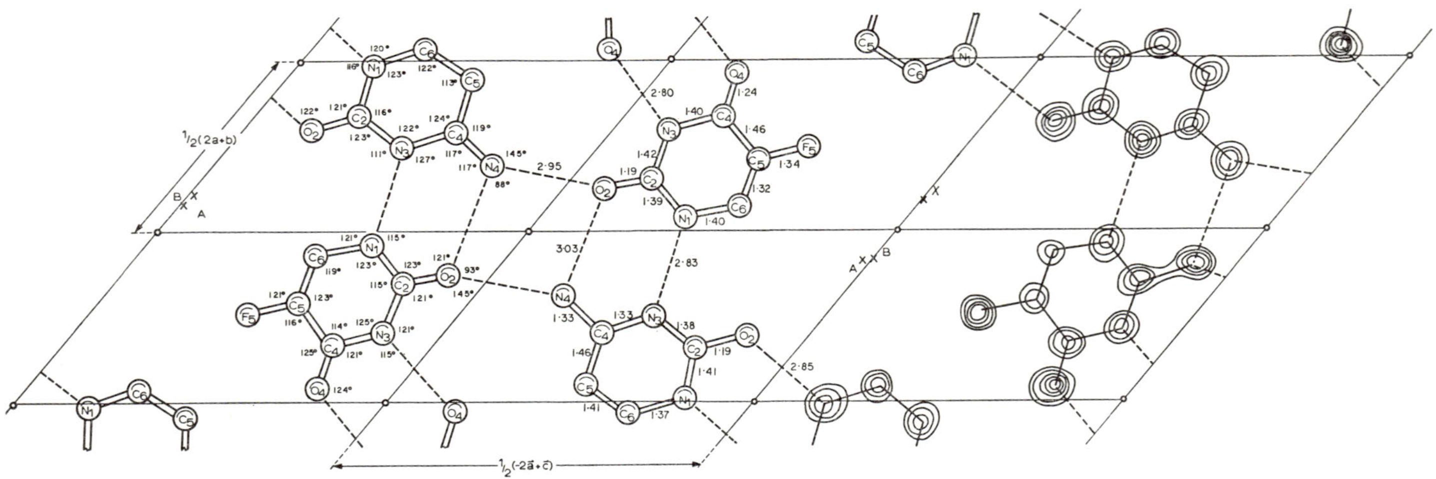

FIG. 23. The $1\bar{2}2$ section of the electron density map of the intermolecular complex cytosine · 5-fluorouracil monohydrate together with a schematic drawing of the molecular structure indicating the bond lengths in Ångstrom units and the bond angles. Hydrogen bonds are indicated by dashed lines. The projected positions of water oxygen atoms A and B are indicated by small ×'s. Taken from Voet and Rich (*84*).

parallel. The data show that in most cases there are only small deviations from coplanarity. The nature of this deviation, whether it is a relative bending or twisting of the molecular planes or a combination of these distortions, is rarely specified. The deviations of the bases from coplanarity is often of the same order of magnitude as the deviations of the atoms of a single base from perfect planarity. Thus it seems reasonable to conclude that these distortions are usually made in response to packing forces in the crystals. However, it is interesting to note that, of the five dihedral angles greater than 10°, all occur in complexes consisting of a trimer of crystallographically unrelated molecules (*41*, *42*, *53b*, *54a,b*).

V. Summary and Conclusions

The object of this review is to document the present status of the X-ray crystallographic studies on the purines and pyrimidines, and to point out the relevance of some of these findings to the chemistry and biology of the nucleic acids. In tabulating the geometry of individual purine or pyrimidine derivatives, we have drawn a number of general conclusions. Thus, for example, all adenine derivatives are generally similar in terms of bond lengths and angles. However, it is also important to note that there are differences in the geometry of the adenine molecule depending upon its environment in the crystal structure. It is often difficult to dissociate the variations seen in the geometry of a particular base in different complexes from the variabilities that are introduced by the fact that different crystal structure analyses have different degrees of reliability. For this reason, we included an introductory discussion of reliability. Reliability factors are listed in Table I to permit an independent assessment of the validity of a particular crystallographic finding. The conclusion to be drawn from the tabular data in the appendices is that the structures of the individual purine and pyrimidine bases retain a fair amount of constancy in a variety of different crystal lattices. However, there are some differences among these molecules that are often real in that the molecule is somewhat responsive to its nearest neighbors in a crystal lattice. The tabular data attempt to quantitate the extent of this variability.

Special attention is directed to the H-bonds that are formed between the purines and pyrimidines. In contrast to the fairly narrow range of covalent-bond lengths that are found within the molecules, there is considerable variation in both the length of H-bonds and their angles, a point that is emphasized in this presentation. Since there is an appreciable range in the geometry of H-bonded interactions, it is likely that a greater variety of stable structures can be formed.

The selectivity of H-bond formation is, of course, the phenomenon that imparts great specificity to nucleic acid interactions. This specificity derives from two sources, one geometrical and the other electronic in nature. The geometrical specificity is seen most clearly in the structure of the polymeric, macromolecular nucleic acids. The electronic specificity is seen in the complementarity that is a feature of the interactions of monomeric purine and pyrimidine derivatives. In solution this is expressed in terms of strong H-bonding between adenine and uracil residues, or guanine and cytosine residues and perhaps weaker bonding between hypoxanthine and adenine derivatives. In the crystal studies, the specificity is expressed by the fact that the majority of the crystalline complexes illustrate the adenine · uracil or the guanine · cytosine pairings.

There is considerable variability in terms of the types of H-bonds that are found between purines and pyrimidines. Although we have attempted to include in this review a complete listing of the different types of H-bonding interactions (Table IV), it is likely that additional types of H-bonding will be found in crystals. An important variable in this connection is the variation in H-bond lengths, since the energy of H-bonding interactions is characterized in part by the length of this bond. The recent extension of our knowledge into the study of the structure of crystalline transfer RNA (*18, 19*) makes it possible that additional types of H-bonding interactions will be found. These may include some unusual codon–anticodon H-bonding interactions. This survey of purine and pyrimidine crystallography underlines the diversity of the H-bonding potentiality of the nucleic acid bases; it is likely that this diversity has been utilized fully by nature in the evolution of the information-transferring system.

Acknowledgments

This work was supported, in part, by a National Institutes of Health Training Grant and by research grants from the National Science Foundation, the American Cancer Society, and the National Aeronautics and Space Administration.

Appendix Tables

APPENDIX I

A. Adenine Bond Distances (Å)

Name \ Bond:	N1–C2	C2–N3	N3–C4	C4–C5	C5–C6	C6–N1	C5–N7	N7–C8	C8–N9	N9–C4	C6–N6	N9–C1′
1. Neutral Molecules												
dAdo (*36*)	1.317	1.326	1.346	1.392	1.414	1.336	1.375	1.307	1.361	1.369	1.331	1.472
9MeAde (*189*)[c]	1.348	1.322	1.338	1.365	1.395	1.348	1.379	1.311	1.354	1.359	1.348	1.468
9MeAde·1MeThy (*37b*)	1.361	1.304	1.347	1.373	1.406	1.355	1.381	1.323	1.363	1.389	1.335	1.453
9EtAde·1MeUra (*38*)	1.30	1.31	1.34	1.35	1.41	1.38	1.41	1.28	1.38	1.38	1.34	1.50
9EtAde·1Me5FUra (*48*)[a]	1.36	1.31	1.39	1.35	1.34	1.39	1.45	1.32	1.40	1.33	1.34	1.49
9EtAde·1Me5BrUra (*46b*)	1.33	1.32	1.37	1.35	1.40	1.36	1.40	1.31	1.38	1.36	1.32	1.46
9MeAde·1Me5BrUra (*47*)[a]	1.41	1.33	1.36	1.43	1.39	1.40	1.40	1.38	1.45	1.34	1.40	1.53
5′Br5′dAdo·Riboflavin·3H_2O (*40*)[a]	1.24	1.37	1.44	1.45	1.43	1.43	1.45	1.23	1.31	1.32	1.29	1.45
Ado·8BrUrd (*39b*)	1.34	1.31	1.36	1.37	1.39	1.30	1.36	1.29	1.37	1.34	1.39	1.51
$(8Br9EtAde)_2$·Phenobarbital (*42*) I	1.348	1.322	1.326	1.361	1.418	1.336	1.390	1.306	1.334	1.386	1.331	1.472
$(8Br9EtAde)_2$·Phenobarbital (*42*) II	1.325	1.311	1.351	1.360	1.392	1.358	1.402	1.266	1.369	1.362	1.340	1.485
9EtAde·5iPr5BrAllylBarb. (*41*)[c]	1.375	1.296	1.365	1.392	1.371	1.366	1.431	1.265	1.393	1.356	1.372	1.483
9Et2,6Am_2Pur·$(1Me5FUra)_2$ (*55*)[a]	1.43	1.20	1.37	1.33	1.46	1.35	1.32	1.33	1.39	1.41	1.40	1.46
Average	1.332	1.315	1.349	1.365	1.404	1.346	1.388	1.297	1.365	1.370	1.341	1.479
σ	0.022	0.008	0.011	0.014	0.012	0.027	0.018	0.021	0.016	0.018	0.023	0.021
2. Rings Protonated on Position N1												
5′-AMP (*43*)	1.368	1.312	1.341	1.403	1.448	1.362	1.364	1.328	1.398	1.377	1.312	1.492
3′-AMP·2H_2O (*44*)	1.349	1.306	1.353	1.381	1.401	1.363	1.384	1.312	1.368	1.355	1.319	1.477
Average	1.358	1.309	1.347	1.392	1.424	1.362	1.374	1.320	1.383	1.366	1.315	1.485
σ	0.013	0.004	0.008	0.016	0.033	0.001	0.014	0.011	0.021	0.016	0.005	0.011
3. Rings Protonated on Positions N1 and N7												
9MeAde·2HBr (*45a,c*)	1.36	1.35	1.37	1.40	1.38	1.39	1.37	1.33	1.35	1.37	1.30	1.38

B. GUANINE BOND DISTANCES (Å)

1. Neutral Molecules

Name \ Bond:	N1–C2	C2–N3	N3–C4	C4–C5	C5–C6	C6–N1	C5–N7	N7–C8	C8–N9	N9–C4	C2–N2	C6–O6	N9–C1′
9EtGua·1MeCyt (*58b*)	1.388	1.326	1.357	1.388	1.412	1.400	1.393	1.302	1.376	1.375	1.335	1.233	1.485
9EtGua·1Me5FCyt (*58b*)	1.375	1.336	1.362	1.363	1.427	1.405	1.395	1.320	1.381	1.381	1.336	1.223	1.467
dGuo·5BrdCyd (*59b*)[a]	1.40	1.38	1.35	1.34	1.50	1.37	1.34	1.29	1.43	1.45	1.36	1.25	1.43
Average	1.381	1.331	1.359	1.375	1.419	1.402	1.394	1.311	1.378	1.378	1.335	1.228	1.476
σ	0.009	0.007	0.004	0.018	0.011	0.004	0.001	0.013	0.004	0.004	0.001	0.007	0.013
2. Rings Protonated on Position N7													
9MeGua·HBr (*61*)	1.42	1.33	1.37	1.32	1.43	1.38	1.39	1.34	1.40	1.39	1.33	1.18	1.41
Gua·HCl·2H_2O (*62b*)	1.374	1.318	1.345	1.377	1.414	1.391	1.378	1.322	1.335	1.375	1.339	1.237	—
Average	1.397	1.324	1.357	1.348	1.422	1.385	1.384	1.331	1.367	1.382	1.334	1.208	1.410
σ	0.033	0.008	0.018	0.040	0.011	0.008	0.008	0.013	0.046	0.011	0.006	0.040	—

C. ADDITIONAL PURINE BOND DISTANCES (Å)

Name \ Bond:	N1–C2	C2–N3	N3–C4	C4–C5	C5–C6	C6–N1	C5–N7	N7–C8	C8–N9	N9–C4	C2–O2	C6–O6	N9–C1′
Purine (*64*)	1.349	1.324	1.336	1.398	1.385	1.330	1.375	1.337	1.311	1.379	—	—	—
Caffeine (*66*)	1.42	1.35	1.42	1.32	1.44	1.36	1.41	1.32	1.34	1.31	1.19	1.26	—
Caffeine·5-Chlorosalicylic acid (*65*)	1.397	1.392	1.373	1.361	1.441	1.413	1.388	1.343	1.341	1.361	1.211	1.208	—
Theophylline·5-Chlorosalicylic acid (*67*)	1.404	1.359	1.376	1.370	1.421	1.387	1.373	1.332	1.334	1.354	1.215	1.217	—
Theophylline (*68*)	1.40	1.35	1.37	1.37	1.41	1.38	1.34	1.31	1.31	1.33	1.19	1.19	—
Ino-5′-P [Na salt] (*76*)	1.38	1.27	1.30	1.33	1.44	1.36	1.40	1.43	1.33	1.39	—	1.22	1.46
1,3,7,9-Me_4Uric (*69*)	1.369	1.387	1.356	1.349	1.412	1.405	1.386	1.353	1.407	1.369	1.209	1.237	1.480
1,3,7,9-Me_4Uric · Pyrene (*71b*)	1.34	1.45	1.35	1.40	1.42	1.42	1.42	1.36	1.44	1.35	1.25	1.24	1.47
Uric acid (*74*)	1.367	1.382	1.356	1.360	1.411	1.397	1.387	1.359	1.376	1.360	1.223	1.233	—
6SIno (*77*) I	1.365	1.285	1.348	1.383	1.414	1.392	1.362	1.331	1.366	1.391	—	—	1.437
6SIno (*77*) II	1.349	1.318	1.359	1.381	1.440	1.367	1.388	1.319	1.377	1.374	—	—	1.468
9EtHyp·5FUra (*78*)	1.37	1.29	1.37	1.36	1.42	1.40	1.38	1.32	1.38	1.38	—	1.24	1.46
7*H*-Tetrazolo[5,1-*i*]-purine·H_2O (*80*)	1.39	1.29	1.36	1.40	1.41	1.36	1.37	1.33	1.33	1.36	—	—	—

APPENDIX I (*Continued*)

D. CYTOSINE BOND DISTANCES (Å)

Name \ Bond:	N1–C2	C2–N3	N3–C4	C4–C5	C5–C6	C6–N1	C2–O2	C4–N4	N1–C1′
1. Neutral Molecules									
Cytosine (*82*)	1.374	1.364	1.337	1.424	1.342	1.357	1.234	1.330	—
Cytosine·H_2O (*83*)	1.376	1.354	1.351	1.432	1.348	1.361	1.260	1.332	—
Cytidine (*86*)	1.379	1.361	1.335	1.420	1.345	1.368	1.246	1.333	1.497
1MeCyt·9EtGua (*58b*)	1.401	1.349	1.345	1.419	1.345	1.362	1.233	1.337	1.468
1Me5FCyt·9EtGua (*58b*)	1.403	1.360	1.334	1.446	1.336	1.345	1.235	1.324	1.477
5BrdCyd·dGuo (*59b*)[a]	1.37	1.33	1.35	1.47	1.35	1.44	1.35	1.29	1.51
Cyt·5FUra·H_2O (*84*)	1.41	1.38	1.33	1.46	1.41	1.37	1.19	1.33	—
1MeCyt·5FUra (*85*)	1.40	1.34	1.34	1.43	1.37	1.36	1.26	1.28	1.43
Average	1.392	1.358	1.339	1.433	1.357	1.360	1.237	1.324	1.468
σ	0.015	0.013	0.007	0.015	0.026	0.008	0.024	0.020	0.028
2. Rings Hemiprotonated on Position N3									
Cytosine-5-acetic acid (*88*)	1.351	1.366	1.353	1.427	1.363	1.365	1.250	1.323	
3. Rings Protonated on Position N3									
3′-CMP(orthorhombic) (*91*)	1.401	1.382	1.339	1.420	1.351	1.350	1.201	1.323	1.475
3′-CMP(monoclinic) (*92*)	1.389	1.403	1.350	1.403	1.365	1.361	1.207	1.315	1.485
1MeCyt·HBr (*45b*)[a]	1.37	1.34	1.33	1.39	1.36	1.36	1.23	1.35	1.48
Average	1.395	1.392	1.344	1.411	1.358	1.355	1.204	1.319	1.480
σ	0.008	0.015	0.008	0.012	0.010	0.008	0.004	0.006	0.007

E. URACIL BOND DISTANCES (Å)

Name \ Bond:	N1–C2	C2–N3	N3–C4	C4–C5	C5–C6	C6–N1	C2–O2	C4–O4	N1–C1′
1. Neutral Molecules									
Uracil (*93b*)	1.371	1.376	1.371	1.430	1.340	1.358	1.215	1.245	—
1-Methyluracil (*94*)	1.38	1.38	1.38	1.42	1.35	1.37	1.23	1.25	1.46
5-Nitrouracil·H_2O (*95*)	1.384	1.370	1.377	1.453	1.358	1.334	1.199	1.224	—
2,4-Dithiouracil (*102*)[b]	1.342	1.406	1.358	1.414	1.365	1.360	—	—	—
5-Fluorodeoxyuridine (*98*)	1.394	1.377	1.373	1.433	1.331	1.365	1.198	1.235	1.481
5-Bromouridine (*97a,b*)	1.40	1.37	1.39	1.43	1.36	1.35	1.23	1.23	1.49
5-Bromodeoxyuridine (*97a,b*)	1.39	1.39	1.38	1.49	1.34	1.42	1.19	1.19	1.40

5-Iododeoxyuridine (*99*)	1.37	1.38	1.36	1.49	1.34	1.37	1.23	1.21	1.49
4-Thiouridine (*116*)[b]	1.40	1.37	1.37	1.44	1.34	1.37	1.20	—	1.46
S-S(5SdUrd)$_2$ (*115*)[b] 1	1.373	1.388	1.392	1.450	1.353	1.354	1.209	1.241	1.505
S-S(5SdUrd)$_2$ (*115*)[b] 2	1.394	1.392	1.384	1.437	1.368	1.363	1.204	1.230	1.502
Urd-5′-P[Ba salt] (*96*)	1.39	1.41	1.40	1.41	1.38	1.38	1.22	1.22	1.44
5FOro[Rb salt]·H_2O (*100*)[a]	1.35	1.39	1.37	1.44	1.33	1.37	1.22	1.23	—
Dihydrouracil (*101*)[b]	1.34	1.39	1.38	1.50	1.51	1.47	1.24	1.22	—
(Ura)$_2$[*cis-syn* photodimer] (*112*)[b] 1	1.336	1.390	1.359	1.499	1.543	1.441	1.225	1.217	—
(Ura)$_2$[*cis-syn* photodimer] (*112*)[b] 2	1.332	1.397	1.364	1.499	1.536	1.437	1.230	1.220	—
1MeUra·9EtAde (*38*)	1.35	1.37	1.36	1.45	1.40	1.38	1.25	1.25	1.50
1Me5FUra·9EtAde (*48*)	1.38	1.37	1.34	1.43	1.33	1.31	1.22	1.24	1.52
1Me5BrUra·9EtAde (*46a,b*)	1.36	1.37	1.39	1.42	1.30	1.36	1.21	1.21	1.49
1Me5BrUra·9MeAde (*47*)[a]	1.42	1.37	1.37	1.45	1.35	1.39	1.23	1.28	1.56
5BrUrd·Ado (*39a,b*)	1.37	1.33	1.39	1.45	1.35	1.34	1.25	1.23	1.53
(1Me5FUra)$_2$·9Et2, 6Am$_2$Pur (*55*)[a]									
N7 bonding	1.34	1.40	1.40	1.40	1.31	1.42	1.23	1.19	1.52
N1 bonding	1.38	1.41	1.28	1.46	1.33	1.37	1.14	1.28	1.51
5FUra·Cyt·H_2O (*84*)	1.39	1.42	1.40	1.46	1.32	1.40	1.19	1.24	—
5FUra·1MeCyt (*85*)	1.34	1.40	1.38	1.44	1.31	1.38	1.24	1.23	—
5FUra·9EtHyp (*78*)	1.35	1.39	1.37	1.44	1.33	1.37	1.22	1.23	—
Average	1.374	1.381	1.380	1.444	1.343	1.370	1.219	1.233	1.476
σ	0.019	0.022	0.013	0.024	0.026	0.022	0.020	0.023	0.028

2. Neutral Thymine Molecules

Name \ Bond:	N1–C2	C2–N3	N3–C4	C4–C5	C5–C6	C6–N1	C2–O2	C4–O4	C5–CM	N1–C1′
Thymine·H_2O (*104*)	1.355	1.361	1.391	1.447	1.349	1.382	1.234	1.231	1.503	—
1-Methylthymine (*105*)	1.379	1.379	1.375	1.432	1.346	1.383	1.214	1.237	1.497	1.470
5Et6MeUra (*107*)	1.360	1.363	1.381	1.455	1.356	1.377	1.240	1.240	1.503	—
5′-TMP[Ca salt] (*106*)	1.365	1.377	1.414	1.452	1.310	1.367	1.247	1.218	1.526	1.466
1MeThy·9MeAde (*37b*)	1.376	1.378	1.377	1.422	1.333	1.382	1.207	1.238	1.510	1.456
Photodimers:										
Thy$_2$(*trans-anti*) (*109*)[b]	1.334	1.390	1.357	1.508	1.547	1.440	1.227	1.211	1.526	—
(1,3Me$_2$Thy)$_2$ (*cis-anti*) (*111*) 1	1.344	1.406	1.377	1.503	1.529	1.449	1.218	1.213	1.513	1.459
(1,3Me$_2$Thy)$_2$ (*cis-anti*) (*111*) 2	1.342	1.426	1.378	1.509	1.533	1.431	1.218	1.203	1.522	1.462
Average	1.367	1.372	1.388	1.442	1.339	1.378	1.228	1.223	1.508	1.464
σ	0.010	0.009	0.016	0.014	0.018	0.007	0.017	0.009	0.011	0.007

APPENDIX I (*Continued*)

3. Uracil Derivatives Protonated on O4

1-Methyluracil·HBr (*118*)	1.38	1.38	1.40	1.34	1.35	1.33	1.23	1.28	

F. Barbiturate Bond Distances (Å)

1. Neutral Molecules

Name \ Bond:	N1–C2	C2–N3	N3–C4	C4–C5	C5–C6	C6–N1	C2–O2	C4–O4	C6–O6
Barbituric acid·$2H_2O$ (*120*)	1.378	1.356	1.388	1.503	1.501	1.367	1.217	1.211	1.219
Barbituric acid (*121*)	1.36	1.35	1.37	1.49	1.47	1.39	1.23	1.20	1.19
Alloxan·$3H_2O$ (*123*)	1.38	1.38	1.37	1.53	1.53	1.37	1.21	1.22	1.22
5-Oxobarbituric acid (*124*)	1.39	1.39	1.36	1.52	1.52	1.36	1.21	1.22	1.22
Alloxan (*125*)	1.37	1.37	1.37	1.54	1.54	1.38	1.22	1.21	1.21
Alloxantin·$2H_2O$ (*126*)	1.38	1.37	1.39	1.55	1.52	1.38	1.23	1.19	1.21
Dilituric acid (*127*)	1.37	1.36	1.35	1.41	1.45	1.38	1.22	—	1.22
Dilituric acid·$3H_2O$ (*128*) A	1.35	1.35	1.38	1.44	1.44	1.37	1.25	1.26	1.24
Dilituric acid·$3H_2O$ (*128*) B	1.36	1.36	1.38	1.44	1.43	1.39	1.22	1.23	1.23
Violuric acid·H_2O (*129*)	1.36	1.38	1.38	1.47	1.50	1.37	1.22	1.21	1.17
Veronal (*130*) I	1.367	1.379	1.367	1.523	1.514	1.381	1.210	1.209	1.211
Veronal (*130*) II	1.379	1.379	1.366	1.514	1.514	1.366	1.196	1.219	1.219
Amytal (*130*) I	1.366	1.370	1.361	1.520	1.511	1.378	1.211	1.218	1.207
Amytal (*130*) IIA	1.363	1.370	1.363	1.519	1.519	1.372	1.213	1.217	1.205
Amytal (*130*) IIB	1.355	1.383	1.361	1.520	1.519	1.369	1.206	1.217	1.205
5-(6′Br3′Et2′Me-Benzimidazolium)-barbiturate (*131*) A	1.360	1.339	1.380	1.405	1.400	1.421	1.256	1.246	1.211
5-(6′Br3′Et2′Me-Benzimidazolium)-barbiturate (*131*) B	1.361	1.350	1.421	1.388	1.403	1.425	1.238	1.248	1.211
Phenobarbital·$(8Br9EtAde)_2$ (*42*)	1.371	1.371	1.362	1.523	1.534	1.359	1.207	1.215	1.203
5iPr5BrAllylBarb.·9EtAde (*41*)[c]	1.411	1.363	1.399	1.554	1.528	1.348	1.245	1.175	1.239
Dialuric acid·H_2O (*132*)[a]	1.35	1.35	1.37	1.42	1.36	1.35	1.22	1.23	—
Average	1.368	1.367	1.372	1.483	1.486	1.381	1.221	1.222	1.212
σ	0.010	0.014	0.021	0.058	0.048	0.018	0.015	0.018	0.015

2. Negatively Charged Molecules

Barb.[NH_4 salt] (*122*)	1.36	1.37	1.40	1.42	1.41	1.40	1.23	1.25	1.23
Veronal [K salt] (*133*)[a]	1.44	1.27	1.41	1.50	1.58	1.33	1.20	1.20	1.15

G. ADDITIONAL PYRIMIDINE BOND DISTANCES (Å)

Name \ Bond:	N1–C2	C2–N3	N3–C4	C4–C5	C5–C6	C6–N1
Pyrimidine (*135*)	1.33	1.34	1.35	1.41	1.38	1.36
2Am4Me6ClPyr (*136*)	1.30	1.34	1.32	1.40	1.40	1.28
4Am2, 6Cl_2Pyr (*137*)	1.304	1.314	1.332	1.400	1.347	1.331
5Br4, 6Am_2Pyr (*137*)	1.36	1.33	1.33	1.37	1.36	1.36
2-(4′-Amino-5′-aminopyrimidyl)-1-penten-4-one (*138b*)	1.31	1.33	1.34	1.40	1.37	1.36
Isocytosine (*139b*):						
[N1 protonated]	1.357	1.333	1.363	1.438	1.331	1.358
[N3 protonated]	1.330	1.369	1.375	1.422	1.356	1.350
Hydrolyzed thiamine-PP (*140*)	1.379	1.290	1.351	1.442	1.342	1.364
Thiamine·HCl (*141*)	1.333	1.306	1.367	1.434	1.354	1.362
Thiamine pyrophosphate (*142*)	1.42	1.27	1.37	1.46	1.35	1.36
Tetrahydroxypyrimido[5,-4*d*]-pyrimine[Na_2 salt] (*143*)	1.38	1.39	1.37	1.47	1.35	1.36
1,3,10-Trimethylisoalloxazonium iodide (*144*)	1.37	1.33	1.37	1.54	1.41	1.37
Riboflavin·HBr·H_2O (*145*)	1.37	1.39	1.38	1.48	1.42	1.35
Riboflavin·5′Br5′dAdo·3H_2O (*40*)	1.50	1.36	1.37	1.56	1.29	1.34
10-Methylisoalloxazine·HBr·2H_2O (*146b*)	1.408	1.379	1.388	1.489	1.422	1.364
8-Azaguanine (*147b*)	1.334	1.379	1.385	1.430	1.383	1.351
Xanthazol·H_2O (*148*)	1.35	1.36	1.39	1.44	1.39	1.34

[a] Not included in the average or σ due to large uncertainties in the bond distances.
[b] Not included in the average or σ due to lack of similarity with other molecules in its group.
[c] Not included in the average or σ.

APPENDIX II

A. Adenine Bond Angles (°)

1. Neutral Molecules

Name \ Angle:	6–1–2	1–2–3	2–3–4	3–4–5	4–5–6	5–6–1	3–4–9	6–5–7	5–7–8	7–8–9	8–9–4	9–4–5	4–5–7	1–6–N6	5–6–N6	4–9–1′	8–9–1′
dAdo (*36*)	119.8	128.8	111.0	126.9	115.4	118.1	128.3	133.9	104.4	113.2	106.8	104.8	110.7	119.2	122.7	123.2	129.6
9MeAde (*189*)[c]	119.8	126.5	112.4	126.6	117.2	117.4	128.7	131.6	104.2	112.0	107.9	104.7	111.2	116.9	125.7	123.8	128.3
9MeAde·1MeThy (*37b*)	116.8	130.9	109.9	127.4	116.9	118.0	127.1	132.1	104.4	112.9	106.2	105.5	111.0	119.0	123.0	125.6	128.2
9EtAde·1MeUra (*38*)	119	130	111	127	118	115	127	132	105	113	106	106	110	120	124	128	127
9EtAde·1Me5FUra (*48*)[a]	116	131	110	124	122	117	126	130	105	110	107	110	108	116	127	130	123
9EtAde·1Me5BrUra (*46b*)	118	131	110	125	119	117	127	131	105	112	106	107	110	118	125	128	126
9MeAde·1Me5BrUra (*47*)[a]	122	128	122	125	119	118	129	132	108	105	112	106	109	114	127	—	—
5′Br5′dAdo·Riboflavin·$3H_2O$ (*40*)[a]	128	127	113	119	124	108	134	130	100	123	103	107	106	126	125	123	134
Ado·8BrUrd (*39b*)	117	127	113	127	117	120	129	134	105	113	106	107	109	118	122	127	128
$(8Br9EtAde)_2$·Phenobarbital (*42*) I	119.7	128.0	111.1	127.5	117.3	116.3	126.8	131.3	101.8	116.4	104.7	105.7	111.4	120.0	123.7	125.3	130.0
$(8Br9EtAde)_2$·Phenobarbital (*42*) II	119.6	128.3	111.2	126.5	117.6	116.6	126.7	132.2	102.8	116.1	104.1	106.8	110.2	119.2	124.2	125.3	130.2
9EtAde·5iPr5BrAllylBarb. (*41*)[c]	118.2	129.5	110.7	125.7	118.9	116.8	125.6	133.9	104.6	115.7	103.8	108.7	107.2	118.6	124.5	124.3	131.9
9Et2,6Am_2Pur·(1Me-5FUra)$_2$ (*55*)[a]	108	140	106	129	114	120	125	134	106	111	102	106	112	119	121	129	129
Average	118.6	129.1	111.0	127.0	117.3	117.3	127.4	132.4	103.8	113.8	105.7	106.1	110.3	119.1	123.5	126.1	128.4
σ	1.4	1.5	1.0	0.4	0.9	1.6	0.9	1.1	1.2	1.4	0.7	0.7	0.8	0.7	0.7	1.6	1.1

2. Rings Protonated on Position N1

Name \ Angle:	6–1–2	1–2–3	2–3–4	3–4–5	4–5–6	5–6–1	3–4–9	6–5–7	5–7–8	7–8–9	8–9–4	9–4–5	4–5–7	1–6–N6	5–6–N6	4–9–1′	8–9–1′
5′-AMP (*43*)	122.8	125.7	112.3	128.5	115.6	114.7	127.4	131.9	104.7	111.8	107.1	104.1	112.3	121.5	123.7	123.8	128.6
3′-AMP·$2H_2O$ (*44*)	123.3	125.9	111.6	127.2	118.2	113.7	127.3	130.6	103.3	113.4	106.6	105.4	111.2	120.2	126.1	125.2	128.0
Average	123.0	125.8	111.9	127.8	116.9	114.2	127.3	131.2	104.0	112.6	106.8	104.7	111.7	120.8	124.9	124.5	128.3
σ	0.4	0.1	0.5	0.9	1.8	0.7	0.1	0.9	1.0	1.1	0.4	0.9	0.8	0.9	1.7	1.0	0.4

3. Rings Protonated on Positions N1 and N7

Name \ Angle:	6–1–2	1–2–3	2–3–4	3–4–5	4–5–6	5–6–1	3–4–9	6–5–7	5–7–8	7–8–9	8–9–4	9–4–5	4–5–7	1–6–N6	5–6–N6	4–9–1′	8–9–1′
9MeAde·2HBr (*45a,c*)	124	127	107	129	120	111	131	130	109	104	116	100	111	124	124	116	127

B. Guanine Bond Angles (°)

1. Neutral Molecules

Name \ Angle:	6–1–2	1–2–3	2–3–4	3–4–5	4–5–6	5–6–1	3–4–9	6–5–7	5–7–8	7–8–9	8–9–4	9–4–5	4–5–7	1–2–N2	3–2–N2	1–6–O6	5–6–O6	4–9–1′	8–9–1′
9EtGua·1MeCyt (*58b*)	125.2	123.3	112.2	128.6	118.9	111.7	125.6	130.7	104.1	114.0	105.8	105.8	110.4	115.9	120.8	119.7	128.6	125.5	128.7
9EtGua·1Me5FCyt (*58b*)	125.3	123.3	112.3	128.3	119.6	111.0	124.5	129.4	103.5	114.0	104.6	107.1	110.8	116.2	120.5	120.5	129.1	125.5	129.8
dGuo·5BrdCyd (*59*)[a]	125	123	109	135	115	112	121	129	104	113	102	104	115	118	118	122	125	131	124
Average	125.2	123.3	112.2	128.4	119.2	111.3	125.0	130.0	103.8	114.0	105.2	106.4	110.6	116.0	120.6	119.8	128.8	125.5	129.2
σ	0.1	0.0	0.1	0.2	0.5	0.5	0.8	0.9	0.4	0.0	0.8	0.9	0.3	0.2	0.2	0.2	0.4	0.0	0.8

2. Rings Protonated on Position N7

Name \ Angle:	6–1–2	1–2–3	2–3–4	3–4–5	4–5–6	5–6–1	3–4–9	6–5–7	5–7–8	7–8–9	8–9–4	9–4–5	4–5–7	1–2–N2	3–2–N2	1–6–O6	5–6–O6	4–9–1′	8–9–1′
9MeGua·HBr (*61*)	126.5	121.5	112.3	128.3	122.0	109.1	119.3	129.5	106.5	110.9	102.4	111.6	107.8	117.7	120.8	120.2	130.8	132.5	123.6
Gua·HCl·2H_2O (*62b*)	125.6	123.4	112.8	127.6	119.9	110.8	126.3	132.7	108.2	109.6	108.6	106.2	107.4	116.0	120.6	120.3	128.9	—	—
Average	126.0	122.4	112.5	127.9	120.9	109.9	122.8	131.1	107.3	110.2	105.5	108.9	107.6	116.8	120.7	120.2	129.8	132.5	123.6
σ	0.6	1.3	0.4	0.5	1.5	1.2	4.9	2.3	1.2	0.9	4.4	3.8	0.3	1.2	0.1	0.1	1.3	—	—

C. Additional Purine Bond Angles (°)

Name \ Angle:	6–1–2	1–2–3	2–3–4	3–4–5	4–5–6	5–6–1	3–4–9	6–5–7	5–7–8	7–8–9	8–9–4	9–4–5	4–5–7	1–2–O2	3–2–O2	1–6–O6	5–6–O6	4–9–1′	8–9–1
Purine (*64*)	117.5	128.5	113.4	123.1	118.5	119.1	127.3	136.0	106.5	114.1	104.6	109.6	105.5	—	—	—	—	—	—
Caffeine (*66*)	127.6	112.9	122.8	121.4	119.9	115.8	127.2	132.8	103.4	112.3	105.7	110.9	107.2	128.7	124.1	118.4	126.4	—	—
Caffeine·5Chlorosalicylic acid (*65*)	127.8	116.1	119.9	121.9	123.4	110.6	126.5	130.5	105.4	113.6	103.4	111.6	106.0	122.3	121.6	122.9	12.74	—	—
Theophylline·5-Chloro-salicylic acid (*67*)	125.9	117.8	119.7	121.2	122.8	112.5	127.5	131.7	106.5	113.2	103.7	111.3	105.3	120.6	121.5	121.2	126.3	—	—
Theophylline (*68*)	126.0	117.8	119.4	121.8	122.6	112.5	125.5	133.1	106.7	114.3	102.0	112.7	104.2	120.8	121.8	121.2	126.5	—	—
Ino-5′-P [Na salt] (*76*)	126	123	114	127	122	107	127	126	103	108	110	107	111	121	132	—	—	124	126
1,3,7,9-Me_4Uric (*69*)	126.3	117.3	118.5	124.0	121.1	112.9	128.3	130.1	108.4	106.7	108.3	107.7	108.9	121.4	121.3	121.3	125.8	131.0	120.7
1,3,7,9-Me_4Uric·Pyrene (*71b*)	130	118	115	125	122	110	131	126	106	107	111	104	112	125	125	126	116	133	115
Uric acid (*74*)	128.8	116.0	118.4	124.1	121.2	111.3	128.3	130.8	108.1	107.1	109.2	107.5	108.0	122.6	121.4	122.3	126.4	—	—
6SIno (*77*) 1	124.1	125.5	112.2	127.9	118.7	111.6	127.1	129.0	103.1	114.1	105.0	105.0	112.3	—	—	—	—	121.7	126.7
6SIno (*77*) 2	126.8	123.9	112.7	126.5	119.5	110.5	127.5	129.5	103.6	113.7	105.7	106.0	111.0	—	—	—	—	122.1	127.3
9EtHyp·5FUra (*78*)	125	125	111	129	119	110	125	130	104	113	106	106	111	—	—	121	129	126	128
7*H*-Tetrazolo[5,1-*i*]-purine·H_2O (*80*)	125	121	116	128	115	115	128	134	103	114	107	105	111	—	—	—	—	—	—

APPENDIX II (*Continued*)

D. CYTOSINE BOND ANGLES (°)

Name \ Angle:	6–1–2	1–2–3	2–3–4	3–4–5	4–5–6	5–6–1	1–2–O2	3–2–O2	3–4–N4	5–4–N4	2–1–C1′	6–1–C1′
1. Neutral Molecules												
Cytosine (*82*)	122.7	118.1	119.4	122.0	117.3	120.1	119.8	122.2	118.2	119.9	—	—
Cytosine·H_2O (*83*)	121.3	120.1	118.9	122.0	117.1	121.6	118.4	121.5	118.7	120.2	—	—
Cytidine (*86*)	120.9	119.7	119.4	121.6	118.2	120.1	119.2	121.2	116.8	121.6	117.2	122.0
1MeCyt·9EtGua (*58b*)	119.9	119.0	120.5	121.4	117.2	121.8	117.5	123.4	118.4	120.3	118.7	121.4
1Me5FCyt·9EtGua (*58b*)	120.6	118.4	122.2	118.4	119.3	121.0	118.7	122.9	121.6	120.0	117.1	121.7
5BrdCyd·dGuo (*59b*)[a]	121	121	122	116	124	115	117	122	116	128	119	119
Cyt·5FUra·H_2O (*84*)	123	116	122	124	113	122	121	123	117	119	—	—
1MeCyt·5FUra (*85*)	120	119	121	121	117	122	118	123	118	120	120	121
Average	121.2	118.6	120.5	121.5	117.0	121.2	118.9	122.5	118.3	120.1	118.2	121.5
σ	1.2	1.3	1.3	1.7	2.0	0.8	1.2	0.8	1.6	0.8	1.4	0.4
2. Rings Hemiprotonated on Position N3												
Cytosine-5-acetic acid (*88*)	122.0	117.6	121.9	120.7	115.6	122.1	121.3	121.1	117.3	121.9		
3. Rings Protonated on Position N3												
3′-CMP (orthorhombic) (*91*)	122.3	113.4	126.0	118.3	117.2	122.7	124.5	122.1	119.9	121.8	117.1	120.4
3′-CMP (monoclinic) (*92*)	122.0	114.0	124.5	118.5	118.4	123.6	123.9	122.0	120.6	120.9	118.2	119.6
1MeCyt·HBr (*45b*)[a]	114	119	127	117	116	127.5	122	119	123	120	122	124
Average	122.1	113.7	125.3	118.4	117.8	123.1	124.2	122.0	120.2	121.3	117.6	120.0
σ	0.2	0.4	1.1	0.1	0.8	0.6	0.4	0.1	0.5	0.6	0.8	0.6

E. URACIL BOND ANGLES (°)

Name \ Angle:	6–1–2	1–2–3	2–3–4	3–4–5	4–5–6	5–6–1	1–2–O2	3–2–O2	3–4–O4	5–4–O4	2–1–C1′	6–1–C1′
1. Neutral Molecules												
Uracil (*93b*)	122.7	114.0	126.7	115.5	118.9	122.3	123.7	122.0	119.2	125.3	—	—
1-Methyluracil (*94*)	121	116	126	115	120	122	122	122	118	127	117	122
5-Nitrouracil·H_2O (*95*)	122.5	114.8	127.8	112.6	120.8	121.5	122.4	122.8	119.5	127.9	—	—
2,4-Dithiouracil (*102*)[b]	126.6	117.4	125.5	117.4	118.9	119.2	—	—	—	—	—	—
5-Fluorodeoxyuridine (*98*)	122.4	114.9	128.0	112.6	122.7	120.3	123.4	122.8	123.3	124.1	117.0	120.2

5-Bromouridine (*97a,b*)	122.0	114.4	127.8	112.8	121.4	121.3	123.8	121.7	121.4	125.8	117.5	120.1
5-Bromodeoxyuridine (*97a,b*)	117.9	116.6	129.1	112.4	119.6	124.2	121.5	121.8	122.0	125.5	120.7	121.1
5-Iododeoxyuridine (*99*)	123	116	125	116	118	121	125	119	121	121	118	119
4-Thiouridine (*116*)[b]	120	115	127	115	119	123	122	122	—	—	121	118
S-S(5SdUrd)$_2$ (*115*)[b] {1	122.7	114.6	126.8	114.0	119.5	122.3	124.1	121.4	119.2	126.8	116.8	120.5
S-S(5SdUrd)$_2$ (*115*)[b] {2	122.7	113.3	127.5	115.0	118.7	122.5	124.4	122.3	119.8	125.2	118.3	118.6
Urd-5′-P [Ba salt] (*96*)	124	112	125	116	120	122	124	124	118	125	116	120
5FOro [Rb salt]·H_2O (*100*)[a]	122	117	125	115	124	117	120	123	122	123	—	—
Dihydrouracil (*101*)[b]	121	116	127	114	113	110	124	120	121	125	—	—
1MeUra·9EtAde (*38*)	123	116	126	117	117	121	123	121	120	123	118	119
1-Me5FUra·9EtAde (*48*)	121	116	126	113	122	121	122	122	122	122	117	122
1Me5BrUra·9EtAde (*46a,b*)	122	115	127	112	124	120	124	121	121	128	113	125
1Me5BrUra·9MeAde (*47*)[a]	121	115	126	117	119	121	—	—	118	—	121	118
5BrUrd·Ado (*39a,b*)	122	118	126	113	121	120	120	122	122	125	117	122
(1Me5FUra)$_2$·9Et2,6Am$_2$ Pur (*55*)[a]												
N7 bonding	124	114	124	115	123	117	127	119	119	126	115	121
N1 bonding	127	109	131	112	124	115	127	124	126	122	115	118
5FUra·Cyt·H_2O (*84*)	123	115	125	114	123	119	123	121	121	125	—	—
5FUra·1MeCyt (*85*)	121	117	125	114	121	121	122	121	120	126	—	—
5FUra·9EtHyp (*78*)	122	116	125	114	122	121	123	121	121	125	—	—
Average	122.0	115.4	126.4	114.1	120.7	121.2	122.9	121.6	120.5	125.3	117.1	120.9
σ	1.4	1.5	1.4	1.6	1.9	1.2	1.3	1.2	1.5	1.8	2.0	1.9

2. Neutral Thymine Molecules

Name \ Angle:	6–1–2	1–2–3	2–3–4	3–4–5	4–5–6	5–6–1	1–2–O2	3–2–O2	3–4–O4	5–4–O4	4–5–CM	6–5–CM	2–1–C1′	6–1–C1′
Thymine·H_2O (*104*)	122.8	115.2	126.3	115.6	118.2	121.8	122.7	122.1	118.3	126.1	119.0	122.8	—	—
1-Methylthymine (*105*)	120.6	115.4	126.3	116.1	118.3	123.3	123.3	121.3	120.9	123.9	119.3	122.4	118.2	121.2
5Et6MeUra (*107*)	123.8	115.2	125.5	116.5	118.0	120.7	122.6	122.1	119.7	123.7	116.6	125.4	—	—
5′-TMP [Ca salt] (*106*)	121.4	115.8	125.7	113.4	120.3	123.3	123.2	120.8	117.7	128.8	124.5	115.1	117.6	121.0
1MeThy·9MeAde (*37b*)	120.7	115.0	126.3	115.7	119.0	123.2	124.3	120.7	119.9	124.4	117.3	123.6	119.1	120.1
Photodimers:														
Thy$_2$(*trans-anti*) (*109b*)[b]	125.7	116.8	127.8	117.7	115.2	116.0	123.5	119.7	120.7	121.5	110.5	115.2	—	—
(1,3Me$_2$Thy)$_2$(*cis-anti*) (*111*)[b] {1	121.6	117.4	126.3	117.8	112.7	115.7	123.2	119.5	121.5	120.7	109.1	117.1	119.2	116.9
(1,3Me$_2$Thy)$_2$(*cis-anti*) (*111*)[b] {2	121.7	116.9	125.8	117.7	112.1	116.3	123.9	119.2	120.8	121.5	110.1	118.3	117.9	117.5
Average	121.9	115.3	126.0	115.5	118.8	122.5	123.2	121.4	119.1	125.4	119.3	121.9	118.3	120.8
σ	1.4	0.3	0.4	1.2	0.9	1.2	0.7	0.7	1.0	2.1	3.1	4.0	0.8	0.6

APPENDIX II (*Continued*)

3. Uracil Derivatives Protonated on O4

1-Methyluracil·HBr (*118*)	123.0	116.2	120.1	120.4	119.2	121.1	123.8	120.0	110.1	129.5	—	—	—	—

F. BARBITURATE BOND ANGLES (°)

1. Neutral Molecules

Name \ Angle:	6–1–2	1–2–3	2–3–4	3–4–5	4–5–6	5–6–1	1–2–O2	3–2–O2	3–4–O4	5–4–O4	5–6–O6	1–6–O6
Barbituric·$2H_2O$ (*120*)	124.8	118.0	125.5	117.4	115.8	118.5	119.3	122.6	118.7	123.8	121.9	119.6
Barbituric acid (*121*)	128	115	125	120	114	116	123	121	118	121	122	122
Alloxan·$3H_2O$ (*123*)	125	118	125	118	114	118	121	121	122	120	120	122
5-Oxobarbituric (*124*)	126	117	126	116	118	116	121	121	123	120	120	123
Alloxan (*125*)	127	117	126	117	114	116	121	122	121	122	122	122
Alloxantin·$2H_2O$ (*126*)	124	118	125	114	110	118	120	122	123	122	121	122
Dilituric acid (*127*)	130	116	124	119	120	114	123	122	—	—	126	120
Dilituric·$3H_2O$ (*128*) A	125	117	125	116	121	116	122	121	117	127	129	115
Dilituric·$3H_2O$ (*128*) B	125	117	125	116	121	116	122	121	117	127	129	115
Violuric·H_2O (*129*)	128	116	126	116	119	114	124	120	120	124	124	121
Veronal (*130*) I	126.4	116.4	126.3	118.4	114.1	118.2	122.0	121.6	121.1	121.8	122.2	119.7
Veronal (*130*) II	126.2	116.2	126.2	118.4	113.5	118.5	122.4	121.4	120.1	121.5	121.5	119.7
Amytal (*130*) I	126.7	116.3	126.2	118.4	113.8	118.0	122.8	120.9	119.8	121.7	121.8	120.2
Amytal (*130*) IIA	127.0	115.9	126.2	118.8	113.5	118.3	123.1	121.5	119.7	121.7	122.0	119.7
Amytal (*130*) IIB	126.7	116.3	126.2	118.4	113.8	118.0	122.8	120.9	119.8	121.7	121.8	120.2
5-(6′Br3′Et2′Me-benzimidazolium)-barbiturate (*131*) A	124.6	117.0	125.0	114.0	125.3	113.7	121.8	121.0	117.7	128.1	125.9	120.3
5-(6′Br3′Et2′Me-benzimidazolium)-barbiturate (*131*) B	124.4	117.3	124.7	116.1	123.1	114.1	119.7	122.9	119.5	124.2	127.4	118.4
Phenobarbital·$(8Br9EtAde)_2$ (*42*)	124.7	118.3	123.5	118.6	110.3	117.8	121.0	120.7	118.0	123.0	122.1	120.0
5iPr5BrAllylBarb.·9EtAde (*41*)[e]	125.3	118.3	125.9	116.1	115.9	118.3	121.3	120.4	122.9	121.0	122.3	119.4
Average	126.1	116.7	125.4	117.3	116.5	116.7	121.7	121.6	119.6	123.0	123.3	120.0
σ	1.5	1.0	0.8	1.6	4.3	1.3	1.3	0.9	1.9	2.3	2.6	1.9

2. Negatively Charged Molecules

Barb. [NH_4 salt] (*122*)	128	115	125	120	114	116	123	121	118	121	122	122
Veronal [K salt] (*133*)[a]	132	116	123	122	113	112	120	123	117	121	122	126

G. ADDITIONAL PYRIMIDINE BOND ANGLES (°)

Name \ Angle:	6–1–2	1–2–3	2–3–4	3–4–5	4–5–6	5–6–1
Pyrimidine (*135*)	115.2	128.2	115.1	122.5	116.3	122.7
2Am4Me6ClPyr (*136*)	114	127	113	129	105	132
4Am2,6Cl_2Pyr (*137*)	110.2	132.3	115.0	119.5	115.4	127.4
5Br4,6Am_2Pyr (*137*)	112	130	114	122	120	122
2-(4′-Amino-5′-aminopyrimidyl)-1-penten-4-one (*138b*)	115.7	127.0	116.8	120.8	117.0	122.2
Isocytosine (*139b*):						
[N1 protonated]	120.2	121.8	119.7	118.7	119.1	120.5
[N3 protonated]	115.9	121.9	123.3	114.8	118.2	125.9
Hydrolyzed thiamine PP (*140*)	118.7	122.9	120.8	118.4	118.9	120.2
Thiamine·HCl (*141*)	120.7	122.9	118.8	120.5	116.8	119.6
Thiamine pyrophosphate (*142*)	125	116	121	123	116	125
Tetrahydroxypyrimido[5,4-*d*]-pyrimidine [Na_2 salt] (*143*)	121.6	118.6	122.5	117.6	120.0	119.6
Riboflavin·HBr·H_2O (*145*)	122	119	123	116	117	121
Riboflavin·5′Br5′dAdo·3H_2O (*40*)	109	116	131	111	112	139
10-Methylisoalloxazine·HBr·2H_2O (*146b*)	123.5	116.3	126.1	115.0	119.4	119.4
8-Azaguanine (*147b*)	112.4	123.2	126.2	110.7	119.6	128.0
Xanthazol·H_2O (*148*)	119	119	127	112	121	123

[a] Not included in the average or σ due to large uncertainties in the bond angle.

[b] Not included in the average or σ due to the lack of similarity with other molecules in its group.

[c] Not included in the average or σ.

APPENDIX III: Deviations from Planarity (in Å × 10³)

A. Adenine

Derivative		rms[c]	N1	C2	N3	C4	C5	C6	N7	C8	N9	N6	C1′
9EtAde·1Me5BrUra		12	24	−2	−6	−12	−13	2	−3	17	11	−11	−8
Ado·5BrUrd		15	7	−9	−11	16	−29	21	−10	4	11	22[a]	76[a]
$(8Br9EtAde)_2$· Phenobarbital	I	9	16	2	−8	−3	−2	2	15	−1	−4	−17	0
	II	63	−19	−55	−21	−15	−7	−32	−56	−71	−28	102	137

B. Guanine

Derivative	rms	N1	C2	N3	C4	C5	C6	N7	C8	N9	N2	O6	C1′
9EtGua·1MeCyt	11	9	1	−15	5	−19	10	−11	9	−11	27[a]	47[a]	2[a]
9EtGUa·1Me5FCyt	7	6	3	−5	−8	−12	4	−4	8	7	37[a]	52[a]	−28[a]
9EtGua·HBr	29	18	−2	−15	30	−40	2	5	18	69	−2	10	43
Gua·HCl·H_2O	13	12	−6	−6	−14	−15	−7	3	19	0	24	14	—

C. Cytosine

Derivative	rms	N1	C2	N3	C4	C5	C6	O2	O4	C1′
Cytosine	5	4	1	−7	7	−1	−4	25[a]	18[a]	—
Cytidine	4	−53[a]	2	0	−4	6	−4	13[a]	33[a]	30[a]
3′-CMP (monoclinic)	10	5	6	−5	−14	−17	4	−7	37[a]	120[a]
1MeCyt·9EtGua	17	24	−20	−4	22	−16	−6	−34[a]	56[a]	79[a]

D. Uracil and Thymine

Derivative	rms	N1	C2	N3	C4	C5	C6	O2	O4	C1′	CM
Uracil	4	5	1	−7	−2	−4	1	7	1	—	—
1Me5BrUra·9EtAde	25	29	−20	0	4	19	4	−63	12	−29[a]	—
5BrUrd·Ado	14	−6	−18	12	−16	−7	−8	24	−4	−3[a]	—
5′-TMP [Ca salt]	8	−12	−4	−3	7	8	9	−101[a]	−64[a]	−50[a]	−61[a]

E. Barbiturate

Derivative	rms	N1	C2	N3	C4	C5	C6	O2	O4	O6
Barbituric acid	3	5	0	−4	−2	−198[a]	3	−4	115[a]	72[a]
Alloxan·$3H_2O$	12	14	−18	14	−5	214[a]	−5	−87[a]	−139[a]	−139[a]
Alloxan	10	−20	−30[a]	10	10	230[a]	−20	−120[a]	−230[a]	−20[a]
Alloxantin·$2H_2O$	20	−20	−50[a]	20	−30	350[a]	30	−80[a]	−340[a]	−190[a]
Veronal I	25	24	−19	−15	39	−33	3	−67[a]	99[a]	−15[a]
Veronal II	4	8	0[b]	8	−7	−7	0[b]	0[b]	−28[a]	−28
Amytal I	38	−31	11	41	−64	42	1	17[a]	−190[a]	14[a]
Amytal II A	32	−30	−12	33	−53	−34	4	34[a]	132[a]	2[a]
Amytal II B	28	30	6	33	−44	−21	14	21[a]	−122[a]	27[a]
Phenobarbital·$(8Br9EtAde)_2$	154	−7	16	2	1	365	10	15	−201	−202[a]

[a] Atoms not included in calculating the least squares plane.
[b] Atoms that must lie in the least squares plane to symmetry considerations.
[c] The root mean square deviation from planarity of the atoms used to calculate the least squares plane.

References

1. J. D. Watson and F. H. C. Crick, *Nature* **171,** 737 (1953).
2. J. D. Watson and F. H. C. Crick, *Nature* **171,** 964 (1953).
3. F. H. C. Crick, *J. Mol. Biol.* **19,** 548 (1966).
4. R. Langridge, H. R. Wilson, C. W. Hooper, M. H. F. Wilkins, and L. D. Hamilton, *J. Mol. Biol.* **2,** 19 (1960).
5. S. Arnott, M. H. F. Wilkins, L. D. Hamilton, and R. Langridge, *J. Mol. Biol.* **11,** 391 (1965).
6. W. Fuller, M. H. F. Wilkins, H. R. Wilson, L. D. Hamilton, and S. Arnott, *J. Mol. Biol.* **12,** 60 (1965).
7. D. A. Marvin, M. H. F. Wilkins, and L. D. Hamilton, *Acta Cryst.* **20,** 663 (1966).
8. M. H. F. Wilkins, A. R. Stokes, and H. R. Wilson, *Nature* **171,** 738 (1953).
9. R. E. Franklin and R. G. Gosling, *Nature* **171,** 740 (1953).
10. S. Arnott, F. Hutchinson, M. Spencer, M. H. F. Wilkins, W. Fuller, and R. Langridge, *Nature* **211,** 227 (1966).
11. W. Fuller, F. Hutchinson, M. Spencer, and M. H. F. Wilkins, *J. Mol. Biol.* **27,** 507 (1967).
12. S. Arnott, M. H. F. Wilkins, W. Fuller, and R. Langridge, *J. Mol. Biol.* **27,** 525 (1967).
13. S. Arnott, M. H. F. Wilkins, W. Fuller, and R. Langridge, *J. Mol. Biol.* **27,** 535 (1967).
14. S. Arnott, M. H. F. Wilkins, W. Fuller, J. H. Venable, and R. Langridge, *J. Mol. Biol.* **27,** 549 (1967).
15. S. Arnott, W. Fuller, A. Hodgson, and I. Prutton, *Nature* **220,** 561 (1968).
16. R. F. Steiner and R. F. Beers, "Polynucleotides." Elsevier, Amsterdam, 1961.
17. D. R. Davies, *Ann. Rev. Biochem.* **36,** 321 (1967).
18. S.-H. Kim and A. Rich, *Science* **162,** 1381 (1968); F. Cramer, F. v.d. Haar, W. Saenger, and E. Schlimme, *Angew. chem.* **7,** 895 (1968); B. Vold, *Biochem. Biophys. Res. Commun.* **35,** 222 (1969).
19. A. Hampel, M. Labanauskas, P. G. Connors, L. Kirkegard, U. L. RajBhandary, P. B. Sigler, and R. M. Bock, *Science* **162,** 1384 (1968).
20. W. Fuller and A. Hodgson, *Nature* **215,** 817 (1967).
21. J. A. Lake and W. W. Beeman, *J. Mol. Biol.* **31,** 115 (1968).
22. L. Pauling and R. B. Corey, *Arch. Biochem. Biophys.* **65,** 164 (1956).
23. J. Kendrew and M. Perutz, *Ann. Rev. Biochem.* **26,** 327 (1957).
24. M. Spencer, *Acta Cryst.* **12,** 59 (1959).
25. M. Spencer, *Acta Cryst.* **12,** 66 (1959).
26. A. Rich and D. W. Green, *Ann. Rev. Biochem.* **30,** 93 (1961).
27. J. Kraut, *Ann. Rev. Biochem.* **34,** 247 (1965).
28. M. Sundaralingam, *J. Am. Chem. Soc.* **87,** 599 (1965).
29. M. Sundaralingam and L. H. Jensen, *J. Mol. Biol.* **13,** 930 (1965).
30. M. Sundaralingam, *Biopolymers,* 1, 821 (1969).
31. A. E. V. Haschemeyer and A. Rich, *J. Mol. Biol.* **27,** 369 (1967).
32. B. Craven and N. Tamberg, "Library of Data for Pyrimidine, Purine, Barbiturate and Related Crystal Structures." Crystallography Laboratory and Knowledge Availability Center, Univ. of Pittsburgh, Pittsburgh, Pennsylvania (1968).
33. "The International Tables for X-Ray Crystallography," Vol. I. The Kynoch Press, Birmingham, England, 1952.
34. A. J. C. Wilson, *Acta Cryst.* **3,** 397 (1950).

35. B. M. Craven and W. J. Takei, *Acta Cryst.* **17,** 415 (1964).
36. D. G. Watson, D. J. Sutor, and P. Tollin, *Acta Cryst.* **19,** 111 (1965).
37a. K. Hoogsteen, *Acta Cryst.* **12,** 822 (1959).
37b. K. Hoogsteen, *Acta Cryst.* **16,** 907 (1963).
38. F. S. Mathews and A. Rich, *J. Mol. Biol.* **8,** 89 (1964).
39a. A. E. V. Haschemeyer and H. M. Sobell, *Proc. Natl. Acad. Sci. U.S.* **50,** 872 (1963).
39b. A. E. V. Haschemeyer and H. M. Sobell, *Acta Cryst.* **18,** 525 (1965).
40. D. Voet and A. Rich, in preparation.
41. D. Voet and A. Rich, in preparation.
42. S.-H. Kim and A. Rich, *Proc. Natl. Acad. Sci. U.S.* **60,** 402 (1968).
43. J. Kraut and L. H. Jensen, *Acta Cryst.* **16,** 79 (1963).
44. M. Sundaralingam, *Acta Cryst.* **21,** 495 (1966).
45a. R. F. Bryan and K. Tomita, *Nature* **192,** 812 (1961).
45b. R. F. Bryan and K. Tomita, *Acta Cryst.* **15,** 1174 (1962).
45c. R. F. Bryan and K. Tomita, *Acta Cryst.* **15,** 1179 (1962).
46a. L. Katz, K. Tomita, and A. Rich, *J. Mol. Biol.* **13,** 340 (1965).
46b. L. Katz, K. Tomita, and A. Rich, *Acta Cryst.* **21,** 754 (1966).
47. Yu. G. Baklagina, M. V. Vol'kenshtein, and Yu. D. Krondraskev, *Zh. Strukturnoi Khim.* **7,** 399 (1966).
48. K. Tomita, L. Katz, and A. Rich, *J. Mol. Biol.* **30,** 545 (1967).
49. K. Watenpaugh, J. Dow, L. H. Jensen, and S. Furberg, *Science* **159,** 206 (1968).
50. E. Shefter, M. Barlow, R. Sparks, and K. Trueblood, *J. Am. Chem. Soc.* **86,** 1872 (1964).
51a. E. Sletten, *Chem. Commun.* 1119 (1967).
51b. T. Kishi, M. Muroi, T. Kusaka, M. Nishikawa, K. Kamiya, and K. Mizuno, *Chem. Commun.* 852 (1967).
52a. H. M. Sobell, *J. Mol. Biol.* **18,** 1 (1966).
52b. F. Mazza, H. M. Sobell, and G. Kartha, *J. Mol. Biol.* **43,** 407 (1969).
53a. S. S. Tavale, T. D. Sakore, and H. M. Sobell, *J. Mol. Biol.* **43,** 375 (1969).
53b. T. D. Sakore, S. S. Tavale, and H. M. Sobell, *J. Mol. Biol.* **43,** 361 (1969).
53c. T. D. Sakore and H. M. Sobell, *J. Mol. Biol.* **43,** 77 (1969).
54a. L. L. Labana and H. M. Sobell, *Proc. Natl. Acad. Sci. U.S.* **57,** 460 (1967).
54b. T. D. Sakore, H. M. Sobell, and F. Mazza, *J. Mol. Biol.* **43,** 385 (1969).
55. R. Chandross and A. Rich, in preparation.
56. C. E. Bugg, U. T. Thewalt, and R. E. Marsh, *Biochem. Biophys. Res. Commun.* **33,** 436 (1968).
57a. J. M. Broomhead, *Acta Cryst.* **1,** 324 (1948).
57b. W. Cochran, *Acta Cryst.* **4,** 81 (1951).
58a. E. J. O'Brien *J. Mol. Biol.* **7,** 107 (1963).
58b. E. J. O'Brien, *Acta Cryst.* **23,** 92 (1967).
59a. A. E. V. Haschemeyer and H. M. Sobell, *Nature* **202,** 969 (1964).
59b. A. E. V. Haschemeyer and H. M. Sobell, *Acta Cryst.* **19,** 125 (1965).
60. H. M. Sobell, K. Tomita, and A. Rich, *Proc. Natl. Acad. Sci. U.S.* **49,** 885 (1963).
61. H. M. Sobell and K. Tomita, *Acta Cryst.* **17,** 126 (1964).
62a. J. Iball and H. R. Wilson, *Nature* **198,** 1193 (1963).
62b. J. Iball and H. R. Wilson, *Proc. Roy. Soc.* **A288,** 418 (1965).
63. J. M. Broomhead, *Acta Cryst.* **4,** 92 (1951).
64. D. G. Watson, R. M. Sweet, and R. E. Marsh, *Acta Cryst.* **19,** 573 (1965).
65. E. Shefter, *J. Pharm. Sci.* **57,** 1163 (1968).
66. D. J. Sutor, *Acta Cryst.* **11,** 453 (1958).

67. E. Shefter, *J. Pharm. Sci.* in press.
68. D. J. Sutor, *Acta Cryst.* **11,** 83 (1958).
69. D. J. Sutor, *Acta Cryst.* **16,** 97 (1963).
70. P. De Santis, E. Giglio, and A. M. Liquori, *Nature* **188,** 46 (1960).
71a. P. De Santis, E. Giglio, A. M. Liquori, and A. Ripamonti, *Nature* **191,** 900 (1961).
71b. A. Damiani, P. De Santis, E. Giglio, A. M. Liquori, R. Puliti, and A. Ripamonti, *Acta Cryst.* **19,** 340 (1965).
72. A. Damiani, E. Giglio, A. M. Liquori, R. Puliti, and A. Ripamonti, *J. Mol. Biol.* **20,** 211 (1966).
73. A. Damiani, E. Giglio, A. M. Liquori, R. Puliti, and A. Ripamonti, *J. Mol. Biol.* **23,** 113 (1967).
74. H. Ringertz, *Acta Cryst.* **20,** 397 (1966).
75. N. Nagashima and Y. Itaka, *Acta Cryst.* **B24,** 1136 (1968).
76. S. T. Rao and M. Sundaralingam, *J. Am. Chem. Soc.* **91,** 1210 (1969).
77. E. Shefter, *J. Pharm. Sci.* **57,** 1157 (1968).
78. S.-H. Kim and A. Rich, *Science* **158,** 1046 (1967).
79. W. M. Macintyre and R. F. Zahrobsky, *Z. Krist.* **119,** 226 (1963).
80. J. P. Glusker, D. van der Helm, W. E. Love, J. A. Minkin and A. L. Patterson, *Acta Cryst.* **B24,** 329 (1968).
81. R. Srinivasan and R. Chandrasekharan, *Acta Cryst.* **B24,** 1698 (1968).
82. D. L. Barker and R. E. Marsh, *Acta Cryst.* **17,** 1581 (1964).
83. G. A. Jeffrey and Y. Kinoshita, *Acta Cryst.* **16,** 20 (1963).
84. D. Voet and A. Rich, *J. Am. Chem. Soc.* **91,** 3069 (1969).
85. S.-H. Kim and A. Rich, *J. Mol. Biol.* **42,** 87 (1969).
86. S. Furberg, C. S. Petersen, and Chr. Romming, *Acta Cryst.* **18,** 313 (1965).
87. F. S. Mathews and A. Rich, *Nature* **201,** 179 (1964).
88. R. E. Marsh, R. Bierstadt, and E. L. Eichhorn, *Acta Cryst.* **15,** 310 (1962).
89. E. Shefter, *Science* **160,** 1351 (1968).
90. J. A. Carrabine and M. Sundaralingam, *Chem. Commun.* p. 746 (1968).
91. M. Sundaralingam and L. H. Jensen, *J. Mol. Biol.* **13,** 914 (1965).
92. C. E. Bugg and R. E. Marsh, *J. Mol. Biol.* **25,** 67 (1967).
93a. G. S. Parry, *Acta Cryst.* **7,** 313 (1954).
93b. R. F. Stewart and L. H. Jensen, *Acta Cryst.* **23,** 1102 (1967).
94. D. W. Green, F. S. Mathews, and A. Rich, *J. Biol. Chem.* **237,** 3573 (1962).
95. B. M. Craven, *Acta Cryst.* **23,** 376 (1967).
96. E. Shefer and K. N. Trueblood, *Acta Cryst.* **18,** 1067 (1965).
97a. J. Iball, C. H. Morgan, and H. R. Wilson, *Nature* **209,** 1230 (1966).
97b. J. Iball, C. H. Morgan, and H. R. Wilson, *Proc. Roy. Soc.* **A295,** 320 (1967).
98. D. R. Harris and W. M. Macintyre, *Biophys. J.* **4,** 203 (1964).
99. N. Camerman and J. Trotter, *Acta Cryst.* **18,** 203 (1965).
100. W. M. Macintyre and M. Zirakzadeh, *Acta Cryst.* **17,** 1305 (1964).
101. D. Rohrer and M. Sundaralingam, *Chem. Commun.* p. 746, (1968).
102. E. Shefter and H. G. Mautner, *J. Am. Chem. Soc.* **89,** 1249 (1967).
103. E. Shefter, M. N. G. James, and H. G. Mautner, *J. Pharm. Sci.* **55,** 643 (1966).
104. R. Gerdil, *Acta Cryst.* **14,** 333 (1961).
105. K. Hoogsteen, *Acta Cryst.* **16,** 28 (1963).
106. K. N. Trueblood, P. Horn, and V. Luzzati, *Acta Cryst.* **14,** 965 (1961).
107. G. N. Reeke, Jr., and R. E. Marsh, *Acta Cryst.* **20,** 703 (1966).
108. J. R. Einstein, J. L. Hosszu, J. W. Longworth, R. O. Rahn, and C. H. Wei, *Chem. Comm.* p. 1063 (1967).

109a. N. Camerman, S. C. Nyburg, and D. Weinblum, *Tetrahedron Letters* **42,** 4127 (1967).
109b. N. Camerman and S. C. Nyburg, *Acta Cryst.* **B25,** 388 (1969).
110. N. Camerman and A. Camerman, *Science* **160,** 1451 (1968).
111. N. Camerman, D. Weinblum, and S. C. Nyburg, *J. Am. Chem. Soc.* **91,** 982 (1969).
112. E. Adman, M. P. Gordon, and L. H. Jensen, *Chem. Commun.* 1019 (1968).
113. I. L. Karle, S. Y. Wang, and A. J. Varghese, *Science* **164,** 183 (1969).
114. P. Tollin, H. R. Wilson, and D. W. Young, *Nature* **217,** 1148 (1968).
115. E. Shefter, M. P. Kotick, and T. J. Bardos, *J. Pharm. Sci.* **56,** 1293 (1967).
116. W. Saenger and K. H. Scheit, *Angew. Chem.* **81,** 121 (1969).
117. E. Shefter and T. I. Kalman, *Biochem. Biophys. Res. Commun.* **32,** 878 (1968).
118. H. M. Sobell and K. Tomita, *Acta Cryst.* **17,** 122 (1964).
119. C. L. Coulter, *Science* **159,** 888 (1968).
120. G. A. Jeffrey, S. Ghose, and J. O. Warwicker, *Acta Cryst.* **14,** 881 (1961).
121. W. Bolton, *Acta Cryst.* **16,** 166 (1963).
122. B. M. Craven, *Acta Cryst.* **17,** 282 (1964).
123. D. Mootz and G. A. Jeffrey, *Acta Cryst.* **19,** 717 (1965).
124. W. Bolton, *Acta Cryst.* **17,** 147 (1964).
125. C. Singh, *Acta Cryst.* **19,** 759 (1965).
126. C. Singh, *Acta Cryst.* **19,** 767 (1965).
127. W. Bolton, *Acta Cryst.* **16,** 950 (1963).
128. B. M. Craven, S. Martinez-Carrera, and G. A. Jeffrey, *Acta Cryst.* **17,** 891 (1964).
129. B. M. Craven and Y. Mascarenhas, *Acta Cryst.* **17,** 407 (1964).
130. E. A. Vizzini, Ph.D. thesis, Univ. of Pittsburgh, 1968.
131. B. W. Mathews, *Acta Cryst.* **18,** 151 (1965).
132. W. Bolton, *Acta Cryst.* **19,** 1051 (1965).
133. P. J. Berthou, B. Rérat, and C. Rérat, *Acta Cryst.* **18,** 768 (1965).
134. M. Calas and J. Martinez, *Compt. Rend. Acad. Sci.* **265C,** 631 (1967).
135. P. J. Wheatley, *Acta Cryst.* **13,** 80 (1960).
136. C. J. B. Clews and W. Cochran, *Acta Cryst.* **1,** 4 (1948).
137. C. J. B. Clews and W. Cochran, *Acta Cryst.* **2,** 46 (1949).
138a. N. F. Yannoni and J. Silverman, *Nature* **202,** 484 (1964).
138b. J. Silverman and N. F. Yannoni, *Acta Cryst.* **18,** 756 (1965).
139a. J. F. McConnell, B. D. Sharma, and R. E. Marsh, *Nature* **203,** 399 (1964).
139b. B. D. Sharma and J. F. McConnell, *Acta Cryst.* **19,** 797 (1965).
140. I. L. Karle and K. Britts, *Acta Cryst.* **20,** 118 (1966).
141. J. Kraut and H. J. Reed, *Acta Cryst.* **15,** 747 (1962).
142. J. Pletcher and M. Sax, *Science* **154,** 1331 (1966).
143. M. Brufani, G. Casini, G. Giacomello, and A. Vaciago, *J. Chem. Soc.* 721 (1966).
144. P. Kierkkegaard, R. Norrestam, P. Werner, A. Ehrenberg, L. E. G. Eriksson, and F. Müller, *Chem. Commun.* 288 (1967).
145. N. Tanaka, T. Ashida, Y. Sasada, and M. Kakudo, *Bull. Chem. Soc. Japan* **40,** 1739 (1967).
146a. C. J. Fritchie, Jr., and B. L. Trus, *Chem. Commun.* 1486 (1968).
146b. C. J. Fritchie, Jr., and B. L. Trus, *Acta Cryst.* in press.
147a. W. M. Macintyre, *Science* **147,** 507 (1965).
147b. W. M. Macintyre, P. Singh, and M. S. Werkema, *Biophys. J.* **5,** 697 (1965).
147c. J. Sletten, E. Sletten, and L. H. Jensen, *Acta Cryst.* **B24,** 1692 (1968).
148. W. Nowacki and H. Bürki, *Z. Krist.* **106,** 339 (1955).
149. W. C. Hamilton, "Statistics in Physical Science." Ronald Press, New York, 1964.

150. A. Pullman and B. Pullman, *Bull. Soc. Chem. France* 766 (1958).
151. A. Pullman and B. Pullman, *Bull. Soc. Chem. France* 594 (1959).
152. C. Singh, *Acta Cryst.* **19,** 861 (1965).
153. L. Pauling, "The Nature of the Chemical Bond." Cornell Univ. Press, Ithaca, New York, 1960.
154. C. A. Coulson, "Valance." Oxford Univ. Press (Clarendon), London and New York, 1961.
155. W. M. Macintyre, *Biophys. J.* **4,** 495 (1964).
156. H. DeVoe and I. Tinoco, Jr., *J. Mol. Biol.* **4,** 500 (1962).
157. E. Shefter, *J. Pharm. Sci.* **57,** 350 (1968).
158. S. C. Wallwork, *J. Chem. Soc.* 494 (1961).
159. "Interatomic Distances," Special Publ. No. 18. The Chemical Society, London, 1965.
160a. C. A. Coulson and U. Danielsson, *Arkiv. Fysik.* **8,** 239 (1954).
160b. C. A. Coulson and U. Danielsson, *Arkiv. Fysik.* **8,** 245 (1954).
161. S. Bratŏz, *Advan. Quant. Chem.* **3,** 209 (1967).
162. B. Pullman and A. Pullman, *Biochim. Biophys. Acta* **36,** 343 (1959).
163. T. A. Hoffman and J. Ladik, *Advan. Chem. Phys.* **7,** 84 (1964).
164. R. E. Marsh, *in* "Structural Chemistry and Molecular Biology" (A. Rich and N. Davidson, eds.), Freeman, San Francisco, 1968.
165. J. Donohue, *J. Phys. Chem.* **56,** 502 (1952).
166. W. Fuller, *J. Phys. Chem.* **63,** 1705 (1959).
167. G. N. Ramachandran, C. Ramakrishnan, and V. Sasisekharan, *J. Mol. Biol.* **7,** 95 (1963).
168. J. Donohue, *in* "Structural Chemistry and Molecular Biology" (A. Rich and N. Davidson, eds.), Freeman, San Francisco, 1968.
169. D. J. Sutor, *Nature* **195,** 68 (1962).
170. W. C. Hamilton and J. A. Ibers, "Hydrogen Bonding in Solids." Benjamin, New York, 1968.
171. W. H. Baur, *Acta Cryst.* **19,** 909 (1965).
172. G. C. Pimentel and A. L. McClellan, "The Hydrogen Bond." Reinhold, New York, 1960.
173. A. Rich, D. R. Davies, F. H. C. Crick, and J. D. Watson, *J. Mol. Biol.* **3,** 71 (1961).
174. J. R. Fresco, *J. Mol. Biol.* **1,** 106 (1959).
175. B. Pullman, P. Claverie, and J. Caillet, *Proc. Natl. Acad. Sci. U.S.* **55,** 904 (1966).
176. R. Langridge and A. Rich, *Nature* **198,** 725 (1963).
177. K. A. Hartman and A. Rich, *J. Am. Chem. Soc.* **87,** 2033 (1965).
178a. J. Donohue, *Proc. Natl. Acad. Sci. U.S.* **42,** 60 (1956)
178b. J. Donohue and K. Trueblood, *J. Mol. Biol.* **2,** 363 (1960).
179 G. Felsenfeld, D. R. Davies, and A. Rich, *J. Am. Chem. Soc.* **79,** 2023 (1957).
180. B. Pullman and J. Caillet, *Theoret. Chim. Acta* **8,** 223 (1967)
181. D. Voet, S.-H. Kim, and A. Rich, unpublished results.
182. V. Sasisekharan and P. B. Sigler, *J. Mol. Biol.* **12,** 296 (1965).
183. Y. Kyogoku, R. C. Lord, and A. Rich, *Science* **154,** 518 (1966).
184. J. Pitha, R. N. Jones, and P. Pithova, *Can. J. Chem.* **44,** 1045 (1966).
185. L. Katz and S. Penman, *J. Mol. Biol.* **15,** 220 (1966).
186. Y. Kyogoku, R. C. Lord, and A. Rich, *Biochim. Biophys. Acta* **179,** 10 (1969).
187a. D. Söll, J. D. Cherayil, and R. M. Bock, *J. Mol. Biol.* **29,** 97 (1967).
187b. D. Söll and U. L. RajBhandary, *J. Mol. Biol.* **29,** 113 (1967).
188. J. Donohue, *Arch. Biochem. Biophys.* **128,** 591 (1968).
189. R. F. Stewart and L. H. Jensen, *J. Chem. Phys.* **40,** 2071 (1964).

190. P. O. P. Ts'o, I. S. Melvin, and A. C. Olsen, *J. Am. Chem. Soc.* **85,** 1289 (1963).
191. O. Jardetzky, *Biopolym. Symp.* **1,** 501 (1964).
192. A. D. Broom, M. P. Schweizer, and P. O. P. Ts'o, *J. Am. Chem. Soc.* **89,** 3612 (1967).
193. R. M. Hamilin, Jr., R. C. Lord, and A. Rich, *Science* **148,** 1734 (1965).
194. Y. Kyogoku, R. C. Lord, and A. Rich, *J. Am. Chem. Soc.* **89,** 496 (1967).
195. R. R. Shoup, H. T. Miles, and E. D. Becker, *Biochem. Biophys. Res. Commun.* **23,** 194 (1966).
196. E. Küchler and J. Derkosch, *Z. Naturforsch.* **21b,** 209 (1966).
197. Y. Kyogoku, R. C. Lord, and A. Rich, *Proc. Natl. Acad. Sci. U.S.* **57,** 250 (1967).
198. H. Tuppy and E. Küchler, *Biochim. Biophys. Acta* **80,** 669 (1964).
199. Y. Kyogoku, R. C. Lord, and A. Rich, *Nature* **218,** 69 (1968).
200. A. Rich, *Nature* **181,** 521 (1958).
201a. R. C. Lord and R. E. Merrifield, *J. Chem. Phys.* **21,** 166 (1963).
201b. K. Nakamoto, M. Margoshes, and R. E. Rundle, *J. Am. Chem. Soc.* **77,** 6480 (1955).
202a. J. Donohue, *J. Mol. Biol.* **41,** 291 (1969).
202b. J. Donohue, *Science* **165,** 1091 (1969).

Author Index

Numbers in parentheses are reference numbers and indicate that an author's work is referred to although his name is not cited in the text. Numbers in italics show the page on which the complete reference is listed.

A

Abelson, J., 24(16), *51*, 59(6), 61(6), 68(6), 73(6), 74(6), *84*
Abrams, R., 94(92), 95(92), 96, *115*
Achmatowicz, S., 63(31), *84*
Adams, A., 29(55, 56, 57), *52*
Adelberg, E., 122(9), *142*
Adman, E., 189(112), 249(112), *263*
Adriaens, E. L., 62, *84*
Agarwal, R. P., 95(100), *115*
Ahmed, Z., 91(26), *114*
Alepa, F. P., 108(218), *118*
Alexander, M., 32(89), *53*
Allan, P. W., 99(159), 104(200), 106(200), 110(159), *117*, *118*
Allan, R., 92(46), *114*
Allard, C., 90(18, 19), *113*
Allende, C. C., 3(10), 9(10), 17(65), *21*, *22*
Allende, J. E., 3(10), 9(10), 17(65), *21*, *22*, 128(33), 129(53), *142*, *143*
Allfrey, V., 32(93), *53*
Ames, B. N., 24, 25, *51*
Anders, M., 29(62, 64, 65), *52*
Anderson, E. P., 103(189), 110(189), *117*
Anderson, M. B., 47, *55*
Anderson, W. F., 27, *51*
Andrews, P., 92(51), *114*
Andronova, L. G., 150(47, 48), 171(48), *180*
Anisimova, V. M., 146(18), *179*
Anthony, D. D., 32(90), *53*
Anthony, R. S., 24(6), *51*
Antón, D., 25(23), *51*
Antonovich, E. G., 146(15, 18, 19, 20), *179*
Apgar, J., 11(42), *22*, 32(87, 96), *53*, 59 (76), 86
Apirion, D., 42(129), *54*
Appleton, M. D., 102(174), 107(174), *117*
Arata, G., 157(73), *181*
Arceneaux, J. L., 41(128), *54*
Arlinghaus, R., 121(4), *142*
Armstrong, D. J., 33(99), *53*, 59(35), 60(35), 68(35), 83(35), *85*
Arnott, S., 185(5, 6, 10, 12, 13, 14, 15), 237, *260*
Ashida, T., 190(145), 251(145), 257(145), *263*
Ashton, D. M., 106(211), 108(222), *118*
Askari, A., 93(57, 59), *114*
Atherly, A. G., 7, 16, *21*, *22*
Atkinson, D. E., 90(8), 107(8), *113*
Atkinson, M. R., 93(73, 77), 95(104), 97(119, 120), 98(119, 120), 104(119, 195, 199), 105(208), 106(209, 210, 213), 112(246), *115*, *116*, *118*, *119*
Axel, R., 49, *55*
Axelrod, V. D., 12(45), *22*, 28(44), *52*, 59(78), *86*

B

Bacharova, R., 32(93), *53*
Baczynskyj, L., 59(3), 60(3), *84*
Baer, H. P., 91(32), 92(32, 54), *114*
Baev, A. A., see Bayev, A. A.
Baguley, B. C., 29(66), 30(76), 31, 39(120), *52*, *53*, *54*, 59(10), 61(10), 73(10), *84*
Bakanova, V. A., 149(39), 150(39), 151 (39), 152(39), 159(39), *180*
Baker, B. R., 92(40), *114*
Baklagina, Yu. G., 188(47), 189(47), 221(47), 230(47), 233(47), 234, 246 (47), 252(47), 255(47), *261*
Baldwin, A. N., 4, 17, 19(13), *21*, 24(8), *51*
Baliga, B. S., 30(79), 49, *53*, *55*
Balis, M. E., 97(132), 104(193), 105(193),

116, *117*, 146(12, 14), 161(12, 74), 167(12), 168(12), *179*, *181*
Bamman, B., 93(72), *115*
Barbata, G., 34(107), *54*
Bardos, T. J., 189(115), 219(115), 249 (115), 255(115), *263*
Barker, D. L., 189(82), 224(82), 225(82), 248(82), 254(82), *262*
Barlow, M., 155(70), *181*, 188(50), 189 (50), 204(50), *261*
Barnett, W. E., 24(108, 109), 34(108, 109), *54*, 140(86), *143*
Barrell, B. G., 13(49), *22*
Bauer, C., 92(50), *114*
Baur, W. H., 220, *264*
Bayev, A. A., 12, *22*, 28(44), *52*, 59(78), *86*, 146(9), *179*
Beck, W. S., 102, *117*
Becker, A., 147(28), *179*
Becker, E. D., 208(195), *265*
Beckwith, J. R., 25(20), *51*
Beeman, W. W., 185(21), *260*
Beers, R. F., 185(16), *260*
Behal, F. J., 91(25), *114*
Beijerinck, M. W., 132(57), *143*
Beletzkaya, I. P., 151(61), *181*
Bell, F. E., 6, 7, 16, *21*, *22*
Bendich, A., 146, 150(54), 151(54), 161 (11, 12), 166(11), 167, 168, *179*, *180*
Benes, J., 76(57), *85*
Benke, P. J., 109(232), *119*
Bennett, E. L., 101(166), *117*
Bennett, L. L., 98(139, 147), 99(147, 159), 104(200), 106(200, 212, 215), 110 (159, 240), *116*, *117*, *118*, *119*
Benzer, S., 24, *51*
Berenbaum, M. C., 104(197), *118*
Berg, P., 2(4), 3(4, 9, 11), 4, 7, 8(25), 9(9), 11, 13, 17, 19(13), *21*, *22*, 24 (8, 9), *51*
Bergel, F., 92(39), *114*
Berger, C. A., 78(68, 69), *85*
Bergkvist, R., 146(2), *179*
Bergmann, F., 2(4), 3(4), 11(39), *21*, *22*
Bergquist, P. L., 39(120), *54*
Berkowitz, W. F., 150(54), 151(54), 167, *180*
Berne, R. M., 103(185), *117*
Berry, A. J., 93(68), 94(68), *114*
Berthou, P. J., 190(133), 205(133), 251 (133), 257(133), *263*
Bertles, J. F., 102, *117*
Beyer, R. E., 147(24), 171(24), *179*
Bezdeck, M., 150(47), *180*
Bieber, A. L., 98(135), *116*
Bieber, L. L., 171(92), *181*
Bieber, S., 108(227), *119*
Biemann, K., 33(98), *53*, 60(5), *84*
Bierstadt, R., 189(88), 204(88), 225(88), 226(88), 248(88), 254(88), *262*
Biesele, J. J., 79(72), *86*
Biggs, D. R., 138(74), *143*
Binkley, F., 146(7), *179*
Birjukova, N. V., 146(16), *179*
Björk, W., 91(22), *113*
Blair, D. G. R., 97, 110(239), *116*, *119*
Blakeslee, J., 78(71), 79(71), *86*
Blakley, R. L., 90, *113*
Bloch, A., 79(73), *86*, 103(192), 105(192), *117*
Bloch, K., 71, *85*
Blow, D. M., *187*
Bluestein, H. G., 17, *22*
Boardman, N. K., 123(14), 134(14), 135(14), *142*
Bock, R. M., 185(19), 241(187a), 244(19), *260*, *264*
Bogdanov, A. A., 145, 146(15, 16, 18, 19, 20, 21), 161(21), *179*
Boissonnas, R. A., 10, *21*
Bolton, W., 190(121, 124, 127, 132), 199 (121), 228(127, 132), 229(124), 250 (121, 124, 127, 132), 256(121, 124, 127), *263*
Boozer, Ch. E., 151(64), 156(64), *181*
Borek, E., 29, 30(71, 72, 73, 78, 79), 31(72, 73), 49(148), *52*, *53*, *55*, 74, *85*
Borenfreund, E., 146(12), 161(12), 167 (12), 168(12), *179*
Boyd, R. F., 34, *54*
Boyer, P. D., 171(85, 91, 92), *181*
Boyle, J. T. A., 33(99), *53*, 59(35), 60(35), 68(35), 83(35), *85*
Brady, T. G., 92, *114*
Bratŏz, S., 206(161), 207, *264*
Bravo, M., 128(33), *142*
Brawerman, G., 123(21), 134(60), 138(80), 139(60), *142*, *143*
Bray, R. C., 92(39), *114*

Brega, A., 138(84), *143*
Brenner, B. M., 91(28), *114*
Brenner, S., 24(16), 25(20), *51*, 59(6), 61(6), 68(6), 73(6), 74(6), *84*
Bridger, W. A., 171(85), *181*
Brink, J. J., 105(206), *118*
Britts, K., 190(140), 205(140), 251(140), 257(140), *263*
Brockman, R. W., 96(111), 98, 103(143, 189), 104(140), 106(215), 110(143, 189, 235, 238), *115*, *116*, *117*, *118*, *119*
Broom, A. D., 67(37), *85*, 207(192), *265*
Broomhead, J. M., 188(57a, 63), 204(63), 221(57a, 63), 224(63), *261*
Brown, A. D., 146(3), *179*
Brown, D. H., 24(108, 109), 34(108, 109), *54*, 140(86), *143*
Brown, G. B., 79(72), *86*
Brown, J. W., 3(70), 19(70), *22*
Brox, L. W., 98(137), *116*
Brufani, M., 190(143), 251(143), 257 (143), *263*
Bryan, R. F., 188(45a, 45c), 189(45a, 45b), 204(45a, 45b, 45c), 246(45a, 45c), 248(45b), 252(45a, 45c), 254 (45b), *261*
Buchanan, J. M., 90(13), 94(85), 95, 97 (113), *113*, *115*, *116*
Buck, C. A., 43(133, 134, 135), 44(133, 134), 45(133), 47(134, 135), 48(134, 135), *54*
Buckley, W. B., 78, *85*
Bucovaz, E. T., 14(59), *22*
Bürki, H., 190(148), 251(148), 257(148), *263*
Bugg, C. E., 188(56), 189(92), 204(92), 224(56), 248(92), 254(92), *261*, *262*
Burger, R., 91(31), 93(64), 94(64), 103(31), *114*
Burrows, W. J., 33(99), *53*, 59(35), 60(35), 68(35), 83(35), *85*
Butler, E. E., 132(58), *143*

C

Caicuts, M., 44(136), 45, *54*
Caillet, J., 221(175), 227(175), 234(175), 235(180), *264*
Calas, M., 190(134), *263*
Caldwell, I. C., 97(129), 99(156, 158), 105(158), 108(224), 110(158), *116*, *117*, *118*
Calendar, R., 3(11), 4(11), 7, *21*, 24(9), *51*
Camerman, A., 189(110), *263*
Camerman, N., 189(99, 109a, 109b, 110, 111), 227(109a, 109b), 249(99, 109a, 109b, 111), 255(99, 109b, 111), *262*, *263*
Canning, L., 125(31), 126(31), 140, *142*
Cantero, A., 90(18, 19), *113*
Cantoni, G. L., 11, 17(65), 18(67), *22*, 32(88), *53*
Capecchi, M. R., 24(14), *51*
Capra, J. D., 30(68, 69), *52*
Carbon, J. A., 31, 32, *53*
Carrabine, J. A., 189(90), *262*
Carraway, K. L., 63(31), 77(63), *84*, *85*
Casini, G., 190(143), 251(143), 257(143), *263*
Cassio, D., 10, *21*
Cavé, A., 62(25), *84*
Ceccarini, C., 34(107), *54*
Celis, J., 3(10), 9(10), *21*
Center, M. S., 91(25), *114*
Cha, S., 95(99, 108), 96(99), *115*
Chakraburtty, K., 7(22), 14(22), 15(62), 16(22, 62), 17(22), 19(22), *21*, *22*
Chalmers, A. H., 104(195), *118*
Chambers, R. W., 12, *22*, 28(48), *52*
Champagne, M., 146(13), *179*
Champe, S. P., 24, *51*
Chandrasekharan, R., 189(81), *262*
Chandross, R., 188(55), 189(55), 224(55), 230(55), 232(55), 233(55), 234(55), 246(55), 249(55), 252(55), 255(55), *261*
Chang, F. N., 129(50), *143*
Chang, S. H., 33(100), *53*, 59(39, 41), 60(39), 68(39), 70(41), *85*
Chanley, J. D., 151(63), *181*
Chapeville, F., 7, 18, *21*, *22*
Charalampous, F. C., 95(103), *115*
Chassy, B. M., 92(55), *114*
Chaykin, S., 90(10), 111, *113*
Chen, C-M., 60(45), 62(24), 71, 72(45), 75(24), *84*, *85*
Cherayil, J. D., 241(187a), *264*
Cherry, J. H., 47, *55*
Chheda, G. B., 59(3), 60(3), *84*
Chrispeels, M. J., 34, *54*

Chuang, H. Y. K., 16, *22*
Chuguev, I. I., 12(45), *22*
Chumley, S., 106(215), 110(238, 240), *118*, *119*
Chvojka, L., 76(57), *85*
Ciferri, O., 121(7), 125(30), 126(32), 127 (30, 43, 44), 128(30, 34), 129(32), 130, 131, 133(7), 134(7), 138(34, 83), 140(7), *142*, *143*, *144*
Clark, B. F. C., 13(48), *22*, 24(16), *51*, 59(80, 81), *86*
Clark, V. M., 164(77), *181*
Clark-Walker, G. D., 138(74, 75, 76), *143*
Claverie, P., 221(175), 227(175), 234(175), *264*
Claybrook, J. R., 76(61), *85*
Clews, C. J. B., 190(136, 137), 228(137), 251(136, 137), 257(136, 137), *263*
Clouse, J. A., 93(75), *115*
Cochran, W., 188(57b), 190(136, 137), 204(57b), 221(57b), 228(137), 251 (136, 137), 257(136, 137), *261*, *263*
Cohen, S. S., 105(207), *118*
Comb, D. G., 45(138), *54*
Conklin, R. A., 171(93), *182*
Connors, P. G., 185(19), 244(19), *260*
Conway, T. W., 129(53), *143*
Cooper, R. A., 171(89, 90), *181*
Copeland, V., 98(141), *116*
Corey, R. B., 185, 196, 213, 219, 237(22), *260*
Cory, J. G., 92(53), *114*
Cory, S., 13(48), *22*, 59(80, 81), *86*
Coulson, C. A., 199(154), 206(160a, 160b) *264*
Coulter, C. L., 189(119), *263*
Cramer, F., 11, 15, *22*, 185(18), 244(18), *260*
Craven, B. M., 186, 189(95), 190(122, 128, 129), 195(35), 205(122), 227(95), 228 (122, 128, 129), 248(95), 250(128, 129), 251(122), 254(95), 256(128, 129), 257(122), *260*, *261*, *262*, *263*
Crick, F. H. C., 24(5), 26(5), *51*, 183(1, 2), 184(3), 204(173), 213(1), 221(173), 230, 232, 241, *260*, *264*
Csonka, L., 60(82), *86*
Cundliffe, E., 138(73), *143*
Cunningham, B., 93(63), 94(63), *114*
Currie, R., 92(39), 102(176), *114*, *117*

D

Damiani, A., 188(71b, 72, 73), 202(71b, 72, 73), 247(71b), 253(71b), *262*
Daniel, V., 12, *22*, 28(39, 40), 39, *52*
Danielsson, U., 206(160a, 160b), *264*
Daves, B. D., 138(68), *143*
David, H., 60(82), *86*
Davidson, E. H., 32(93), *53*
Davidson, J. D., 110(241), *119*
Davidson, J. N., 100(164), *117*
Davie, E. W., 3(7), 8(26), *21*, 127(40, 41), *142*
Davies, D. R., 185(17), 204(173), 221 (173), 230(179), 233(179), *260*, *264*
Davies, J., 138(67, 68), *143*
Davies, P. J., 112(244), *119*
Debavadi, C. S., 96(111), *115*
De Carli, L., 128(34), 138(34, 84), *142*, *143*
Decker, K., 98(150), *116*
de Lamirande, G., 90, *113*
DeLuca, M., 11, 19(69), *21*, *22*, 171(91), *181*
DeMoss, J. A., 8(24), *21*
Derkosch, J., 208(196), *265*
De Santis, P., 188(70, 71a, 71b), 202(71a, 71b), 247(71b), 253(71b), *262*
Desseaux, B., 40(123), *54*
Desvages, G., 62(27), *84*
Deutscher, M. P., 13(53), 18(53), *22*
De Verdier, C. H., 95(98), *115*
DeVoe, H., 202, 207(156), *264*
Deyrup, J. A., 62, *84*
Dieckmann, M., 2(4), 3(4), 11(39), *21*, *22*
Dietrich, C. P., 2(2), *21*
Dixon, G. J., 106(215), *118*
Doepner, H., 11(43), *22*
Doi, R. H., 32, 40(121, 122), 41(126), *53*, *54*
Donaldson, G. R., 104(202), 107(202), *118*
Donohue, J., 196, 219, 220, 229, 237, *264*, *265*
Dorizzi, M., 18, *22*
Dorsey, J. K., 74, *85*
Doudoroff, M., 122(9), *142*
Dounce, A., 146(17), *179*
Dove, W. F., 23(2), *51*
Dow, J., 188(49), *261*

Downey, K. M., 127(41), *142*
Drigin, Yu. F., 146(21), 161(21), *179*
Drummond, G. I., 91(32), 92(32, 54), *114*
Dube, S. K., 13(48), *22*, 59(80, 81), *86*
Dudkin, S. M., 175(97), *182*
Dütting, D., 59(8), 60(5), 61(8), 63(8, 34), 73(8), *84*, *85*
Dulmadge, E. A., 106(215), *118*
Duncan, E. L., 91(32), 92(32), *114*
Dunn, D. B., 57, 63(32), *84*, *85*
Dutting, D., 33(98), *53*
Dvilansky, A., 101(173), *117*
Dwyer, M. R., 138(79), *143*

E

Earhart, C. F., 33, *53*
Ebner, K. E., 171(91), *181*
Eckermann, G., 104(199), *118*
Edmonds, M., 94(92), 95(92), 96(92), *115*
Ehrenberg, A., 190(144), 251(144), *263*
Eichhorn, E. L., 189(88), 204(88), 225(88), 226(88), 248(88), 254(88), *262*
Eigner, E. A., 3, 4, 6(16, 17), 19(13a), *21*
Eigner, J., 146(22), *179*
Einstein, J. R., 189(108), 227(108), *262*
Eisenstadt, J. M., 123(21), 134(60), 138(80), 139(60), *142*, *143*
Ejiiri, S., 128(35), 132(35), *142*
Elder, A., 146(14), *179*
Elion, G. B., 95(106), 96(106, 109), 98(134), 103(190, 191), 104(134, 191), 108(228), *115*, *116*, *117*, *119*
Elliott, D. C., 112(246), *119*
Elson, D., 89(4), *113*
Engelbrecht, T., 76(58), *85*
Engelhardt, D. L., 24(15), *51*
Engstrom, L., 171(86), *181*
Ennis, H. L., 138(81), *143*
Epler, J. L., 24(109), 34(109), *54*
Epstein, H. T., 139(85), *143*
Eriksson, L. E. G., 190(144), 251(144), *263*
Erwin, M. J., 97(113), *116*
Evans, H. J., 147(26), 171(26), *179*
Everett, G. A., 11(42), *22*, 32(87, 96), *53*, 59(76), *86*

F

Fagan, V. M., 74(50), *85*
Falaschi, A., 138(84), *143*
Farber, E., 49, *55*
Faulkner, R. D., 59(41), 70(41), *85*
Favorova, O. O., 19, *22*, 34(111), *54*
Feageson, E., 151(63), *181*
Feldbush, T., 79(75), *86*
Feldmann, H., 33(98), *53*, 59(8), 60(5), 61(8), 63(8, 34), 73(8), *84*, *85*, 147, *179*
Felicetti, L., 132(55, 56), 134(56), 140(55), *143*
Felsenfeld, G., 230(179), 233(179), *264*
Fink, G. R., 25(23), *51*
Fittler, F., 33(102), *53*, 60(43), 67(38), 68, 71, 72(38, 47), 73(47), 74(38), *85*, *86*
Flaks, J. G., 97(113), *116*
Fleissner, E., 29(63), *52*
Fodov, I., 28(**44**), *52*
Foft, J. W., 28(37, 38), 38(37, 38), *52*
Fontenelle, L. J., 101(170), *117*
Fox, J. J., 79, *86*, 103(192), 105(192), *117*
Francki, R. I. B., 123(14), 134(14), 135(14), *142*
Franklin, J. E., 93(57), *114*
Franklin, R. E., 185(9), *260*
Franz, H. E., 98(150), *116*
Franz, M., 98(150), *116*
Fredericksen, S., 92(56), *114*
Frei, E., 98(155), 99(155), *117*
Freidkin, M., 94(83), 95(105), *115*
Fresco, J. R., 29(55, 56, 57), *52*, 221(174), *264*
Fridovich, I., 91(35, 36), *114*
Friedman, R., 108(227), *119*
Friedman, S. M., 138(69), *143*
Friedrich, A., 76(57), *85*
Friedrichs, B., 91(30), 101(30), *114*
Fritchie, C. J., Jr., 190(146a, 146b), 251(146a), 257(146b), *263*
Fritzson, P., 90, 91(20), *113*
Fujii, T., 63(30), 77(63), *84*, *85*
Fujimoto, A., 171(93), *182*
Fujiwara, Sh., 157(73), *181*
Fukugawa, Y., 96(110), *115*
Fuller, W., 185(6, 10, 11, 12, 13, 14, 15, 20), 219, 237, *260*, *264*
Furberg, S., 188(49), 189(86), 248(86), 254(86), *261*, *262*
Furth, J. J., 32(89), *53*

G

Gabrio, B. W., 95(96), *115*
Gadd, R. E. A., 97(118, 124), *116*
Gale, G. O., 76(62), *85*
Gallo, R., 79(74), *86*
Galston, A. W., 112(244), *119*
Garrison, A. W., 151(64), 156(64), *181*
Gartland, W. J., 29(50, 54), 41(127), 42(127), *52*, *54*
Gasior, E., 121(5), *142*
Gassen, H. G., 61(7a), *84*
Gatica, M., 3(10), 9(10), *21*
Gatinskaya, L. G., 150(55), 151(55), 152 (55, 72), 154(55), 155(72), 162(72), 163(72), 166(72), *180*, *181*
Gefter, M. L., 33, *53*, 68, 69, *85*, 147(28), *179*
Genuth, S. M., 8(24), *21*
George, H., 11(37), *21*
Gerdil, R., 189(104), 227(104), 249(104), 255(104), *262*
Geroch, M., 29(56), *52*
Ghose, S., 190(120), 228(120), 250(120), 256(120), *263*
Giacomello, G., 190(143), 251(143), 257 (143), *263*
Gibor, A., 135(62), 139(62), *143*
Giglio, E., 188(70, 71a, 71b, 72, 73), 202 (71a, 71b, 72, 73), 247(71b), 253 (71b), *262*
Gilbert, W., 129(47), 138(67), *143*
Gilbert, W. R., Jr., 106(211), *118*
Gillis, J., 92(54), *114*
Ginsberg, T., 59(10), 61(10), 73(10), *84*
Glebov, R. N., 149(31), 150(31), *179*
Glišin, M. V., 32(91), *53*
Glišin, V. R., 32(91), *53*
Glusker, J. P., 189(80), 247(80), 253(80), *262*
Goehler, B., 32, 41(126), *53*, *54*
Gold, M., 29(62, 64, 65), 30(80), *52*, *53*
Goldfinger, S., 102(177), 108(177), *117*
Goldthwait, D. A., 32(90), *53*
Goodman, H. M., 24(16), *51*, 59(6), 61(6), 68(6), 73(6), 74, *84*
Goodwin, T. W., 71, *85*
Gordon, M. P., 189(112), 249(112), *263*
Gorini, L., 138(67, 68), *143*
Gosling, R. G., 185(9), *260*
Gots, J. S., 96(112), *116*
Gotto, A. M., 94(86), 95(86), *115*
Gould, B. J., 95(98), *115*
Goutaul, R., 62(25), *84*
Goutier, R., 100(164), *117*
Grace, J. T., Jr., 30(75), 31(75), *53*, 78, 79, *86*
Graham, D., 138(79), *143*
Grandi, M., 121(7), 126(32), 129(32), 130 (7, 32), 131(7), 133(7), 134(7), 140(7), *142*
Granger, G. A., 43(133, 134), 44(133, 134), 45(133), 47(134), 48(134), *54*
Granick, S., 135(62), 139(62), *143*
Green, D. W., 185(26), 189(94), 226(94), 227(94), 248(94), 254(94), *260*, *262*
Green, R. H., 93(61), *114*
Grieve, A., 138(79), *143*
Griffin, A. C., 125(31), 126(31), 140(88), *142*, *144*
Grillo, R. S., 78, *86*
Grimm, W. A. H., 63, *84*
Grollman, A. P., 129(52), 138(82), *143*
Gromova, E. S., 149(44), 150(44), 152(44), 156(44), 158(44), 159(44), *180*
Gros, D., 46(142), *54*
Gros, F., 14, 17, *22*, 123(22), *142*
Gross, H. J., 33(100), *53*, 59(39), 60(39), 68(39), *85*
Gross, P. R., 24, 32(92, 94), *51*, *53*
Grünberger, D., 127(38), *142*
Gurin, S., 103(186), *117*
Gurova, G. I., 150(52), 151(52), 152(72), 155(72), 162(72), 163(72), 166(72), *180*, *181*
Gussin, G. N., 24(14), *51*

H

Haab, W., 102(174), 107(174), *117*
Haar, F. v.d., 11(43), 15(61), *22*, 185(18), 244(18), *260*
Haavik, C. A., 97(114), *116*
Hacker, B., 79(75), *86*
Hageman, E., 146(6), *179*
Hahn, G. A., 3(70), 19(70), *22*
Hakala, M. T., 78(71), 79(71), *86*
Hall, R. H., 30(75), 31(75), 33(97, 102, 103), *53*, 57, 59(3), 60(3, 4, 33, 43, 45, 82), 62(4, 24, 28), 63(4, 33), 65(4), 66(4, 33), 67(38), 68, 71, 72(38, 45,

47), 73(47), 74(38), 75, 76(33, 53, 62), 78(71), 79(71), 81, 82(2), *84, 85, 86*
Hall, Z. W., 170(82), *181*
Halman, M., 151(62), *181*
Halvorson, H. O., 125(28), 128(28), *142*
Hamanda, M., 96(110), *115*
Hamilin, R. M., Jr., 208(193), *265*
Hamilton, L. D., 185(4, 5, 6, 7), 237(4, 5, 6, 7), *260*
Hamilton, W. C., 196(149), 219(170), 220, *263, 264*
Hampel, A., 185(19), 244(19), *260*
Hampton, A., 79, *86*
Hamzi, H. Q., 63(31), 77(63), *85*
Hancock, R. L., 30(74), *53*
Hansen, K., 98(144), 99(144), *116*
Hansen, R. G., 146(6), 146(8), *179*
Harada, F., 28(45), 33(100), *52, 53*, 59 (39), 60(39), 68(39), *85*
Hardy, J., 29(58), *52*
Harris, D. R., 189(98), 248(98), 254(98), *262*
Harris, G., 146(4), *179*
Harris, H., 110(240a), *119*
Hartenstein, R. C., 91(35), *114*
Hartman, K. A., 204(177), 226, *264*
Hartman, P. E., 24, 25, *51*
Hartman, S. C., 90(13), *113*
Haschemeyer, A. E. V., 185(31), 188(39a 39b, 59a, 59b), 189(39a, 39b, 59a, 59b), 198(31), 230(39a, 39b), 233(39a, 39b), 237(59a, 59b), 238(59a, 59b), 239, 246(39b), 247(59b), 248(59b), 249(39a, 39b), 252(39b), 253(59a, 59b), 254(59a), 255(39a, 39b), *260, 261*
Hase, E., 146(10), *179*
Hatfield, D., 44(136), 45, *54*
Hattman, S., 38(118), *54*
Hausmann, R., 30(80), *53*
Hay, J., 27, 28(35), 39(35), *51*
Hayward, R. S., 32(84), *53*
Hecht, L., 11, *22*
Hecht, S. M., 33(99), *53*, 59(35), 60(35), 68(35), 77(65), 83(35), *85*
Heinemeyer, C., 13(51), 15(51), *22*
Hele, P., 13, *22*
Helgeson, J. P., 61(11), 62(23), 77(63), *84, 85*
Helinski, D. R., 24(12), *51*
Henderson, A. M., 95(106), 96(106), *115*
Henderson, J. F., 97, 98, 99(156, 158), 100, 101, 102, 103(188), 105(158), 106 (214), 107(217), 108(220, 224), 109, 110(158), *116, 117, 118, 119*
Hendley, D. D., 33(105), *53*
Heppel, L. A., 90(15), 91(33), *113, 114*
Heredia, C. F., 125(28), 128(28), *142*
Herrington, K. A., 97(133), 98(133), *116*
Hershko, A., 101(169, 172), 102, 108(223), 109(223), *117, 118*
Hershko, C., 108(223), 109(223), *118*
Heyman, T., 40, *54*
Hill, D. L., 98(147), 99(147), 106(212), *116, 118*
Hirsch, G. P., 33(105), 34(105), *53*
Hirsh, D. I., 3, 9, *21*
Hitchings, G. H., 95(106), 96(106, 109), 103(190), 104(201), 108(227, 228), *115, 117, 118, 119*
Ho, D. H. W., 98(155), 99(155), *117*
Hoagland, M. B., 2(1), *21*, 147, *179*
Hodgson, A., 185(15, 20), 237(15), *260*
Hoffman, T. A., 206(163), *264*
Hofschneider, P. H., 38(118), *54*
Holland, J. J., 43, 44(133, 134), 45(133), 47(134, 135), 48, *54*
Holley, R. W., 11, *22*, 32(87, 96), *53*, 59(76), *86*
Holmes, K. C., *187*
Holmsen, H., 98(148, 149), *116*
Hoogsteen, K., 188(37a, 37b), 189(37a 37b, 105), 219(105), 227(105), 230 (37a, 37b), 232, 246(37b), 249(37b, 105), 252(37b), 255(37b, 105), *261, 262*
Hooper, C. W., 185(4), 237(4), *260*
Horecker, B. L., 91(33), *114*
Hori, A., 91(29), *114*
Hori, M., 97, *116*
Horn, P., 189(106), 249(106), 255(106), *262*
Horowitz, N. H., 42, 43(131, 132), *54*
Hoskinson, R. M., 59(41), 68, 70(41), *85, 86*
Hosszu, J. L., 189(108), 227(108), *262*
Howell, B. A., 93(75), *115*
Howell, L. G., 91(36), *114*
Howell, R. R., 108(218), *118*

Hsu, W. T., 28(37, 38), 38(37, 38), *52*
Huang, M., 138(74), *143*
Hudson, P. B., 95(97), *115*
Huennekens, F. M., 95(96), *115*
Hultquist, D. E., 171(91), *181*
Hung, L., 31, 32(81), *53*
Hung, P. P., 28, 39(41), *52*
Hurwitz, J., 29(62, 64, 65), 30(80), 32(89), *52*, *53*, 91(33), *114*, 147(28), *179*
Hutchinson, F., 185(10, 11), 237(11), *260*
Hutchison, D. J., 96(111), 110(236), *115*, *119*
Hutner, S., 136(64), 138(70), *143*

I

Iaccarino, M., 11, *21*
Iball, J., 188(62a, 62b), 189(97a, 97b), 204(62a, 62b), 224(62a, 62b), 247 (62b), 248(97a, 97b), 253(62b), 255 (97a, 97b), *261*, *262*
Ibers, J. A., 219(170), 220, *264*
Igarashi, R. T., 40(122), *54*
Ilan, J., 46, *54*, 128(36), *142*
Imahori, K., 3(57), 14(57), *22*
Imura, N., 28(48), *52*
Ingram, V. M., 32(86), 45(137), *53*, *54*
Ipata, P. L., 91, *113*
Ishida, T., 29(52, 53, 54, 59), *52*
Ishikura, H., 61(7), *84*
Itaka, Y., 188(75), *262*
Itano, H. A., 24, *51*
Itoh, R., 90(16, 17), *113*
Iwakata, S., 79(86), *86*
Izak, G., 101(173), *117*

J

Jacob, F., 23, *51*
Jacobson, K. B., 34, *54*
Jakovcic, S., 108(219), *118*
James, H. L., 14, *22*
James, M. N. G., 189(103), 227(103), *262*
Janot, M., 62, *84*
Jardetzky, O., 138(71), *143*, 207(191), *265*
Jeffrey, G. A., 189(83), 190(120, 123, 128), 199(123), 224(83), 225(83), 228(120, 123, 128), 248(83), 250(120, 123, 128), 254(83), 256(120, 123, 128), *262*, *263*
Jensen, L. H., 151(65, 68), 155(69), *181*, 185(29), *187*, 188(43, 49, 189), 189 (91, 93b, 112), 190(147c), *197*, 204 (43, 91), 219(91), 226(93b), 227(93b), 246(43, 189), 248(91, 93b), 249(112), 251(147c), 252(43, 189), 254(91, 93b), 257(147c), *260*, *261*, *262*, *263*, *264*
Jesensky, C., 30(68), *52*
Jones, D. S., 31, 32(81), *53*
Jones, J. W., 67(37), *85*
Jones, O. W., 108(222), *118*
Jones, R. N., 208(184), *264*
Jorgensen, S., 102, *117*
Josan, V., 92, *114*
Josse, J., 146(22), *179*
Julian, G. R., 138(71), *143*
Junga, I. G., 104(203), *118*
Juodka, B. A., 149(41, 42, 43), 150(41, 42, 43, 53), 151(41, 42, 43, 53), 152(43, 53), 153(53), 156(43), 158(43), 167 (41, 53), 169(80), *180*, *181*

K

Kakudo, M., 190(145), 251(145), 257(145), *263*
Kalckar, H. M., 91(37), 94(83), 95(105), *114*, *115*
Kalle, G. P., 96(112), *116*
Kalman, T. I., 189(117), *263*
Kamiya, K., 188(51b), *261*
Kan, J., 27(29), 36(29), 37(117), 40(117), *51*, *54*
Kaneko, I., 40(121, 122), 41(126), *54*
Kano-Sueoka, T., 25, 26(25), 27(29, 31), 28(49), 35, 36, 37(31, 116), 38(31), 40(116), 41(127), 42(127), *51*, *52*, *54*
Karle, I. L., 189(113), 190(140), 205(140), 251(140), 257(140), *263*
Karnofsky, D. A., 104(194), *117*
Kartha, G., 188(52b), 189(52b), 221(52b), 230(52b), 235(52b), *261*
Katz, L., 188(46a, 46b, 48), 189(46a, 46b, 48), 208(185), 218(46a), 230(46a, 46b, 48), 233(48), 234, 239(48), 246 (46b, 48), 249(46a, 46b, 48), 252(46b, 48), 255(46a, 46b, 48), *261*, *264*
Katze, J. R., 12(46a), *22*
Kaye, A. M., 30(77), 31, *53*
Kazarinova, L. Y., 28(44), *52*
Keller, E. B., 2(1), *21*, 24(6), *51*

Kelley, G. G., 98(141), *116*
Kelley, W. N., 97(121, 131), 98(136, 137, 138), 107(217), 108, 109(131, 138, 217, 231), *116*, *118*, *119*
Kelmers, A. D., 27(28), 36(28), *51*
Kende, H., 112(245), *119*
Kendrew, J., 185(23), *260*
Kerr, D. S., 32(90), *53*
Kevin, S. P., 78(69), *85*
Khanina, M. F., 146(9), *179*
Khein, Kh. Ya., 78(67), *85*
Khorana, H. G., 68, *86*, 175(98), *182*
Kierkkegaard, P., 190(144), 251(144), *263*
Kim, B. K., 95, 96, *115*
Kim, S.-H., 185(18), 188(42), 189(78, 85), 190(42), 192(85), 199(42), 212(42), 221(42), 227(78, 85, 181), 230(42), 235(42), 236, 238(78, 85), 240, 241 (85), 243(42), 244(18), 246(42), 247 (78), 248(85), 249(78, 85), 250(42), 252(42), 253(78), 254(85), 255(78, 85), 256(42), *260*, *261*, *262*, *264*
Kimball, A. P., 98(145), 99(145), *116*
Kimura, F., 28(45), 33(100), *52*, *53*, 59(39), 60(39), 68(39), *85*
Kingdon, H. S., 3(7), *21*
Kinoshita, Y., 189(83), 224(83), 225(83), 248(83), 254(83), *262*
Kirkegard, L., 185(19), 244(19), *260*
Kiselev, L. L., 11, 19, *22*, 34(111), *54*
Kishi, T., 188(51b), *261*
Kizer, D. E., 93(74, 75), *115*
Klämbt, D., 61(12, 13), *84*
Klein, G., 110(240a), *119*
Klenow, H., 94(79), 98(144), 99(144), *115*, *116*
Kline, L. K., 60(43), 67, 71(43), 72(47), 73(47), 74(38), *85*
Klinenberg, J. R., 102(177), 108(177, 218), *117*, *118*
Klink, F., 130(54), 132(54), *143*
Knight, P. R., 104(195), *118*
Kobayashi, M., 28(46, 47), *52*
Kobayashi, T., 128(35), 132(35), *142*
Koblet, H., 46(142), *54*
Konigsberg, W., 12(46a), *22*
Konigsberger, V. V., 8(26), *21*
Korn, E. D., 94(85), 95, *115*
Kornberg, A., 88, 105(208), *113*, *118*, 147 (23), *179*
Kornberg, H. L., 171(89, 90), *181*
Korngold, G. C., 146(12), 161(12), 167 (12), 168(12), *179*
Kosalopoff, G. M., 6(19), *21*
Koshimizu, K., 61, *84*
Kotick, M. P., 189(115), 219(115), 249 (115), 255(115), *263*
Koyama, G., 96(110), *115*
Kraemer, K., 32(92), *53*
Krajewska, E., 127(42), *142*
Kraut, J., 151(65), *181*, 185(27), 188(43), 190(141), 204(43), 205(141), 246(43), 251(141), 252(43), 257(141), *260*, *261*, *263*
Krebs, H. A., 94, 105(81), 107(81), *115*
Krenitsky, T. A., 94(93), 95(93, 106), 96, 98(134), 104(134), *115*, *116*
Krim, M., 146(12), 161(12), 167(12), 168 (12), *179*
Krishnan, P. S., 92, *114*
Krishnaswamy, P. R., 3(8), 9(8), *21*
Krondraskev, Yu. D., 188(47), 189(47), 221(47), 230(47), 233(47), 234(47), 246(47), 249(47), 252(47), 255(47), *261*
Krone, W., 127(39), *142*
Krueckel, B. J., 101(166), *117*
Krug, R., 32(89), *53*
Krutilina, A. I., 59(78), *86*
Krygier, V., 99(162), *117*
Küchler, E., 208(196), 209(198), *265*
Küntzel, H., 123(15), 134(15, 61), 135 (15), 138(61), 139(61), *142*, *143*
Kukhanova, M. K., 19, *22*
Kulayeva, O. N., 78(67), *85*
Kuliev, P., 146(16), *179*
Kull, F. J., 34, *54*
Kumar, S., 92(41, 42), *114*
Kung, H., 59(9), 61(9), 73(48), *84*, *85*
Kuo, T., 11, *21*
Kusaka, T., 188(51b), *261*
Kusaki, T., 61(20), *84*
Kwan, C. N., 42, *54*
Kyogoku, Y., 208(183, 194), 209(183, 197), 210, 212(197, 199), 241(186), *264*, *265*

L

Labana, L. L., 188(54a), 189(54a), 224 (54a), 230(54a), 233(54a), 234(54a), 243(54a), *261*
Labanauskas, M., 185(19), 244(19), *260*
Labouesse, B., 18, *22*
Ladik, J., 206(163), *264*
Lagerkvist, U., 9(29), 13(29), 14, *21*, *22*
Lajtha, L. G., 100, 103(187), *117*
Lake, J. A., 185(21), *260*
Landy, A., 59(6), 61(6), 68(6), 73(6), 74(6), *84*
Langridge, R., 185(4, 5, 10, 12, 13, 14), 204(176), 226, 237(4, 5, 12, 13, 14), *260*, *264*
Lapidot, A., 151(62), *181*
Larsson, A., 90(12), *113*
Laskowski, M., 89(5), 91(22), *113*
Laster, W. R., 104(205), 107(205), *118*
Lazdunski, C., 171(88), *181*
Lazdunski, M., 171(88), *181*
Lazzarini, R. A., 40, 41, *54*
Leboy, P. S., 30(77), 31, *53*
Leder, P., 68, *86*, 127(45), *143*
Lederberg, S., 129(46, 48), *143*
Lederberg, V., 129(48), *143*
Ledneva, R. K., 149(40), 150(40), 162(76), 172(94, 95), 174(96), 175(97), *180*, *181*, *182*
Lee, J. C., 45(137), *54*
Lee, L. W., 13(52), *22*
Lee, Y. P., 93, 94, *114*, *115*
Leff, J., 139(85), *143*
Legault-Demare, J., 40(123), *54*
Lehman, I. R., 89(6), *113*, 170(82), *181*
Lemaire, G., 18, *22*
Lemoine, F., 10(33), *21*
Lengyel, P., 12(46a), *22*, 121(3), 132(3), *142*
Lennarz, W. J., 2(3), *21*
Leonard, N. J., 33(99), *53*, 59(35), 60(35), 61(11), 62, 63, 68(35), 77(63, 65), 83(35), *84*, *85*
LePage, G. A., 98(145), 99, 100, 101, 102, 104(203), 105(206), *116*, *117*, *118*
Letham, D. S., 61, *84*, 112, *119*
Li, L., 59(78), *86*
Libenson, L., 94(92), 95(92), 96(92), *115*
Lilley, R. M., 106(213), *118*
Lindahl, T., 29(55, 56, 57), *52*
Lindberg, B., 98(144), 99(144), *116*
Linnane, A. W., 138(74, 75, 76, 77), *143*
Linsmaier, E., 76(55), *85*
Lipmann, F., 3, 8, *21*, 121(1, 2), 122, 123 (24), 125(8, 27), 126(2, 8), 128(36), 129(49), 130(1, 2), 132(2, 56), 134 (56), *141*, *142*, *143*, 171(84), *181*
Lipsett, M. N., 32, *53*
Liquori, A. M., 188(70, 71a, 71b, 72, 73), 202(71a, 71b, 72, 73), 247(71b), 253 (71b), *262*
Lisowski, J., 91(27), *114*
Littauer, U. Z., 12, *22*, 28(39, 40), 30, 39(39, 40), *52*
Loening, U. E., 123(18), *142*
Leoppky, R. N., 63(31), 77(63), *84*, *85*
Loftfield, R. B., 3, 4, 6, 19(13a), *21*
Logan, D. M., 62(24), 75(24, 52), *84*, *85*
London, I. M., 101(168, 171), *117*
Longworth, J. W., 189(108), 227(108), *262*
Lord, R. C., 207(201a), 208(183, 193, 194), 209(183, 197), 210(197), 212(197, 199), 241(186), *264*, *265*
Love, W. E., 189(80), 247(80), 253(80), *262*
Lowenstein, J. M., 91(31), 93(63, 64, 65, 66), 94, 103(31), *114*, *115*
Lowy, B. A., 101(168, 171), *117*
Lubin, M., 138(81), *143*
Lucas-Lenard, J., 121(2), 123(24), 126(2), 130(2), 132(2), *141*, *142*
Luce, J. K., 98(155), 99(155), *117*
Lukens, L. N., 97(133), 98(133), *116*
Luzzati, V., 189(106), 249(106), 255(106), *262*
Lyn, G., 147(28), *179*

M

McCalla, D. R., 76(59), *85*
McCaman, R. E., 94(84), *115*
McClellan, A. L., 220, *264*
McCollister, R. J., 106(211), *118*
McComb, R. B., 103, *117*
McConnell, J. F., 190(139a, 139b), 203 (139a, 139b), 228(139a, 139b), 229 (139a, 139b), 251(139b), 257(139b), *263*
McDonald, I. R. C., 61(16), *84*

McElroy, W. D., 11(36), 19(69), *21*, *22*, 91(34), *114*
Mach, B., 46, *54*
Macindoe, H. M., 123(17), *142*
Macintyre, W. M., 189(79, 98, 100), 190 (147a, 147c), 202(155), 248(98), 249 (100), 254(98), 255(100), *262*, *263*, *264*
McLennan, B. D., 60(82), 62(24), 75(24, 52), *84*, *85*, *86*
McQuillen, K., 138(73), *143*
Macrae, A. R., 164(77), *181*
Madison, J. T., 11(42), *22*, 32(96), *53*, 59 (9, 76), 61(9), 73(48), *84*, *85*, *86*
Mager, J., 93(58), 97(125), 101(169, 172, 173), 102(172), 108(223), 109(223), *114*, *116*, *117*, *118*
Maggio, R., 34(107), *54*
Maguire, M. H., 92(48), *114*
Maitra, U., 30(80), *53*
Makman, M. H., 18, *22*, 32(88), *53*
Maling, B. D., 24(12), *51*
Malkin, L. I., 32(92), *53*
Mandel, H. G., 103(188), *117*
Mandel, L. R., 29, *52*
Mandel, M., 139(85), *143*
Mandel, P., 94(80), 105(80), *115*
Marcker, K. A., 59(80, 81), *86*
Margoshes, M., 207(201b), *265*
Marker, K. A., 13(48), *22*
Marquisee, M., 11(42), *22*, 32(96), *53*, 59(76), *86*
Marsh, R. E., 188(56, 64), 189(82, 88, 92, 107), 190(139a), 203(139a), 204(88, 92), 219, 224(56, 82), 225(82, 88), 226, 227(107), 228(139a), 229(139a), 247(64), 248(82, 88, 92), 249(107), 253(64), 254(82, 88, 92), 255(107), *261*, *262*, *263*, *264*
Martin, D. M. G., 66, 67, 82(36), *85*
Martin, T. E., 129(51), *143*
Martinez, J., 190(134), *263*
Martinez-Carrera, S., 190(128), 228(128), 250(128), 256(128), *263*
Marvin, D. A., 185(7), 237, *260*
Mascarenhas, Y., 190(129), 228(129), 250 (129), 256(129), *263*
Matamola, M., 3(10), 9(10), *21*
Mathews, B. W., 190(131), 228(131), 250 (131), 256(131), *263*
Mathews, F. S., 188(38), 189(38, 87, 94), 225(87), 226(87, 94), 227(94), 230 (38), 233(38), 246(38), 248(94), 249 (38), 252(38), 254(94), 255(38), *261*, *262*
Matsubara, S., 61(20), *84*
Matsuhashi, M., 2(2), *21*
Matthaei, J. H., 123(20), *142*
Mautner, H. G., 189(102, 103), 227(102, 103), 248(102), 254(102), *262*
Mazen, A., 146(13), *179*
Mazza, F., 188(52b, 54b), 189(52b, 54b), 221(52b), 224(54b), 230(52b, 54b), 232(54b), 233(54b), 234(54b), 235 (52b), 243(54b), *261*
Mehler, A. H., 3(12, 28, 35), 4(12, 14, 15), 6(15, 18), 7, 8(28), 9(28), 10(35), 13(54), 14(22), 15(62), 16(22, 62), 17(22), 18(18), 19(18, 22, 69), *21*, *22*
Meikle, A. W., 94(86), 95(86), *115*
Meister, A., 3(8), 9(8), 11(37), *21*
Melnikova, V. I., 150(50, 51), 151(50, 51), 166(50), *180*
Melvin, I. S., 207(190), *265*
Mendicino, J., 93(62), *114*
Merrifield, R. E., 207(201a), *265*
Merrill, S. H., 11(42), *22*, 32(96), *53*, 59(76), *86*
Metz, E. N., 108(229), *119*
Michelson, A. M., 175(99), *182*
Mihara, S., 146(10), *179*
Milanesi, G., 125(30), 127(30), 128(30), *142*
Miles, H. T., 208(195), *265*
Miller, C. O., 61(15, 19), 76, *84*, *85*
Miller, C. S., 103(186), *117*
Miller, H. R., 97(121), *116*
Miller, J., 97(131), 108(226), 109(131, 226, 231), *116*, *118*, *119*
Miller, M. J., *144*
Miller, O. J., 110(240a), *119*
Miller, R. L., 98(135), *116*
Minkin, J. A., 189(80), 247(80), 253(80), *262*
Mirsky, A. E., 32(93), *53*
Mirzabekov, A. D., 28(44), *52*, 59(78), *86*
Mitchinson, J. M., 129(46), *143*
Mitra, S. K., 3(28), 4(14), 6(18), 7(22), 8(28), 9(28), 13(54), 14(22), 16, 17 (22), 18(18), 19(18, 22), *21*, *22*

Mitsui, A., 90(16, 17), *113*
Mitsui, T., 61(20), *84*
Mittelman, A., 30(75), 31, *53*
Miura, G. A., 76, *85*
Miyazaki, M., 59(77, 79), *86*
Mizuno, K., 188(51b), *261*
Mizutani, T., 59(79), *86*
Moldave, K., 121(5, 6), *142*
Momparler, R. L., 99(162), *117*
Monod, J., 23, *51*
Monro, R. E., 137(66), 138(66), *143*
Monseur, X. G., 62, *84*
Montgomery, J. A., 99(159), 106(215), 110(159, 240), *117, 118, 119*
Moon, H. M., 121(3), 132(3), *142*
Moore, A. E., 79(72), *86*
Mootz, D., 190(123), 199(123), 228(123), 250(123), 256(123), *263*
Mora, G., 3(10), 9(10), *21*
Morgan, C. H., 189(97a, 97b), 248(97a, 97b), 255(97a, 97b), *262*
Morrison, J. C., 14(59), *22*
Morton, R. K., 95(104), 106(209, 210), *115, 118*
Mothes, K., 77, *85*
Moyer, R. W., 171(85), *181*
Müller, F., 190(144), 251(144), *263*
Muench, K. H., 29(51), *52*
Muntz, J. A., 93(60, 62), *114*
Murakami, M., 59(77), *86*
Murase, M., 96(110), *115*
Murashige, T., 76(56), *85*
Muroi, M., 188(51b), *261*
Murray, A. W., 91(30), 93(73, 77), 97(119, 120, 123, 126, 127, 128, 130), 98, 99 (146), 101(30), 104(119, 202), 106 (213), 107(202), 112(242), *114, 115, 116, 118, 119*
Mushinski, J. F., 46, *54*

N

Nagashima, N., 188(75), *262*
Najjar, V. A., 171(83), *181*
Nakamoto, K., 207(201b), *265*
Nass, M. M. K., 24(11), *51*
Nathans, D., 122, 125(8), 126(8), *142*
Nedbye, V. K., 149(43), 150(43), 151(43), 152(43), 153(43), 156(43), 158(43), 169(80), *180, 181*
Neidhardt, F. C., 33, 34, *53, 54*
Nesbitt, J. A., III, 2(3), *21*
Newton, A., 25(20, 21), *51*
Nicholls, P. B., 112(242), *119*
Nirenberg, M. W., 29(54), 37(116, 117), 40 (116, 117), *52, 54*, 68, *86*, 123(20), 127(45), *142, 143*
Nishikawa, M., 188(51b), *261*
Nishimura, S., 28, 33(100), *52, 53*, 59(39), 60(39), 61, 68(39), *84, 85*
Nishizuka, Y., 121(1), 130(1), *141*
Nobel, J., 6(17), *21*
Noll, H., 123(15, 16, 19), 134(15), 135(15), *142*
Norrestam, R., 190(144), 251(144), *263*
Norris, A. T., 3(9), 4, 9(9), *21*
Norris, E. R., 91(38), *114*
Novelli, G. D., 8(24), *21*, 25, 27(28), 28, 36, 46, *51, 52, 54*, 123(25), *142*
Nowacki, W., 190(148), 251(148), 257 (148), *263*
Noyes, W. D., 103(187), *117*
Nurk, E., 95(96), *115*
Nyburg, S. C., 189(109a, 109b, 111), 227 (109a, 109b), 249(109a, 109b, 111), 255(109b, 111), *263*

O

O'Brien, E. J., 188(58a, 58b), 189(58a, 58b), 224(58a, 58b), 237(58a, 58b), 238(58a, 58b), 247(58b), 248(58b), 253 (58b), 254(58b), *261*
Obruchnikov, I. V., 149(43), 150(43), 151 (43), 152(43), 153(43), 156(43), 158 (43), *180*
Occolowitz, 59(35), 60(35), 68(35), 83(35), *85*
Ochoa, M., Jr., 138(69), *143*
Ochoa, S., 123(23), 125(26), 126(26), 138 (72), *142, 143, 144*
O'Connell, W., 92, *114*
Oda, K., 28(45), *52*
Ofengand, E. J., 2(4), 3(4), 11(39), *21, 22*
Ogura, H., 157(73), *181*
Ohta, T., 3(57), 14, *22*
Okamoto, T., 125(29), *142*
Oliver, R., 103(187), *117*
Olivera, B. M., 170(82), *181*
Olomucki, M., 62, *84*

Olsen, A. C., 207(190), *265*
Ono, Y., 121(3), 132(3), *142*
Oparin, A. I., 112(247), *119*
Orsulak, P. J., 102(174), 107(174), *117*
Osborne, D. J., 76(59), 77, *85*
Otey, M. C., 94(91), *115*
Otsuka, H., 146(10), *179*
Overby, L. R., 28, 39(41), *52*
Overgaard-Hansen, K., 102, *117*
Owens, S. L., 6, *21*

P

Papaioannou, R., 98(134), 104(134), *116*
Papas, T. S., 3(35), 4(15), 6(15), 7, 10(35), *21*
Parin, A. V., 11, *22*
Parisi, B., 121(7), 125, 126(32), 127(30, 43, 44), 128(30, 34), 129(32), 130(7, 32), 131(7), 133(7, 59), 134(7), 136 (59), 138(34, 83), 140(7), *142*, *143*, *144*
Parks, R. E., 95(99, 100, 107, 108), 96(99, 107), 97(114), 98(142), *115*, *116*
Parry, G. S., 189(93a), 219(93a), 226(93a), 227(93a), *262*
Patel, N., 46(140), *54*
Paterson, A. R. P., 91(29), 94(87, 88, 89), 96, 97(129), 99(156, 158), 104(198, 204), 105(158), 107(204), 110(158, 237), *114*, *115*, *116*, *117*, *118*, *119*
Pattee, H. H., 112(248), *119*
Patterson, A. L., 189(80), 247(80), 253 (80), *262*
Pauling, L., 185, 196, 198(153), 199(153), 206(153), 213, 219, 229(153), 237(22), *260*, *264*
Pavan, M., 138(84), *143*
Penman, S., 208(185), *264*
Penswick, J. R., 11(42), *22*, 32(96), *53*, 59(76), *86*
Perani, A., 121(7), 125(30), 126(32), 127 (30), 128(30, 34), 129(32), 130(7, 32), 131(7), 133(7), 134(7), 136(63), 137 (63), 138(34), 140(7, 63), *142*, *143*, *144*
Perry, S., 79(74), *86*
Persson, K. Ö., 146(5), *179*
Perutz, M., 185(23), *260*
Peter, J. B., 171(91), *181*
Peterkofsky, A., 30, 32(85), *52*, *53*, 71(44), *85*
Petermann, M., 123(10), *142*
Petersen, C. S., 189(86), 248(86), 254(86), *262*
Pfrogner, N., 92(49, 52), *114*
Philippovich, I. I., 123(13), 134(13), 135 (13), *142*
Philippsen, P., 12(46), *22*
Pierre, K. J., 98((145), 99, *116*, *117*
Pillinger, D., 30(78), *53*
Pimentel, G. C., 220, *264*
Pintner, J. J., 136(64), *143*
Pinto, B., 94(90), 95, *115*
Pitha, J., 208(184), *264*
Pithova, P., 208(184), *264*
Plesner, P., 94(91), *115*
Pletcher, J., 190(142), *263*
Polsky, R., 78, *86*
Pomales, R., 108, *119*
Porter, J. W., 74, *85*
Potter, J., 146(17), *179*
Potter, M., 46, *54*
Poulson, H. E., 102, *117*
Pouyet, J., 146(13), *179*
Preiss, J., 11, *22*
Preobrazhenskaya, N. N., 149(33, 34, 45), 150(34, 45, 57), 152(45, 57, 71), 153 (45, 57, 71, 75), 155(34, 45, 57, 71), 158(71), 162(75, 76), 164(75), 174(96), *180*, *181*, *182*
Price, V. E., 94(91), *115*
Prokofiev, M. A., 146(15, 16, 18, 19, 20, 21), 149(31, 32, 33, 34, 35, 36, 37, 38, 39, 41, 42, 43, 44, 45), 150(31, 32, 34, 35, 36, 37, 38, 39, 41, 42, 43, 44, 45, 46, 47, 48, 49, 50, 51, 52, 53, 55, 56), 151(36, 37, 39, 41, 42, 43, 50, 51, 52, 53, 55, 59), 152(37, 39, 43, 44, 45, 53, 55, 56, 59, 71, 72), 153(43, 45, 53, 59, 71, 75), 154(37, 55, 59), 155(34, 45, 56, 59, 71, 72), 156(43, 44), 157(56), 158(43, 44, 71), 159(37, 39, 44), 161 (21), 162(72, 75, 76), 163(72), 164(75), 165(78), 166(50, 72), 167(41, 53), 169 (79, 80), 171(48, 49), 172(94, 95), 174 (96), 175(97), *179*, *180*, *181*, *182*
Provasoli, L., 136(64), 138(70), *143*
Prutton, I., 185(15), 237(15), *260*
Puliti, R., 188(71b, 72, 73), 202(71b, 72, 73), 247(71b), 253(71b), *262*

Pullman, A., 196(150, 151), 198(150, 151), 206(162), *264*
Pullman, B., 196(150, 151), 198(150, 151), 206(162), 221, 227, 234(175), 235(180), *264*
Pullman, M. E., 171(83), *181*
Purigin, P. P., 149(40), 150(40), 151(59), 152(59), 153(59), 154(59), 155(59), *180*, *181*

R

Racker, E., 103, *117*
Rahn, R. O., 189(108), 227(108), *262*
RajBhandary, U. L., 33(100), *53*, 59(39, 41), 60(39), 68(39), 70(41), *85*, 185 (19), 241(187b), 244(19), *260*, *264*
Ralph, R. K., 39(120), *54*
Ramachandran, G. N., 219, *264*
Ramakrishnan, C., 219(167), *264*
Ramaley, R. F., 171(85), *181*
Rao, S. N., 93(59), *114*
Rao, S. S., 129(52), *143*
Rao, S. T., 188(76), 247(76), 253(76), *262*
Ravel, J. M., 13, 15, *22*
Razin, A., 93(58), 97(125), 101(169, 172, 173), 102(172), *114*, *116*, *117*
Reed, H. J., 190(141), 205(141), 251(141), 257(141), *263*
Reeke, G. N., Jr., 189(107), 227(107), 249 (107), 255(107), *262*
Reese, C. B., 66, 67, 82(36), *85*
Reichard, P., 90(11, 12), *113*
Reid, E., 102(180), *117*
Reis, J. L., 91(26), *114*
Reisner, A. H., 123(17), *142*
Reiss, J., 2(3), *21*
Rendi, R., 125(26), 126(26), 138(72), *142*, *143*
Rérat, B., 190(133), 205(133), 251(133), 257(133), *263*
Rérat, C., 190(133), 205(133), 251(133), 257(133), *263*
Revel, M., 30(70), *52*, 123(22), *142*
Rich, A., 185(18, 26, 31), 188(38, 40, 41, 42, 46a, 46b, 48, 55, 60), 189(38, 46a, 46b, 48, 55, 60, 78, 84, 85, 87, 94), 190 (40, 41, 42), 198(31), 199(42), 204(173, 176, 177), 208(183, 193, 194), 209(183, 197), 210(197), 212(41, 42, 197, 199), 218(46a), 221, 224(55), 225(84, 87), 226, 227(78, 84, 85, 94), 230(38, 40, 41, 42, 46a, 46b, 48, 55, 179), 232(55), 233 (38, 48, 55, 179), 234(40, 46a, 46b, 55), 235(42), 236, 237(41, 181), 238(60, 78, 84, 85), 239(48), 240, 241(84, 186, 200), 242, 243(41, 42), 244(18), 246(38, 40, 41, 42, 46b, 48, 55), 247(78), 248(84, 85, 94), 249(38, 46a, 46b, 48, 55, 78, 84, 85), 250(41, 42), 251(40), 252(38, 40, 41, 42, 46b, 48, 55), 253(78), 254 (84, 85, 94), 255(38, 46a, 46b, 48, 55, 84, 85), 256(41, 42), 257(40), *260*, *261*, *262*, *264*, *265*
Richardson, C. C., 147(27), 170(81), 171 (81), *179*, *181*
Richter, D., 130(54), 132(54), *143*
Richter, J. F. P., 164(77), *181*
Ringertz, H., 188(74), 224(74), 225(74), 247(74), 253(74), *262*
Ripamonti, A., 188(71a, 71b, 72, 73), 202 (71a, 71b, 72, 73), 247(71b), 253(71b), *262*
Rjabova, T. S., 149(31, 32, 38), 150(31, 32, 38, 48, 49), 151(60, 61), 152(60), 169 (79), 171(48, 49), *179*, *180*, *181*
Robins, E., 94(84), *115*
Robins, M. J., 33(97), *53*, 60(4, 33), 62(4, 28), 63(4, 33), 65(4), 66(4, 33), 76(33) *84*, *85*
Robins, R. K., 67(37), 85
Roche, J., 62(27), *84*
Rockwell, M., 92(48), *114*
Rogg, H., 59(10), 61(10), 73(10), *84*
Rogozinska, J. H., 62, *84*
Rohrer, D., 189(101), 202(101), 249(101), 255(101), *262*
Romanko, E. G., 78(67), *85*
Romming, Chr., 189(86), 248(86), 254(86), *262*
Ronca, G., 92(50), *114*
Rosenbloom, F. M., 97(121, 131), 98(136, 137, 138), 107(217), 108(220, 224, 225, 226), 109, *116*, *118*, *119*
Rosenkranz, H. S., 146(11), 161(11), 166 (11), 167(11), 168(11), *179*
Rossi, C. A., 92(50), *114*
Roth, J. R., 25, *51*
Rouget, P., 7, 18, *21*, *22*
Roush, A., 91(38), *114*
Rowe, J., 123(17), *142*

Roy, J. E., 94(94), 95(94), *115*
Roy, J. K., 97(114), *116*
Rozenberg, M. C., 98(148, 149), *116*
Rundle, R. E., 207(201b), *265*
Rundles, R. W., 108(229), *119*
Russell, R. L., 33, *53*, 68, 69, *85*
Rymo, L., 14(56), *22*

S

Saenger, W., 185(18), 189(116), 244(18), 249(116), 255(116), *260*, *263*
Sagan, L., 140(87), *144*
Sakore, T. D., 188(53a, 53b, 53c, 54b), 189 (53a, 53b, 54b), 221(53a, 53c), 224 (54b), 230(53a, 53b, 54b), 232(54b), 233(53b, 54b), 234(54a, 54b), 235(53a, 53b), 241, 243(53b, 54b), *261*
Sala, F., *144*
Salas, M., 123(23), *142*, *144*
Salser, J. S., 146(14), 161(74), *179*, *181*
Sampson, S. D., 99(157), *117*
Samuel, D., 151(62), *181*
Sandrin, E. 10, *21*
Sanger, K. C. S., 92(42), *114*
Santangelo, E., 41, *54*
Sarid, S., 28(39, 40), 39(39, 40), *52*
Sasada, Y., 190(145), 251(145), 257(145), *263*
Sasisekharan, V., 219(167), 237, *264*
Satarova, L. G., 150(46), *180*
Saveliev, E. P., 149(41, 44), 150(41, 44, 53, 56), 151(41, 53, 61), 152(44, 53, 56, 71), 153(53, 71), 155(56, 71), 156(44), 157(56), 158(44, 71), 159(44), 167(41, 53), *180*, *181*
Sawa, T., 96(110), *115*
Sax, M., 190(142), 205(142), 251(142), 257 (142), *263*
Scannell, J. P., 104(201), *118*
Schabel, F. M., 104(205), 106(215), 107 (205), *118*
Schachtschabel, D., 127(39), *142*
Schatz, A., 138(70), *143*
Scheit, K. H., 189(116), 249(116), 255 (116), *263*
Scherberg, N. H., 28(38), 38(38), *52*
Schiff, J. A., 137(65), 139(85), *143*
Schlessinger, D., 42(129), *54*
Schlimme, E., 11(43), 15(61), *22*, 185(18), 244(18), *260*
Schmitz, R. Y., 77(65), *85*
Schnebli, H. P., 98(147), 99, 106(215), 110 (159), *116*, *117*, *118*
Scholtissek, C., 98(151, 152, 153, 154), *116*, *117*
Schulman, L. H., 12, *22*
Schwartz, J. H., 171(84, 87), *181*
Schwartz, R., 104(196), *118*
Schweet, R., 121(4), *142*
Schweizer, M. P., 59(3), 60(3), *84*, 207 (192), *265*
Seeds, N. W., 129(53), *143*
Seegmiller, J. E., 97(121, 131), 98(136, 137, 138), 102(177), 107(217), 108(177, 218, 220, 224, 225, 226), 109(131, 138, 217, 226, 231), *116*, *117*, *118*, *119*
Segal, H. L., 91(28), *114*
Seidel, H., 11(43), *22*
Seno, T., 28(46, 47), *52*
Serov, S., 40(123), *54*
Setlow, B., 93(65, 65, 66), 94, *114*, *115*
Shabarova, Z. A., 149(31, 32, 33, 34, 35, 36, 37, 38, 39, 40, 41, 42, 43, 44, 45), 150 (31, 32, 34, 35, 36, 37, 38, 39, 40, 41, 42, 43, 44, 45, 46, 47, 48, 49, 50, 51, 52, 53, 55, 56, 57), 151(36, 37, 39, 41, 42, 43, 50, 51, 53, 55, 59, 60, 61), 152(37, 39, 43, 44, 45, 53, 55, 56, 57, 59, 60, 71, 72), 153(43, 45, 53, 57, 59, 71, 75), 154(37, 55, 59), 155(34, 45, 56, 57, 59, 71, 72), 156(43, 44), 157(56), 158(43, 44, 71), 159(37, 39, 44), 162(72, 75, 76), 163(72), 164(75), 165(78), 166(50, 72), 167(41, 53), 169(79, 80), 171(48, 49), 172(94, 95), 174(96), 175(97), *179*, *180*, *181*, *182*
Shaeffer, J., 121(4), *142*
Shannon, J. S., 61(16), *84*
Sharma, B. D., 190(139a, 139b), 203(139a, 139b), 228(139a, 139b), 229(139a, 139b), 251(139b), 257(139b), *263*
Sharon, N., 8, *21*
Sharpless, T. K., 92(45, 46), *114*
Shaw, G., 61, *84*
Shearn, A., 42, 43(131, 132), *54*
Sheen, M. R., 95(107), 96(107), *115*
Shefter, E., 151(66), 155(66, 70), *181*, 188 (50, 65, 67, 77), 189(50, 89, 96, 102,

103, 115, 117), 202, 204(50), 219(65, 67, 115), 221(67), 224(67), 225(89), 227(102, 103), 247(65, 67, 77), 248 (102), 249(96, 115), 253(65, 67, 77), 254(102), 255(96, 115), *261*, *262*, *263*, *264*
Shepherd, W. M., 28(35), 39(35), *51*
Shiflet, R. N., 14(59), *22*
Shigeura, H. T., 99(157), *117*
Shimada, I., 3(57), 14(57), *22*
Shimura, K., 128(35), 132(35), *142*
Shinsky, N. G., 149(45), 150(45), 152(45), 153(45), 155(45), *180*
Shirley, B. C., 93(75), *115*
Shive, W., 13(51, 52), 15(51), *22*, 76(61), *85*
Shoshani, T., 101(169, 172), 102(172), *117*
Shoup, R. R., 208(195), *265*
Shugar, D., 89(7), *113*
Sierakowska, H., 89(7), *113*
Sigler, P. B., 185(19), 237, 244(19), *260*, *264*
Sih, C. J., 129(50), *143*
Silberman, H. R., 108(229), *119*
Silbert, D. F., 25(23), *51*
Silverman, J., 190(138a, 138b), 228(138a, 138b), 229(138a, 138b), 251(138b), 257 (138b), *263*
Simpson, A. I., 94(87, 88), 96, 104(204), 107(204), *115*, *118*
Simpson, M. S., 98(139), *116*
Singer, M. F., 11, *22*
Singh, C., 190(125, 126), 197, 199(125, 126), 203, 204, 228(125), 250(125, 126), 256(125, 126), *263*, *264*
Singh, P., 190(147b), *263*
Siperstein, M. D., 74(50), *85*
Sissakian, N. H., 123(13), 134(13), 135(13), *142*
Sjöquist, J. A., 32(86), *53*
Skinner, C. G., 76(61), *85*
Skipper, H. E., 98(139), 104(205), 106 (215), 107(205), *116*, *118*
Skogerson, L., 121(6), *142*
Skoog, F., 33(99), *53*, 59(35), 60(35), 61 (12), 62(23), 63(31), 68(35), 76(55, 56), 77, 83(35), *84*, *85*
Skoultchi, A. S., 121(3), 132(3), *142*
Sletten, E., 188(51a), 190(147c), 251 (147c), 257(147c), *261*, *263*
Sletten, J., 190(147c), 251(147c), 257 (147c), *263*
Smallwood, B. M., 61(17), *84*
Smellie, R. M. S., 100(164), *117*
Smiley, K. L., 93, 94, *114*
Smillie, R. M., 138(79), *143*
Smith, D. E., 94(84), *115*
Smith, D. W. E., 39(119), *54*
Smith, J. D., 24(16), *51*, 59(6), 61(6), 68 (6), 73(6), 74(6), *84*
Smith, L. D., 93(74), *115*
Smith, R. A., 171(93), *182*
So, A. G., 127(40, 41), *142*
Sobell, H. M., 188(39a, 39b, 52a, 52b, 53a, 53b, 53c, 54a, 54b, 59a, 59b, 60, 61), 189(39a, 39b, 52a, 52b, 53a, 53b, 53c, 54a, 54b, 59a, 59b, 60, 118), 204(61, 118), 221(52a, 52b, 53a, 53c), 224(54a, 54b), 230(39a, 39b, 52a, 52b, 53a, 53b, 54a, 54b), 232(54b), 233(39a, 39b, 53b, 54a, 54b), 234(54a, 54b), 235 (52a, 52b, 53a, 53b), 237(59a, 59b), 238(59a, 59b, 60), 239, 241, 243(53b, 54b), 246(39b), 247(59b, 61), 248 (59b), 249(39a, 39b), 252(39b), 253 (59a, 59b, 61), 254(59a), 255(39a, 39b), 256(118), *261*, *263*
Sober, H. A., 32(95), *53*
Söll, D., 12(46a), *22*, 241, *264*
Sokolova, N. I., 149(33, 34, 40), 150(34, 40, 50, 51, 52, 55, 57), 151(50, 51, 52, 55, 59), 152(55, 57, 59, 72), 153(57, 59), 154(55, 59), 155(34, 57, 59, 72), 162(72), 163(72), 166(50, 72), *180*, *181*
Soloviova, A. B., 150(57), 152(57), 153 (57), 155(57), *180*
Sonnenbichler, T., 33(98), *53*, 60(5), *84*
Sorenson, L. B., 108(219), 109(232), *118*, *119*
Sparks, R., 155(70), *181*, 188(50), 189(50), 204(50), *261*
Spasokukotskaya, T. N., 34(111), *54*
Spencer, D., 138(78), *143*
Spencer, M., 185, 196, 219, 237(11), *260*
Srinivasan, P. R., 29(60, 67), 30(71, 72, 73, 79), 31(72, 73), 49(148), *52*, *53*, *55*, 74, *85*, 189(81), *262*
Srivastava, B. I. S., 76(53, 60), *85*
Stadtman, E. R., 90(9), *113*

Staehelin, M., 29(66), 30(76), 31, *52*, *53*, 59(10), 61(10), 73(10), *84*
Stanier, R., 122(9), *142*
Stanley, W., Jr., 123(23), *142*
Starr, J. L., 32(90), *53*
Stasiuk, L., 33(97), *53*, 60(4), 62(4), 63(4), 65(4), 66(4), *84*
Stavy, L., 32(94), *53*
Steiner, R. F., 185(16), *260*
Stent, G. S., 25, 27, 42, *51*, *54*
Stephenson, J., 104(199), *118*
Stephenson, M., 11(40), *22*
Stern, R., 3(12), 4(12), 19(69), *21*, *22*, 30 (70), *52*
Stevens, B. M., 102(180), *117*
Stevens-Clark, J. R., 171(93), *182*
Stewart, P. R., 138(77), *143*
Stewart, R. F., 188(189), 189(93b), 226 (93b), 227(93b), 246(189), 248(93b), 252(189), 254(93b), *262*, *264*
Stokes, A. R., 185(8), *260*
Storck, R., 123(11), 135(11), *142*
Stout, G. H., *187*
Strehler, B. L., 33(105), 34, *53*
Strelitz, R. A., 96(109), *115*
Strominger, J. L., 2(2), *21*
Stuart, A., 59(41), 70(41), *85*
Stulberg, M. P., 27(28), 36(28), *51*
Stumbravichute, Z. A., 151(59), 152(59), 153(59), 154(59), 155(59), *181*
Stutts, P., 96(111), 98(141), *115*, *116*
Stutz, E., 123(16), *142*
Subak-Sharpe, H., 27, 28(35, 36), 39, *51*, *52*
Suelter, C. H., 93, 94 (68, 71), *114*
Sueoka, N., 24, 25, 26(25), 27(29, 31), 28 (49), 29(50, 52, 53, 54, 58, 59), 35, 36, 37(31, 116, 117), 38(31), 40(116, 117), 41(127, 128), *51*, *52*, *54*
Suhadolnik, R. J., 92(53, 55), *114*
Sulkowski, E., 91(22), *113*
Sundaralingam, M., 151(67, 68), 155(69), *181*, 185(28, 29, 30), 188(44, 76), 189 (90, 91, 101), 197, 202(101), 204(44, 91), 219(91), 246(44), 247(76), 248 (91), 249(101), 252(44), 253(76), 254 (91), 255(101), *260*, *261*, *262*
Sundharadas, G., 12, *22*
Sutor, D. J., 188(36, 66, 68, 69), 202(66), 219, 221(36, 68), 224(68), 246(36), 247 (66, 68, 69), 252(36), 253(66, 68, 69), *261*, *262*, *264*
Svetailo, E. N., 123(13), 134(13), 135(13), *142*
Sweet, R. M., 188(64), 247(64), 253(64), *261*
Sweetman, L., 109(233), 110(233, 234), *119*
Switzer, R. L., 147(25), 171(25), *179*
Sypherd, P. S., 47, *55*
Szer, W., 127(42), *142*
Szweykowska, A. M., 63(31), 77(63), *84*, *85*

T

Tachibana, T., 110(240a), *119*
Takeda, M., 129(49), *143*
Takei, W. J., 195(35), *261*
Takemura, S., 59(77, 79), *86*
Takeuchi, T., 96(110), *115*
Talbert, F. D., 76(61), *85*
Tamberg, N., 186, *260*
Tamiya, H., 146(10), *179*
Tanaka, N., 190(145), 251(145), 257(145), *263*
Tarr, H. L. A., 94(94), 95, 99(161), *115*, *117*
Tavale, S. S., 188(53a, 53b), 189(53a, 53b), 221(53a), 230(53a, 53b), 233(53b), 235 (53a, 53b), 243(53b), *261*
Taveres, J. E., 112(245), *119*
Tay, B. S., 106(213), *118*
Taylor, M. M., 123(11), 135(11), *142*
Taylor, M. W., 43(133, 134, 135), 44(133, 134), 45, 47(134, 135), 48, *54*
Terganova, G. V., 146(15, 16, 18, 19, 20), *179*
Tewari, K. K., 92(41, 42), *114*
Thedford, R., 33(97), *53*, 60(4, 33), 62(4), 63(4, 33), 65(4), 66(4, 33), 76(33), *84*, *85*
Thewalt, U. T., 188(56), 224(56), *261*
Thiebe, R., 12(46), *22*, 70, *85*
Thies, G., 61(12), *84*
Thoai, N-V., 62(27), *84*
Thomas, H. J., 106(215), *118*
Thompson, A., 170(81), 171(81), *181*
Thomson, J. R., 98(139), *116*
Thomson, R. Y., 100(164), *117*
Tiboni, O., 128(37), 132(37), 133(37), 134

(37), 136(37), 138(37, 83), *142, 143, 144*
Tillack, T. W., 39(119), *54*
Tinoco, I., Jr., 202, 207(156), *264*
Tobin, N. F., 138(79), *143*
Todd, Lord, 164, *181*
Tollin, P., 188(36), 189(114), 221(36), 246 (36), 252(36), *261, 263*
Tomita, K., 188(45a, 45c, 46a, 46b, 48, 60, 61), 189(45a, 45b, 46a, 46b, 48, 60, 118), 204(45a, 45b, 45c, 61, 118), 218 (46a), 230(46a, 46b, 48), 233, 234(46a, 46b), 238(60), 239(48), 246(45a, 45c, 46b, 48), 247(61), 248(45b), 249(46a, 46b, 48), 252(45a, 45c, 46b, 48), 253 (61), 254(45b), 255(46a, 46b, 48), 256 (118), *261, 263*
Tomozawa, Y., 93(72), *115*
Touster, O., 94(86, 90), 95, *115*
Townsend, L. B., 67(37), *85*
Trass, T. C., 14(59), *22*
Travers, A. A., 35(149), *55*
Trotter, J., 189(99), 249(99), 255(99), *262*
Trueblood, K. N., 151(66), 155(66, 70), *181*, 188(50), 189(50, 96, 106), 204 (50), *229*, 249(96, 106), 255(96, 106), *261, 262, 264*
Trus, B. L., 190(146a, 146b), 251(146a), 257(146b), *263*
Ts'O, P. O. P., 123(12), *142*, 207(190, 192), *265*
Tsuboi, K. K., 95(97), *115*
Tsunakawa, S., 33(98), *53*, 60(5), *84*
Tsushima, K., 90(16, 17), *113*
Tsutsui, E., 30(73), 31, *52*
Tuppy, H., 209(198), *265*

U

Umezawa, H., 96(110), *115*
Unkeless, J. C., 90(21), *113*
Uziel, M., 61(7a), *84*

V

Vaciago, A., 190(143), 251(143), 257(143), *263*
Vail, M. H., 99(159), 110(159, 240), *117, 119*
Vanderheiden, B. S., 94(78), *115*
van der Helm, D., 189(80), 247(80), 253 (80), *262*
Vane, J. R., 100, *117*
Van Etten, J. L., 125(30), 127(30, 44), 128 (30), *142, 143*
Varghese, A. J., 189(113), *263*
Vazquez, D., 137(66), 138(66), *143*
Venable, J. H., 185(14), 237(14), *260*
Venkstern, T. V., 59(78), *86*
Veres, K., 76(57), *85*
Vescia, A., 34(110), *54*
Vitols, E., 90, *113*
Vizzini, E. A., 190(130), 202(130), 228 (130), 250(130), 256(130), *263*
Voet, D., 188(40, 41), 189(84), 190(40, 41), 212(41), 225(84), 226(84), 227(84), 230(40, 41), 234(40), 237(41, 181), 241 (84), 242, 243(41), 246(40, 41), 248 (84), 249(84), 250(41), 251(40), 252 (40, 41), 254(84), 255(84), 256(41), 257(40), *261, 262, 264*
Vold, B. S., 47, *55*
Vol'kenshtein, M. V., 188(47), 189(47), 221 (47), 230(47), 233(47), 234(47), 246 (47), 249(47), 252(47), 255(47), *261*
Voll, M. J., 25(23), *51*
Von Ehrenstein, G., 125(27), *142*
Vorobjov, O. E., 149(35, 36, 37), 150(35, 36, 37, 50), 151(36, 37, 50, 58), 152 (37), 154(37), 159(37, 58), 165(78), 166(50), *180, 181*

W

Wahba, A., 123(23), *142, 144*
Wainfan, E., 30(71, 72), 31(72), *52*
Waldenström, J., 9(29), 13(29), 14(56), *21, 22*
Waller, E., 10(33), *21*
Wallwork, S. C., 202, *264*
Wang, M. C., 104(204), 107(204), *118*
Wang, M. H., 93(76), 94, *115*
Wang, S.-F., 13(51), 15(51), *22*
Wang, S. Y., 189(113), *263*
Ware, G., 76(60), *85*
Warwicker, J. O., 190(120), 228(120), 250 (120), 256(120), *263*
Watanabe, K. A., 79(73), *86*, 103(192), 105(192), *117*
Watenpaugh, K., 188(49), *261*

Waters, L. C., 36, *54*
Watson, D. G., 183(1, 2), 188(36, 64), 213 (1), 221(36), 230, 232, 246(36), 247 (64), 252(36), 253(64), *260*, *261*
Watson, J. D., 204(173), 221(173), *264*
Way, J. L., 98(142), *116*
Webster, L. T., Jr., 3(7), *21*
Webster, R. E., 24(15), *51*
Wehrli, W., 59(10), 61(10), 73(10), *84*
Wei, C. H., 189(108), 227(108), *262*
Weil-Malherbe, H., 93(61), *114*
Weinblum, D., 189(109a, 111), 227(109a), 249(109a, 111), 255(111), *263*
Weinstein, I. B., 49, *55*, 138(69), *143*
Weisblum, B., 129(50), *143*
Weiss, B., 147(27), 170(81), 171(81), *179*, *181*
Weiss, G. B., 28(48), *52*
Weiss, S. B., 28(37, 38), 32(84), 38(37, 38), *52*, *53*
Weissbach, H., 129(53), *143*
Wenkstern, T. V., 146(9), *179*
Werkema, M. S., 190(147b), *263*
Werner, P., 190(144), 251(144), *263*
Wettstein, F. O., 27(30), 42, *51*, *54*
Whang-Peng, J., 79(74), *86*
Wheatley, P. J., 190(135), 251(135), 257 (135), *263*
Whybrew, W. D., 14(59), *22*
Widnell, C. C., 90(21), *113*
Wildman, S. G., 123(14), 134(14), 135(14), *142*
Wilhelm, R. C., 24(15), *51*
Wilken, D., 146(8), *179*
Wilkins, M. H. F., 185(4, 5, 6, 7, 8, 10, 11, 12, 13, 14), 237(4, 5, 6, 7, 11, 12, 13, 14), *260*
Williams, L. S., 34, *54*
Williams, M. K., 101(168, 171), *117*
Wilson, A. J. C., 195(34), *260*
Wilson, A. R., 98(139), *116*
Wilson, D. V., 61(17), *84*
Wilson, D. W., 103(186), *117*
Wilson, H. R., 185(4, 6, 8), 188(62a, 62b), 189(97a, 97b, 114), 204(62a, 62b), 224 (62a, 62b), 237(4, 6), 247(62b), 248 (97a, 97b), 253(62b), 255(97a, 97b), *260*, *261*, *262*, *263*
Wind, E., 101(173), *117*
Winter, T. S., 110(241), *119*
Wintermeyer, W., 12(46), *22*
Wiseman, A., 146(4), *179*
Witkus, E. R., 78(68, 69), *85*
Wolfenden, R., 92(44, 45, 46), 93(72), *114*, *115*
Wong, P. C. L., 97(128), *116*
Wood, H. G., 147(26), 171(26), *179*
Wool, I. G., 129(51), *143*
Worst, P., 110(240a), *119*
Wu, R., 103, *117*
Wyngaarden, J. B., 106(211), 107(216), 108(222), *118*

Y

Yamada, E. W., 94(95), 95(95), *115*
Yamada, Y., 61(7), *84*
Yamane, T., 24, 26, *51*
Yang, S. S., 45(138), *54*
Yang, W. K., 46, *54*
Yaniv, M., 13(49), 14, 17, *22*
Yannoni, N. F., 190(138a, 138b), 228(138a, 138b), 229(138a, 138b), 251(138b), 257(138b), *263*
Yanofsky, C., 24, *51*
Yarus, M., 13, 17, *22*
Yegian, C. D., 27, *51*
Yohn, D. S., 30(75), 31(75), *53*
Young, D. W., 189(114), *263*
Yu, C. T., 24(10), *51*
Yushok, W. D., 103, *117*

Z

Zachau, H. G., 12, *22*, 33(98), *53*, 59(8), 60(5), 61(8), 63(8, 34), 70, 73(8), *84*, *85*, 147, *179*
Zahrobsky, R. F., 189(79), *262*
Zamecnik, P. C., 2(1), 11(40), *21*, *22*
Zamir, A., 11(42), *22*, 32(96), *53*, 59(76), *86*
Zeitlin-Beck, R., 101(169), *117*
Zeldin, M. H., 137(65), *143*
Zillig, W., 127(39), *142*
Zinder, N. D., 24(15), *51*
Zipser, D., 25(20), *51*
Zirakzadeh, M., 189(100), 249(100), 255 (100), *262*

Subject Index

A

Adaptor,
  chemical basis for alteration,
    aminoacyl-transfer ribonucleic acid synthetase and, 33–35
    chain cleavage and, 28–29
    conformational changes and, 29
    methylation and, 29–31
    minor nucleosides and, 32–33
    terminal trinucleotide and, 32
    thiolated bases and, 31–32
Amino acid(s),
  activating enzymes,
    reactions of, 3–4
    transfer ribonucleic acid and, 11–15
  consequences of structure,
    aminoacyl adenylates and, 8–11
    analogs and, 4–6
    competitive inhibition and, 6–8
Aminohydrolases, catalytic and regulatory properties of, 91–94

B

Bond angles (Appendix II), 252–257
Bond distances (Appendix I), 246–251

C

Chemotherapy, purine metabolism and, 103–111

D

Deviations from planarity (Appendix III), 258–259
Differentiation,
  development and, 45–47
  different tissues and organs, 43–45
  growth conditions and, 40–43
  neoplasia and, 47–49
  virus infection and, 35–39
Drugs,
  control of purine metabolism by, 105–107
  conversion to active metabolites, 103–105
  resistance, purine metabolism and, 110–111

G

Gout, purine metabolism and, 107–109
Growth,
  conditions, differentiation and, 40–43
  control, purine metabolism and, 103–111

H

Hydrogen bonding,
  bases with other bases, 229–243
  bases with themselves, 220–229
  geometry of, 213–220
  theory and selective affinity, 205–213

I

N^6-(Δ^2-Isopentenyl)adenosine,
  biological activity of, 76–80
  biosynthesis of, 70–74
  chemical reactions of, 64–67
  enzymatic degradation of, 75–76
  in free state, 61–62
  occurrence of, 60–61
  properties of, 64
  synthesis of, 62–63
  transfer-ribonucleic acid function, 80–83
    base "Y" deletion and, 70
    deletion and, 68–69
    iodination and, 67–68
N^6-(Δ^2-Isopentenyl)-2-methylthioadenosine, synthesis of, 64

L

Lesch-Nyhan syndrome, purine metabolism and, 109–110

M

Molecular structures,
  bond angles and lengths, 196–199
  flatness of rings, 199–203
  location of protons on bases, 203–205

N

Neoplasia, differentiation and, 47–49
Nucleic acids, X-ray crystallography of, 183–186
5′-Nucleotidases, catalytic and regulatory properties of, 90–91
Nucleotide(s), binding of, 18
Nucleotide-peptide(s),
  nomenclature of, 147–148
  properties of bonds,
    conclusions, 169–170
    peptide phosphoesters, 167–169
    peptide phosphoramidates, 151–167
    specific cleavage of, 169
Nucleotide-(P-N)-peptide phosphoramidates,
  bonds, properties of, 151–167
  as models of "active nucleotides,"
    nucleotidyl transferases and, 170–172
    ribonucleoside-5′-phosphoamidase and, 172–178
  synthesis of, 148–151
Nucleotide-(5′ → O)seryl(or threonyl) phosphoesters, bonds, properties of, 167–169
Nucleotide-(P-O)-seryl(or threonyl)peptide phosphoesters, synthesis of, 148–151
Nucleotidyl transferases, nucleotide-peptides and, 170–172

O

Organs, differentiation in, 43–45

P

Protein synthesis,
  heterologous systems *in vitro*, 125–129
  regulation,
    ribosome specificity and cellular organelles, 135–137
  ribosome-specific inhibitors of, 137–139
  role of ribosomes, 122–125
  transfer ribonucleic acid mediated,
    abundancy and, 26–27
    adaptor and, 25–26
    inhibition and, 28
    modulation and, 24–25
    new synthesis and, 27–28
Purine(s),
  interaction with pyrimidines, 205–243
  metabolism, defects in, 107–111
  nucleosides and bases, supply of, 99–103
Purine nucleoside kinases, catalytic and regulatory properties of, 98–99
Purine nucleoside phosphorylases, catalytic and regulatory properties of, 94–96
Purine phosphoribosyltransferases, catalytic and regulatory properties of, 96–98
Pyrimidine(s), interaction with purines, 205–243

R

Ribonucleic acid, transfer,
  amino acid binding and, 15–16
  binding of, 13–15, 16–17
  catalysis and, 17
  control of protein synthesis and, 24–28
  enzyme activity and, 13
  isopentenyladenosine and, 67–70, 80–83
  recognition of, 11–13
Ribonucleoside-5′-phosphoamidase, nucleotide-peptides and, 172–178
Ribosomes,
  role in protein synthesis, 122–125
  specific inhibitors of, 137–139
  specificity,
    cellular organelles and regulation of protein synthesis, 135–137
    characterization of, 129–135

S

Specificity, basis of, 19–20

T

Thymine, similarity to uracil, 205
Tissues, differentiation in, 43–45

U

Uracil, similarity to thymine, 205

V

Variability, similarity and, 18–19
Virus, infection, cell differentiation and, 35–39

X

X-ray crystallography,
  experimental data, 186–195
  nucleic acids and, 183–186